Elementary Approach to Special Relativity

Helmut Günther

Elementary Approach to Special Relativity

With 76 Figures

Fig. 4.1 Source Christina Günther

Springer

Helmut Günther
Berlin, Germany

ISBN 978-981-15-3167-5 ISBN 978-981-15-3168-2 (eBook)
https://doi.org/10.1007/978-981-15-3168-2

Revised edition of "Elementary Theory of Relativity: Lattice-Ether-Symmetry" Revised English Edition of "Grenzgescheindigkeiten und ihre Paradoxa", Teubner-Texte zur Physik Bd. 31, Stuttgart-Leipzig, Teubner-Verlag 1996. Shaker Verlag, Aachen 2000, ISBN3-8265-7104-5

This Springer imprint is published by the registered company Springer Nature Singapore Pte Ltd.
The registered company address is: 152 Beach Road, #21-01/04 Gateway East, Singapore 189721, Singapore

Preface

Light somewhat like sound.
Sound somewhat like light.

Ernst Mach.

Can one really comprehend the Special Theory of Relativity (SRT) the same way we understand the mechanics of automobiles and bicycles? In order to explain Special Relativity, the physicist calls upon the universal constancy of the speed of light. This constancy is an indisputable certainty of the almighty Einsteinian principle of relativity, if applied to MAXWELL'S electrodynamics. Considering this fact, every objection against the theory, however resourceful and shrewd, breaks down miserably. Finally, the inquirer submits and withdraws, unhappy about the fact that he has to leave the elementary cognition of his existence in space and time in the hands of specialists. Do we just have to get used to the merciless consequences of irrefutable principles? Sometimes one gets an uneasy feeling that we could be made to believe that an X is a U. Are we supposed to comfort ourselves with the recognition, comprehension is a process of accustomisation? Should it then make a difference to adapt to either everyday experiences or logical consequences of principles? Are not logical consequences more reliable than everyday experiences? However, what happens if the principles themselves break down?

This book will introduce an alternative representation of EINSTEIN's Special Theory of Relativity. We believe that we can make Special Relativity much more comprehensible. Moreover, surprisingly, we will come across a fundamental relationship between the Special Theory of Relativity and the mechanics of a space lattice.

In all previous formulations, the EINSTEINian special principle of relativity, in one or the other form, is used as the starting point for Special Relativity. In correspondence to this principle, one takes it for granted a priori that all observers independent of their uniform motion to each other measure one and the same propagation velocity of a light signal. Notice that here the definition of simultaneity for the uniform moving frames is assumed with the help of exactly this velocity of light. The rest constitutes unavoidable consequences: Moving clocks go behind, lengths shorten when moving and inertial masses increase.

Here, we will take a completely different approach in formulating the Special Theory of Relativity. We will specifically abandon the process of accepting a priori the validity of any abstract physical principle of the likes of EINSTEIN's principle. In opposition to traditional representations of Special Relativity, we at first suppose a preferred reference system for our observations, and we ask what happens if measuring-rods and clocks are moving with respect to this system.

Nevertheless, if we want to arrive at the covariant relativistic formalism, we cannot however continue without a certain principle of relativity. We need a rule or a regulation to be able to synchronise clocks in a moving frame after this has been done in a supposed preferred 'static' system. It is a most simple and not in the least controversial principle that we use for this, the so-called reciprocity principle, an *elementary principle of relativity*, as we will see it that anticipated should sound like:

If you observe that I have the velocity v, then I observe that you have velocity $-v$.

We demand no more and no less of relativity. The original paper deriving EINSTEIN's Special Theory of Relativity along these lines is published in GÜNTHER [31]. The axiomatic system we use is without any restraint equivalent to the axiomatic system used by EINSTEIN. In Chap. 16, we will show a comparable representation of the different axiomatic approaches to the Special Theory of Relativity. Here too, we will state that our method is in fact just a continuation of the old ideas of H. A. LORENTZ.

However, we want to continue. We want to understand how it can happen that measuring-rods change their length and clocks change their pace? Which familiar and for us comprehensible processes can cause such phenomena? In order to answer this question, we are looking for a physical model, a 'miniatur' of the Special Theory of Relativity' which should be present in reality.

To this end, we will concentrate on using the laws of NEWTONian mechanics on the lattice structure of matter. We will ask ourselves the following question: How can we measure spatial distances and time differences in a lattice if we exclusively use geometrical objects that physically exist in this lattice? Such physical objects that come into question are dislocations. These dislocations can be found in great measure inside of a crystal. Firstly, we search for those physically stabile forms of a dislocation that will be able to provide us with a measure of length. We then secondly continue our search for a stabile oscillating dislocation that will provide us with an oscillation period for a clock. This can be achieved using the so-called sine-GORDON equation, for which we can give a most simple physical explanation. We will discover, building on this procedure, one relativistic effect after another, and at the end, we will arrive at the principle of the universal constancy of critical signal velocity, which is itself defined in a lattice.

With other words, the ideal space lattice represents our model for the vacuum, whereas certain local deviations from the ideal structure existing in this lattice (configurations of dislocations in the crystal) possess those inertial masses with respect to the lattice whose motions within certain boundaries of validity lead to the observation and confirmation of the laws of the Special Theory of Relativity. In this 'miniature version' of SRT, the question about the ether finds its physical explanation in the relationship between the ideal space lattice and its localised structural imperfections. Incidentally, this model may be seen as an additional point of argumentation in favour of a lattice structure of our physical vacuum:

The continuum approximation of a crystalline lattice becomes recognisable as a model for relativistic spacetime.

That is, we will undertake two things within the context of Special Relativity:

1. We will introduce a new axiomatic system for the Special Theory of Relativity that will be completely equivalent to EINSTEIN's axiomatic system and will still consist of simple and understandable statements.
2. We will develop this new axiomatic approach using a crystalline lattice which will claim the largest part of our explanations. In this model, we will immediately see how it can happen that moving rods suffer a length contraction and moving clocks go behind. Following this path, we will in fact discover an actually existing mechanical model, a 'miniature' for the Special Theory of Relativity.[1] Here, we will discover the rational core of the philosophical statement made by E. MACH quoted at the beginning, cf. THIELE [1].

Unlike EINSTEIN's procedure, our approach involves a definition of simultaneity which is independent from the other part of our SRT axiomatics. As a spin-off, the repeated debates on some questions of H. REICHENBACH's philosophy of space and time could be brought to an end. In Chap. 12, we derive the transformation formulas of REICHENBACH's non-covariant absolute simultaneity and we discuss, in the framework of SRT, REICHENBACH's LORENTZ contraction versus EINSTEIN contraction argumentation. Our method for evaluating philosophical questions concerning SRT should be noticed. It is only necessary to have a look at the real existing mechanical model of SRT for immediately 'seeing' the answer.

The first part of our discussion, Chaps. 1–8, serves as a preparation for our investigation of SRT inside of a space lattice. Chapter 3 gives a short summary of the physical statements made by SRT. However, the knowledge of these statements is not required for further reading. In Chaps. 4–6, we will deal in detail with the NEWTONian motion of masses and will investigate the mechanical oscillations and waves more closely. The expert may skim through these passages.

The central part of our explanations, Chaps. 9–21, will be on the kinematics of SRT, with its own specific effects. In Chaps. 22 and 23, we concern ourselves with the questions of dynamics in the SRT, with mass and energy. The core of SRT, the universal constancy of the critical signal velocity, is developed from our axiomatic approach in Chap. 12. The paradox and oddities resulting from the existence of critical velocities that we comprehend using the background of a space lattice can be traced throughout the whole of this book. In Chap. 11, we come across an unknown clock paradox, and in Chap. 17, the famous twin paradox will be

[1]The question of dimensionality of our model is discussed in Chap. 16, p. 141.

described in great detail from different viewpoints. The analysis of the well-known DOPPLER effect in Chap. 18 will bring surprising statements even for the physicist, and in Chap. 19, we give an elementary description for the exact relativistic effect of aberration.

Exact representations of physical context have their price: mathematics. I have tried to write this book in such a manner that it can still be read, even when mathematical equations are taken as granted and just left out. Then, again we do not want to discuss anything that an interested reader cannot himself reproduce mathematically. The frame of small mathematical pocketbooks of the likes of BRONSTEIN/SEMENDJAJEW [1] will seldomly be breached. I have placed a few supplementary explanations in Chaps. 24–30 of the appendix. Here, it is once again the paradox situations that find our interest when we conclusively turn to the tachyon problem considered in a crystal lattice.

This book is thought of as a lecture for physicists, mathematicians and computer scientists and concentrates on the students of these fields. It will also hopefully reach a broad circle of interested readers from the fields of natural sciences and philosophy. I personally hope that engineers will find this book invigorating.

I owe the inspiration for this theme to H. J. TREDER. I am much obliged to H. F. GOENNER, E. KRÖNER and D. E. LIEBSCHER for extended discussions, and I am indebted to J. STACHEL for drawing my attention to the history of the reciprocity principle.

I want to record my gratitude to MR. C. SHERWOOD for his enormous work in a primary formulation of the English text.

I would like to thank particularly MR. N. BEHRENT for imperative technical installations without which the manuscript could not have been written.

Above all, I am deeply grateful to my wife who unbounded patiently read the manuscript and gave me an invaluable list of corrections.

I express my gratitude to the Springer Publishing House, especially to Dr. Loyola d'Silva for realising this book in the present revised version.

I would like to thank the Springer team very much for their understanding cooperation.

Berlin, Germany
November 2019

Helmut Günther

Contents

Chapter 1
The Discovery of the Ether

The discovery of the ether had taken a long time. Once it was finally discovered, one could neither show it, see it, hear it nor touch it. There was nothing. Really discovered?

However, incredible phenomena came to light, curious properties of measuring-rods and clocks. One was repeatedly able to demonstrate these effects more precisely—even the furious spectacle of mass bursting into raging energy during nuclear fission.

Ether or Special Relativity? No question. Does the old ether have to be thrown into the proverbial dustbin where it belongs?

We want to retrieve it from this dustbin. For the sake of its inventors, of whom we know so much about it. This time we will be able to see and touch the ether. We will not see the old ether. That is impossible. We will have to resort to a trick that is used in films. We will organise a double, a double so identical to the original in all aspects that it could be taken to be an identical twin, so that we can correctly state: This is 'the ether'. We will observe in which way the Special Theory of Relativity is valid due to the ether, and we will see the conditions under which Special Relativity fails.

Ernst Mach was the forerunner of the Special Theory of Relativity.

Hendrik Antoon Lorentz continuously got nearer to the theory.

Henri Poincaré prepared it and nearly had it.

Albert Einstein discovered it.

Since then, there have been enthusiastic critics, who today would belong to those that would still construct a perpetual-motion machine. It is nevertheless not unusual that physicists are of a different opinion when discussing the experimental details of the twin paradox. Undisputed is Karl Popper's [77] statement that a physical theory can never be verified. It can only be shown to be false. We can never prove that a physical theory is correct. At the most, we can show where a theory falters and where it no longer applies.

H. Günther, *Elementary Approach to Special Relativity*,
https://doi.org/10.1007/978-981-15-3168-2_1

Could it be possible that those searching for a flaw in the Special Theory of Relativity are in the right?—The Special Theory of Relativity is in itself without flaw. It is as flawless as the laws of geometry. Anyone can verify this using a pair of compasses and a ruler. In D. E. LIEBSCHER's [57] pocketbook about Special Relativity, it is described and explicitly shown in all aspects for everybody. Special Relativity can be falsified in physical terms, if its boundaries of validity are passed, e.g. when considering gravity. We can only experiment with gravitational masses. As is well known, the gravitational field cannot be described from the standpoint of Special Relativity.

Using gravitation does not help us comprehend Special Relativity any better. In the ideal experiments concerning Special Relativity, we can keep all relevant masses as small as possible, so that the gravitational effects stay as small as one likes in comparison with the effects of Special Relativity. Using gravity as a means of explaining the twin paradox in the Special Theory of Relativity has led to much confusion. Besides, it would be unfair to make a difficult theory as we see it, even more difficult by considering and including a gravitational field. (It could only work in the opposite direction. We can concentrate on the General Theory of Relativity after the Special Theory of Relativity including its conception of ether has been fully understood. This in turn includes a generalisation of the special relativistic idea of ether. We will however not discuss this problem here).

Faultless application of Special Relativity has become routine for physicists, and instruction of usage has become so elementary that Special Relativity is now instructed in some grammar schools.

All physical applicable statements about the Special Theory of Relativity can be found in EINSTEIN's papers. EINSTEIN specifically asked himself the question:

What is the ether?
He asked this question in his typical subtle and careful manner. The problem of ether really was the origin of all the excitement. Since then, ether has been banned and removed from all explanations concerning Special Relativity. This was done quite simply, because those mentioning ether were in fear of being branded as critics of SRT.

In theoretical physics, Special Relativity stands for the so-called LORENTZ symmetry of a physical system,[1] as well as the LORENTZ symmetry of measuring-rods and clocks when observing a physical system. The exact meaning of this symmetry will be discussed in Chaps. 13–16, where we will illustrate it. The following explanation about the meaning of LORENTZ symmetry will have to suffice for the meanwhile: As long as this symmetry is valid without restrictions, we do not need to take the ether into account. However, the ether becomes physically important if LORENTZ symmetry breaks down. For the fundamental equations of physics, such as MAXWELL's equa-

[1] For the system of electromagnetic fields with spatial distributed electrical charges and currents, this means that two observers moving towards each other with uniform velocity would observe the same MAXWELL equations. Their results are identical. The question about the mathematical form of MAXWELL's equations for 'moving' observers was one of the fundamental problems a physicist faced, before EINSTEIN discovered the Special Theory of Relativity.

tions, we do not need to reckon with a breaking of the LORENTZ symmetry. There are however phenomena in solids that are characteristic for both the LORENTZ symmetry and the breaking of this symmetry. These phenomena therefore supply us with two important models: A mechanical model describing the Special Theory of Relativity, with which we can describe all relativistic effects, as well as a model of the ether, with which we can realise all these effects (see the end of Chap. 6) and Chap. 13 for the nomenclature of the terminus LORENTZ transformation).

The thesis 'The ether does not exist at all', which EINSTEIN expressively warned against in 1920 and which can be found in many textbooks of this subject, can be proved false inside a solid. In EINSTEIN's speech, 'Ether and Relativity' held on the 5 May 1920 in the Reichsuniversität of Leiden, cf. EINSTEIN [12], he stated, 'More careful reflection teaches us, however, that the special theory of relativity does not compel us to deny ether. We may assume the existence of an ether; only we must give up ascribing a definite state of motion to it,... To deny the ether is ultimately to assume that empty space has no physical qualities whatever. The fundamental facts of mechanics do not harmonize with this view'. (Quoted from the translation of EINSTEIN's speech in EINSTEIN [13]).

We will show how this theoretical conclusion does in fact hit the nail on the head. The thesis of the non-existence of the ether is of the same physical quality as the statement 'a solid does not exist', when considering a certain class of mechanical phenomena in a material atomic lattice. We therefore want to ignore and forget such formulations.

We will be able to develop and show an astounding congruence between the two terms describing the motion in our physical spacetime on the one hand and the motion of certain structures in a space lattice on the other hand. This conformity was probably only foreseen by E. MACH, and it is from this point of view that we will try to understand the quotation found in front of the preface. This statement, if taken literally, cannot be upheld or proven correct. Principally, there can be no Special Relativity for sound (see at the end of Chap. 6). We will, however, be able to show how the MACHian vision is applicable, when observing a certain class of motions in a crystal lattice. In Chap. 18, we will come to a conclusion concerning this question. Due to the fact that ERNST MACH's discourses in mechanics had an enormous influence on ALBERT EINSTEIN, it is said that MACH was one of the pioneers of the Special Theory of Relativity. This, however, was a position that MACH himself could not accept, so strong was his opposition in its later years to EINSTEIN's Special Theory of Relativity. Perhaps our discourse will be able to shed some light on MACH's position.

With the quantum theory came the term 'physical vacuum'. This meant, in the light of the new discoveries made by the quantum structure of motion of matter, that the classical vacuum had to be corrected. The classical vacuum is synonymous with the questionable and suspicious term 'ether'. The term physical vacuum has not answered our question about the ether. Ether is just the old term for the 'classical vacuum'. Here, we will not include the quantum theoretical corrections, or the corrections made by the General Theory of Relativity into our considerations. Nevertheless when analysing the ether model resulting from our considerations of solids, we will stumble across some oddities of motion of ‘quasiparticles’ (particles with respect to the lattice as explained in Chap. 22) through the ether. The same oddities of motion of physical particles through three-dimensional space were introduced by quantum mechanics.

Chapter 2
The Ether and the Wave Equation

For the Greeks ether is 'fine air'. Chemically, ether is a group of especially volatile substances. Ether as gas has a characteristic smell. In the field of physics, ether has for a long time produced an anaesthetic effect—similar to the anaesthetic effect of ether during a medical operation.

Ether concerns itself physically with wave propagation. The first thing we associate with the word wave is probably a wave in water. Mathematically, however, these waves are not so elementary as it seems, since these waves are related to the surface of the water. We are, however, interested in the waves that completely exist inside of a medium (inside of the 'space for the waves'). Examples of these types of waves can be found in acoustics in the form of sound waves and in electrodynamics in the form of radio waves, light waves and X-rays. Even though these waves possess extremely different physical properties, they can be described mathematically in the same way. Wave propagation in either a medium or in space, the transmission of waves, their deflection around obstacles and the transmission of signals through space with the help of waves—the fundamental base for all these phenomena are one single equation, J. LE R. D'ALEMBERT's *wave equation*.

The wave equation is without doubt the common denominator that connects all wave phenomena, regardless of their acoustic or electromagnetic nature. We get a completely different picture with regard to its meaning for our physical comprehension of nature, for a modern foundation of electromagnetic theory s. HEHL, F. W. and OBUKHOV, Y. N. [41].

Up to the turn of the century, to be more precise until AD 1905, mechanics was the fundamental base used for the theoretical description of all physical processes. Mechanical wave phenomena were accepted as fundamental facts. Propagation of sound through air or any other gas, fluid or solid medium was widely researched and understood. One was also convinced that electromagnetic processes functioned in exactly the same way. One was convinced that light waves moved the same way

H. Günther, *Elementary Approach to Special Relativity*,
https://doi.org/10.1007/978-981-15-3168-2_2

through an undiscovered mechanical medium called *ether*, as sound waves moved through the air. The last step was 'only' to experimentally discover this mechanical transport medium for electromagnetic waves. This, however, could not be achieved. The experiments reached new levels of cunning, and the explanations for their negative results became more complex and complicated. The climax of these attempts was reached with the decade long experiments made by A. A. MICHELSON. Using an inventive experimental arrangement dating back to J. C. MAXWELL, here one tried to use the interference of light itself to prove the state of motion of the ether. The idea and the principle construction of the MICHESON experiments will be discussed in Chap. 24. A modern SRT test experiment using the MÖSSBAUER effect is described by D. C. CHAMPENEY, G. R. ISAAK and A. M. KHAN [9], cf. also E. LIEBSCHER [57]. We refer also to M. P. HAUGAN's and C. M. WILL's [39] review article on modern SRT test experiments on the occasion of the hundred year jubilee of the publication of A. A. MICHELSON's and E. W. MORLEY's [65] famous paper in the Am. J. Sci. Here, we wish to remark that the first successful measurements were made by MICHELSON in 1881 in a cellar of the Astrophysical Observatory in Potsdam-Babelsberg; please refer to the documentation of U. BLEYER et al. [5].

No less a person than MAXWELL, the inventor of the theory of electromagnetic phenomena, was for the whole of his life very convinced of the existence of a mechanical ether as a carrier of electromagnetic waves. Even the young EINSTEIN was until 1901 of the same conviction; see A. PAIS [71]. However, all attempts failed. The search for the ether was like a never-ending expedition. The explorer was like a lost man in the desert, who would, in the search for escape keep stumbling across his old footprints. The ether was a physicist's nightmare from which they were awakened in 1905 by EINSTEIN's *Special Theory of Relativity*. This historical date finally liberated MAXWELLian electrodynamics once and for all from the domination of mechanics. It in fact became a self-sufficient theory. Moreover, the laws for the propagation of electromagnetic waves now became the key for a new comprehension about space and time. The principles of electrodynamics became the new fundamental of theoretical physics. The field of acoustics was now in comparison suddenly just a storm in a tea cup. The electromagnetic waves now had a higher priority than the acoustic waves. A scapegoat had to be found and made responsible for the wasted years. The scapegoat was the ether. Even though EINSTEIN [12, 13] in 1920 warned the physicists not to start a witch hunt, they denied it not only its mechanical civic rights of a state of motion, and they also declared the ether for non-existent and even went as far as to remove the term ether from the dictionary.

We will now turn our thoughts to those physical properties that all waves possess, independent of their physical nature be they electromagnetic or acoustic.

The transportation of energy is of central importance for physics and engineering. We can store chemical energy in a certain material form, e.g. coal, petrol, a battery, etc. These materials can be transported to the place where we want to extract the energy. Such transport systems like pipelines or oil transports and transportation via rail or via road need to be controlled and if necessary renewed or repaired.

However, energy is transported by waves in a completely different manner. Each wave is loaded with a certain amount of energy which is transported through a medium (or through space) with a characteristic velocity by specific motions of a medium. During this process, no material is transported. The medium (or space) still has the same form, and it is in the same state after the energy in the wave has 'flowed' past. The transportation of energy with the help of waves can be repeated as often as needed without any side effects. This system needs no servicing to keep it in order. The medium (or space) in which the transportation of energy takes place shows no phenomena of wear. We can send radio waves through the 'ether' as often as we want. The space between the sender and the receiver does not show any signs of damage. The same applies to laser rays. The transmission of electric energy through high-voltage lines is also based on the propagation of electromagnetic waves funnelled along a wire in a certain direction. The motion of these waves along the wire causes absolutely no damage to the wire. Take a conversation between two neighbours as another example. One can talk as much and as long as one likes. After the transmission of mechanical energy with the help of sound waves, the medium through which the waves travelled, here air, returns to its original state.[1]

Let us examine and apply the wave equation on an especially simple system. We will take a straight elastic rod. Here, the simplification is that there is only one special direction, the direction of the rod itself. The propagation of waves through the rod is part of the field of acoustics. If we generate an impulse at one end of the rod (e.g. we hit one of the ends with a hammer), we produce an elastic deformation that moves along the rod with the speed $c = \sqrt{E/\rho}$, the sound velocity. The deformation energy is transported to the other end of the rod. Here, ρ is the mass density of the rod, and the material parameter E is the modulus of elasticity. The constant c is the propagation velocity of an acoustic signal. MORSE code signals that are tapped on the left end of the rod with the length L arrive at the right end of the rod after the time $t = L/c$. The rod itself does not change permanently in form or appearance, unless the rod was bent, deformed or broken by too violent blows of the hammer.

The elastic deflection s out of the mass particle's position of equilibrium at the location x at time t is the function $s = s(x,t)$ for every rod's state of oscillation. In the field of acoustics, one can show that the wave equation of D'ALEMBERT applies

$$\frac{\partial^2}{\partial x^2}s(x,t) - \frac{1}{c^2}\frac{\partial^2}{\partial t^2}s(x,t) = 0 \qquad \text{D'ALEMBERT's wave equation} \qquad (1)$$

with[2]

$$c = \sqrt{\frac{E}{\rho}}\,. \qquad \text{Signal velocity} \quad (2)$$

[1] To be exact, air absorbs some of the sound energy and rises in temperature. Thermodynamic processes play no role in our considerations. We will therefore ignore all friction and scattering phenomena that change the ordered energy of the waves into disordered heat energy.

[2] Notice that we understand the critical velocity by the term 'signal velocity'.

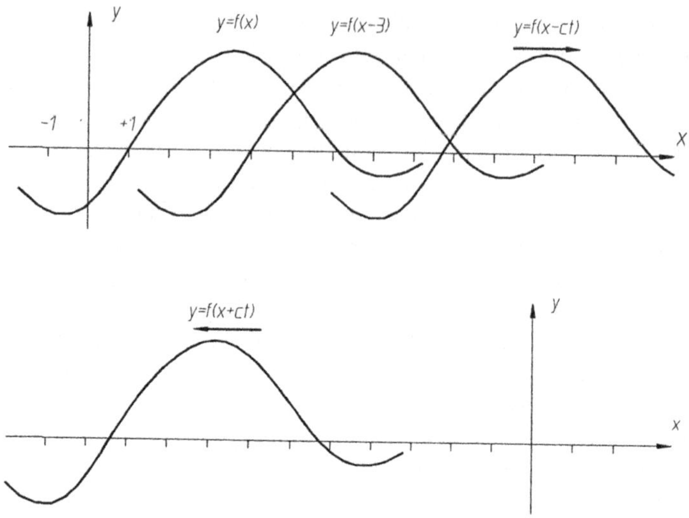

Fig. 2.1 Displacement of a graph of a function $y = f(x)$ by an additive constant. When $a > 0$, the function $f(x - a)$ is in comparison with the function $y = f(x)$ displaced by the value of a to the right. With a constant velocity $c > 0$ after time t, the function $f(x - c\,t)$ is in comparison with the function $y = f(x)$ displaced by the value $a = c\,t$ to the right. For the function $f(x - c\,t)$ with the velocity c, we get the graph of the function moving to the right with velocity c. The graph of the function $f(x + c\,t)$ moves to the left, as indicated by the arrows

The origin of this wave equation will be examined in detail in the next chapter.

The general form of the solutions of (1) is

$$\left.\begin{array}{l} s(x,t) = f(x - c\,t) \\ \text{or} \\ s(x,t) = f(x + c\,t)\,. \end{array}\right\} \qquad \text{General solution of the wave equation} \qquad (3)$$

Here, $f = f(\xi)$ is an arbitrary, twice differentiable function. With $\xi = x - c\,t$ is

$$\frac{\partial f}{\partial x} = \frac{df}{d\xi} \cdot \frac{\partial \xi}{\partial x} = f'(\xi)\,, \quad \frac{\partial f}{\partial t} = \frac{df}{d\xi} \cdot \frac{\partial \xi}{\partial t} = f'(\xi)(-c) \quad \text{hence}$$

$$\frac{\partial^2 f}{\partial x^2} = f''(\xi)\,, \qquad \frac{\partial^2 f}{\partial t^2} = f''(\xi)(c^2)\,. \quad \text{The same applies to } \xi = x + c\,t.$$

For the first case, we get waves moving to the right; in the second case the waves move to the left. The displacement of a left or right moving graph of a function caused by an additive constant in the argument of a function $y = f(x)$ will be made clear in Fig. 2.1.

We will firstly examine an infinite medium. Additional conditions that have to fulfil the solutions of a finite rod have not yet been considered here, e.g. a rod's fixed ends remain motionless, whereas a rod braced in its middle vibrates violently at both of its ends, see Chap. 4.

Harmonic waves are exceedingly important. They are special solutions of the wave equation.

$$\left.\begin{array}{l} s(x,t) = A \cdot \sin(k\,x - \omega\,t) \\ \text{and} \\ s(x,t) = A \cdot \sin(k\,x + \omega\,t)\,. \end{array}\right\} \qquad \text{Harmonic waves} \quad (4)$$

For a constant time t, the elastic deflection s of the mass particle describes a sine function in space with the wavelength $\lambda = 2\pi/k$. The quantity k is the wave number (respectively wave vector for the case of spatial propagation). At a fixed point x, the masses oscillate according to a sine function of time with an oscillation period $T = 1/\nu = 2\pi/\omega$. Frequency multiplied by 2π is called the angular velocity $\omega = 2\pi\,\nu$ of an oscillation. One can register these as pure notes in the field of acoustics (even though they sound terrible).

This function (4) is only a solution of the wave equation if the above condition $s(x,t) = f(x - c\,t)$ is fulfilled,

$$s(x,t) = f(x - c\,t) = A \cdot \sin\left[k\left(x - \frac{\omega}{k}\,t\right)\right] = A\,\sin[k\,(x - c\,t)]\ .$$

The wavelength λ, frequency ν and velocity of propagation c of a wave are connected by the relation $c = \lambda\,\nu = \omega/k$ and or $\omega = c\,k$. However, the propagation velocity c for waves of arbitrary frequencies does not in any case need to have one and the same value, as in our example with the elastic rod. The general dependence of a wavelength on the frequency is called the dispersion relation. We will come back to this in Chap. 4 when discussing the linear chain. If the angular velocity ω is proportional to the wave number k with the unchangeable constant c for the velocity of propagation, one states that the medium (or space) is free of dispersion,

$$c = \lambda\,\nu = \frac{\omega}{k}\,. \qquad \text{Dispersion free medium} \quad (5)$$

The wave's velocity of propagation c is independent of its frequency. This has been fulfilled in near approximation for the field of acoustics and is especially valid for electromagnetic waves in a vacuum. When light travels through a medium, e.g. light through glass or water, it gets dispersed or broken. This can be observed in a spectrum of a glass prism, or in drops of water in a rainbow. The front velocity of an electromagnetic wave in a medium is dependent on its frequency (its colour). In this case, more complicated equations are at work, but we will not take them into consideration.

Taking the dispersion relation $\omega = c\,k$, we can describe the general solution of the wave equation as a superposition of harmonic waves with a variable wave number k,

$$s(x,t) = \sum_k [A(k)\,\sin(k(x - c\,t) + B(k)\,\cos(k(x - c\,t))]\,. \qquad \text{General solution} \quad (6)$$

In other words, the general solution is a superposition of waves with variable wavelengths $\lambda = 2\pi/k$, or waves with variable frequencies $\nu = c/\lambda$. Using suitable superpositions of these waves, we can fulfil the boundary conditions for a finite medium which we will need in Chap. 4.

The basis of all these properties is the aforementioned wave equation which is equally valid for acoustic and electromagnetic processes. For electromagnetic processes, we have to substitute the speed of light c_L in a vacuum for the sound velocity c. We will now have to take a remarkable fact into consideration when dealing with the speed of light: An observer receives a light signal from a light source, which is at rest with respect to him. The light signal travels across the distance between its source and the observer with the speed of light in a vacuum c_L. If there is another observer moving towards the source of light with an arbitrary uniform velocity v and we ask him how fast the light signal moves towards him, we get the following answer, 'the light signal is moving towards me with the speed of light in a vacuum $c_L = 299792458$ m/s'.[3] We receive exactly the same answer from the observer if he moves with velocity v away from the light source. The MICHELSON experiments have removed the last question marks about the correctness of this fact, a fact that is for us astounding, because we expect the quite another result.

Exactly this happens when we observe sound waves. A source of sound S can transmit MORSE code signals with the sound velocity c through air. Assume at first that the source of sound, the medium air and an observer are at rest with respect to each other. Then, the observer measures this sound velocity c for the propagation of the signals. Now an observer moving through the air and towards this source with the velocity v measures the speed of sound coming from the sound source towards him as $c + v$. If he moves away from the source of the sound waves, he measures the speed as $c - v$. This is easily explained for the propagation of sound waves in the medium air which is shown in Fig. 2.2. It also makes absolutely no difference through which transport medium these sound waves travel.[4]

Let us return to our light source. It can send MORSE code signals through space with the speed of light c_L. The question is, how do these light waves reach the observer? Is there a mechanical transport medium, an ether maybe, in which these light waves spread in a similar way as sound through the air? If this is the case, then an observer moving through this ether towards the light source with velocity v would have to measure the speed of the light signal as $c_L + v$ and correspondingly, if moving through the ether away from the source with the velocity v he would have to measure $c_L - v$. In fact, however, in both cases independent of his velocity towards or away from the source, he measures one and the same signal velocity c_L.

This is the problem behind the term ether: If we say that light waves propagate through a mechanical ether, then it is impossible to register the slightest motion or presence of this ether. There is absolutely nothing of the ether to be detected.

---

[3]The inaccuracy of this answer is less than 1 m/s = 3, 6 km/h. In comparison, the speed of sound in air is $c_{air} = 331$ m/s, and the speed of sound through aluminium is $c_{Al} = 5090$ m/s.

[4]Notice however that this result takes the so-called GALILEI transformation as granted, cf. Chaps. 13 and 15 as well as the detailed discussion concerning the definition of relative velocities in Chap. 17.

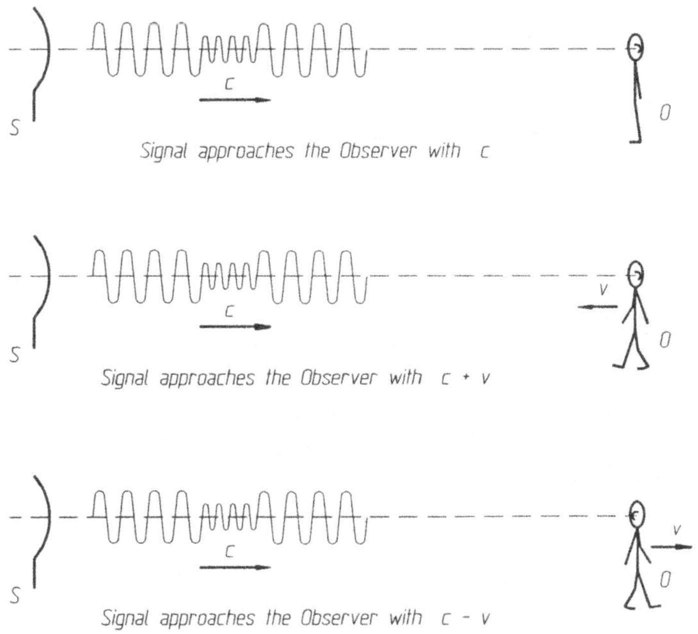

Fig. 2.2 Comparison between the signal velocity, sound waves e.g., that an observer measures when he moves through the transport medium towards or away from the source of the signal with the velocity of either $+v$ or $-v$. Here, it is assumed that the source is at rest with respect to the transport medium, air e.g

MICHELSON's experiments were so detailed and precise that we can safely state that the failure of these experiments was not based on inaccurate measurements. We will, however, explicitly not rush to the hasty conclusion that there is no such thing as an ether.

The fact that these experiments and their enormous importance for the comprehension of light propagation played no key role in EINSTEIN's considerations is surprisingly remarkable. EINSTEIN was inspired by his symmetry principle and his principle of relativity in his search for a solution of the electromagnetic wave problem. Here, we wish to quote this principle. We will deal with its contents in detail later on. In his famous paper, A. EINSTEIN [14] postulated[5]:

'The laws by which the states of physical systems undergo change are not affected, whether these changes of state be referred to the one or the other of two systems of coordinates in translatory motion'. (Quoted from the translation of EINSTEIN's paper in EINSTEIN [15]).

We come across and comprehend this axiom in Chap. 15.

---

[5]Here EINSTEIN uses the term 'coordinate system' instead of the term used in later years, that we also use, 'reference system' or 'frame'.

The result of applying his principle of relativity on MAXWELL's electrodynamics was expressively formulated by EINSTEIN as the 'principle of the constancy of the velocity of light', cf. EINSTEIN [14, 15]:

'Any ray of light moves in the "stationary" system of coordinates with a determined velocity c whether the ray be emitted by a stationary or a moving body. Hence

$$\text{velocity} = \frac{\text{light path}}{\text{time interval}}$$

where time interval is to be taken in the sense of the definition in Chap. 1'.

From this, EINSTEIN deduced the relativistic effects that we will describe in the next chapter. The velocity c is our speed of light c_L, and the 'time interval' is the time display of synchronised clocks (see the next chapter). In Chap. 12, the principle of the universal constancy of the critical signal velocity will be explained and derived using our new axiomatic approach.

Chapter 3
The Physical Elements of the Special Theory of Relativity

In 1905 EINSTEIN [14, 15] in his paper, 'On the electrodynamics of moving bodies', managed to explain all those curious effects that were connected to the propagation of light. This paper was the foundation of EINSTEIN's Special Theory of Relativity, which itself was then the basis for the following theoretical physics. Starting out from a single basic postulate, the EINSTEINian principle of relativity, all those curious effects that were observed in the propagation of light were deducible. This paper therefore implicitly contains the answer about the ether without however explicitly discussing it, cf. EINSTEIN [14, 15]. Here, we will now sketch out using a purely empirical approach the relatively simple contents of Special Relativity.[1]

The negative outcome of all the ether experiments should for the present be formulated with utmost caution.

> There is no instrument of measurement capable of responding to the ether.

This statement is reason enough for us to make our instruments of measurement, our measuring-rods and clocks themselves the objects of our observation. The questions concerning the reference system for our observations which defines the framework will be discussed in the next chapter. The following statement will have to suffice for the moment: Our reference system is the laboratory in which the laws of NEWTONian mechanics are valid. This is a so-called inertial system.

Our measurements now lead us into making two exciting discoveries. All clocks and measuring-rods a physicist can build show the following properties:

---

[1]Today, we could actually follow this path. At the beginning of the century, the necessary experiments could not be carried out with the required precision. In those days, one was forced on to the thin ice of conclusions deduced from a bold theory. In 1905, EINSTEIN achieved this with his paper. One should not however forget that other outstanding theoreticians also tried the same thing, but failed or could not bring their thoughts to a conclusive end. E. MACH, H. A. LORENTZ and H. POINCARÉ still stand as the pioneers of the Special Theory of Relativity. We refer here to the historical depiction of all the facts and developments in A. PAIS' [71] book.

H. Günther, *Elementary Approach to Special Relativity*,
https://doi.org/10.1007/978-981-15-3168-2_3

1. We measure using a measuring-rod of a certain length L_o (e.g. one metre) a distance $\mathbf{X}$ by laying down the measuring-rod as many times as necessary and recording $\mathbf{X} = x\, L_o$. In other words, the measuring-rod L_o fits x times in the distance $\mathbf{X}$, and x is the coefficient of measure for the distance $\mathbf{X}$ referred to the unit of measure L_o. If we now move this measuring-rod at a constant velocity v ms^{-1} along this distance, we observe that the moving measuring-rod has the length

$$L' = L_o \sqrt{1 - \frac{v^2}{c_L^2}} \,. \qquad \text{Length contraction of the moving measuring-rod} \tag{7}$$

The measuring-rod in motion has shortened.

The moving measuring-rod fits x' times into the same distance $\mathbf{X}$, with $x' = x/\sqrt{1 - v^2/c_L^2}$ so that $\mathbf{X} = x'\, L'$. This effect is the so-called LORENTZ contraction or simply length contraction and has today been confirmed without doubt by the countless SRT experiments. It makes no difference if we use the Parisian standard metre or the wavelength λ of the yellow sodium line. If these objects move in reference to our laboratory, we observe this length contraction.

2. We examine a set of identically constructed precision clocks positioned one metre from each other along a distance $\mathbf{X}$. We now have to synchronise these clocks, i.e. start them 'at the same time'. To do this, we send out a signal with a precise velocity v_o ms^{-1} from the first clock U_o and set this clock at 0 s.[2] When the signal reaches the second clock U_1 one metre away, this clock is started at the setting $1/v_o$ s. When the signal reaches the clock U_2 two metres away from the first clock, this clock is started at the setting $2/v_o$ s, etc., until all clocks are ticking synchronically. We now let a further clock, U_v, of the same construction glide past all the other clocks with a constant velocity v ms^{-1}. At the first clock U_o, the moving clock is set to the same setting as U_o, i.e. 0. This starting parameter can be freely chosen. We will now observe that when the moving clock reaches U_1, it has a setting of t'. The static clock U_1 however has the setting t, where

$$t' = t \sqrt{1 - \frac{v^2}{c_L^2}} \,. \qquad \text{Time dilatation of the moving clock} \tag{8}$$

The moving clock goes behind.

This effect is called time dilatation and has also been exactly proven by many experiments. It does not matter if we examine an old 'Nuremburger Ei' or a modern instrument of time measurement, e.g. a caesium atomic clock. We will always observe this time dilatation if the clocks move.

[2] Traditionally since EINSTEIN, one uses the speed of light in a vacuum c_L for this synchronisation procedure (for this ideal experiment). Notice, however, that we only need an arbitrary velocity for this which however is precisely determined.

We can also formulate this effect of time dilatation leaning towards the formulation of the length contraction and in foresight of our upcoming discussion on mechanics as follows. The unit of measure for time measurement is a certain period of oscillation which has the value T_o for the static clocks. For the time **T** that the moving clock U_v needs to move from U_o to U_1, we measure using the static clocks $\mathbf{T} = t\ T_o$. In other words, t is the coefficient of measure of time **T** referred to the unit of measure T_o; t is the number of oscillations. The moving clock oscillates slower. Its unit of measure, the oscillation period T', is in comparison with T_o according to $T' = T_o/\sqrt{(1 - v^2/c_L^2}$ longer, and it is stretched. This is the reason for the term 'time dilatation'. For the time **T** that the moving clock U_v needs to move from U_o to U_1, we measure using the moving clock U_v itself $\mathbf{T} = t'\ T'$. Here, $t' = t\ \sqrt{1 - v^2/c_L^2}$ is the above coefficient of measure of time **T** referred to the unit of measure T'; t' is the number of oscillations of the moving clock U_v.[3]

It is noticeable that all these formulas, even those used for mechanically built measuring-rods and clocks do in fact include the speed of light. What is even more striking is that it is not possible to build any clocks or measuring-rods without it. In Chap. 11, we will have to return to this point. We will observe in the examined phenomena that the above described the behaviour of measuring-rods and clocks is characteristic of a 'participation of an ether'. In the light of this, we will try to understand EINSTEIN's [12, 13] statement 'Only we must be on our guard against ascribing a state of motion to the ether'. This means that the physical properties of our measuring-rods and clocks are in respect to the ether of such a kind that they are not able to respond to the ether.

An important fact has yet to be mentioned. The question 'why' concerning time dilatation and length contraction is not specifically asked in the Special Theory of Relativity. It is only shown that if we measure one and the same velocity for light, then we have to conclude that in principle all measuring-rods and clocks show this behaviour. This does not mean, however, that the question 'why' about these effects is forbidden by the Special Theory of Relativity. However, it is just that all traditional attempts of explanation are excluded by the particular axiomatic structure of this theory. In EINSTEIN's axiomatic approach, all problems of SRT can be reduced to the universal constancy of the speed of light. This is the axiomatic base of the theory. A more simple explanation than the reduction to the axioms does not exist. This is the problem. Is there any point in trying to discover a 'mechanism' that could clearly explain these changes for moving measuring-rods and clocks? We will come back to this question if we succeeded in finding out a real existing physical model, our 'miniatur version' of SRT.

The first direct observation of time dilatation of a 'clock' was achieved in 1938/39, as described in the papers of H. J. IVES [44, 45] and G. J. STILLVELL and G. OTTING's

---

[3]Here, we will inform you of our notation which we will carry on throughout this book. All statements about space and time described using moving measuring-rods and clocks are represented by primed symbols. If measuring-rods and clocks of different velocities are involved, then tildes or roofs are used; see Chap. 17.

[70] paper.[4] We will return to this at the end of Chap. 18. EINSTEIN spoke of this experiment as the 'experimentum crucis' of the Special Theory of Relativity. In comparison, the negative results of the many refined MICHELSON experiments carried out with much effort played only a secondary role in EINSTEIN's conclusions. The results of these experiments 'only' showed that there was no way of measuring an ether wind. The supposed discovery of an ether wind in April of 1921 by D. C. MILLER in the mount Wilson observatory leads to one of EINSTEIN's most famous quotations, 'Subtle is the Lord, but malicious He is not'. He later added the wonderful statement, 'Nature hides her secret because of her essential loftiness, but not by means of ruse'. Here, I once again refer to A. PAIS [71]' book which contains a detailed description of the historic facts concerning the Special Theory of Relativity. The consequence derived from EINSTEIN's Special Theory of Relativity that all our clocks really do behave as strangely as stated by Special Relativity and that this can be measured using our clocks and measuring-rods, which was proven correct by the experiments in 1938/39. The result of these experiments proved that the dilatation of time, EINSTEIN's most astounding statement was correct. There was no room for another interpretation.

For the mathematically interested reader, we recommend EINSTEIN's [14, 15] original paper that even today has not lost any of its beauty and clearness.

Up to now, we have described the so-called kinematic effects of Special Relativity. These effects are at the centre of our interest, and we will concern ourselves with them in Chaps. 9–21. The Special Theory of Relativity makes another spectacular statement. Only three months after his famous paper Einstein [14], in a further paper EINSTEIN [16] came to the conclusion of the inertia of energy, cf. also the translation in EINSTEIN [15]. The results lead to the most public and well-known equation science had ever produced,

$$E = m\,c_L^2. \qquad \text{Energy mass equivalence} \qquad (9)$$

The reason for this interest is without doubt the range of the consequences this equation carried. In certain processes of the nuclear fission, enormous unimaginable amounts of energy are released, as in the uncontrolled explosion of an atomic bomb, or controlled in a nuclear power plant. Another example is nuclear fusion, a process that causes the production of energy in the sun and which can until now only be realised on the earth in the uncontrolled explosion of a hydrogen bomb. EINSTEIN's discovery forces us to unify two properties of masses that were held to be independent of each other: the inertia of a mass and the energy of a mass. Superficial descriptions have sometimes given an incorrect picture of EINSTEIN's energy mass equivalence. Energy does not change into mass or vice versa. Any energy E has an inertial mass m, so that $m = E/c_L^2$; every mass m, the mass of every atom, every single bit of matter is a special form of 'concentrated' energy E, so that $E = m\,c_L^2$. This energy is unbelievably enormous because of the enormous numerical value of the speed

[4] It was the confirmation of the transversal DOPPLER effect. In Chap. 18, we will see that this effect is an immediate expression for time dilatation. Here, the examined clock is an excited atom. Its monochromatic light delivers the frequency standard of the clock.

of light. The consequence of EINSTEIN's equivalence of mass and energy is that this energy, as every other sort of energy, can principally be transformed into any other form of energy, e.g. into heat. The sum of the energy, as well as sum of the masses remain constant in any process. It is however a completely different thing to have one milligram of matter lying on a desk in front of us, or to have the energy of this one milligram changed into heat energy in our environment. We will give an example to illustrate this:

We observe a pot containing 200 000 l of water with a temperature of 0°C, and we then add one milligram of matter to the pot. The pot now contains a total mass of 200 000, 000 001 kg, with a total energy of $E = mc_L^2 = 200\,000,000\,001\,\text{kg} \cdot 9 \cdot 10^{16}\,\text{m}^2\text{s}^{-2} = 200\,000,000\,001 \cdot 9 \cdot 10^{16}$ J.

We will now completely transform the energy of the added milligram into heat energy. Using EINSTEIN's equation, we receive the following heat quantity $E = 10^{-6}\,\text{kg} \cdot 9 \cdot 10^{16}\,\text{m}^2\text{s}^{-2} = 9 \cdot 10^{10}\,\text{Nm} = 9 \cdot 10^{10}$ J, in other words $E = 9 \cdot 10^{10} \cdot 2,4 \cdot 10^{-4}\,\text{kcal} = 2 \cdot 10^7\,\text{kcal} = 200\,000 \cdot 100\,\text{kcal}$.

This amount of energy would be enough to raise the temperature of the 200 000 l of water from 0 °C to its boiling point. The energy needed to raise the temperature of one litre of water by one degree is calculated as one kcal. (We ignore the fact that the energy needed to raise the temperature from 1 °C to 2 °C and from 99 °C to 100 °C is not quite the same amount). The total sum of the energies and masses of the 1 milligram, and the 200 000 l stay the same. Special Relativity teaches us that the mass (the inertia) of the 200 000 l at 100 °C is approximately 1 milligram larger than the mass of 200 000 l (with the same number of molecules) at 0 °C.

Every energy possesses an inertial mass; every mass is a carrier of energy.

It is an immediate expression of the equivalence of mass and energy that a moving mass has to be larger than a static mass. A moving mass possesses not only the energy of a static mass, but also the kinetic energy of its motion. Let us now compare the inertial mass m_o of a static solid, e.g. our milligram on the desk with the inertial mass m that the same solid possesses whilst moving with respect to us with the velocity v. During the process of passing by with the velocity v, we notice that the inertia of our moving mass has increased, i.e. its resistance against acceleration has increased. According to Special Relativity, the following equation is valid

$$m = \frac{m_o}{\sqrt{1 - v^2/c_L^2}}. \tag{10}$$

The moving mass increases.

It follows that an object with a static mass larger than zero can never be accelerated to the speed of light, even if we increase the forces beyond measure. The speed of light is the privilege of the speed of light. A particle moving at the speed of light, e.g. a photon or a neutrino will for an arbitrary observer always retain this speed and will move with the speed of light throughout all eternity.

These phenomena connected to the inertia of masses constitute the dynamic effects of the Special Theory of Relativity. We will concentrate on this in Chaps. 22–23.

An exciting fact of Special Relativity is that we can observe all of these effects in any inertial reference system we choose. None of these effects allows us to give preference to any one of these frames. This is also valid for the phenomena of light propagation, for the propagation of electromagnetic waves. All these phenomena are defined with respect to an inertial system, and all inertial systems are on equal footing. This principle of relativity has stood next to the cradle of physics during the process of evolving into an exact science and was named after GALILEI. I. NEWTON founded his mechanics on this principle. This will be explained in the following chapter in detail. Notice that Eq. (10) is the only relativistic correction to NEWTONian mechanics.

In Chap. 2, we tried to explain the propagation of electromagnetic waves with the help of a mechanical medium. This ended in hopeless conflict with the field of mechanics, where the acoustic phenomena show us the state of motion of its carrier medium (see also Chap. 18). EINSTEIN, however, stuck to his principle of relativity. This relativity was the driving, the motivating force in the discovery of his theory, hence the name, the Theory of Relativity. The different possible axiomatic approaches to the Special Theory of Relativity will be discussed in Chap. 16. For an extended elaboration of the aspects of Special Relativity s. H. GÜNTHER [38] AND V. MÜLLER, the reader also finds a lot of references.

We will now search for in the field of mechanics, the origin of the wave equation which plays a key role in the Special Theory of Relativity.

Chapter 4
Where Does the Wave Equation Come From?

Motion is always motion in reference to something, i.e. a reference system (reference frame) where the observers sit, watch and measure. A mass m moves relative to the reference system of an observer. The observer is per definition considered as resting. If we take two observers moving against each other in two different reference systems, each observer will come to a different conclusion about the motion of the mass m. I sit at my desk on which a mass m is lying. A colleague swivels on a revolving chair next to me. He sees that the mass m, the desk and I are orbiting around him; see Fig. 4.1.

Are both frames, myself sitting at a desk and my colleague on his revolving chair equal to one another? Well, after a few minutes my colleague should start to get dizzy, whereas I do not. If I were more sensitive, I should also get dizzy, because of the rotation of the earth. If my perceptiveness was far greater, the orbiting of the earth around the sun and perhaps the accelerated motion of the complete solar system could cause me dizziness. Special systems of reference are those systems that principally do not cause 'dizziness'. These special systems are called *inertial systems* or *inertial frames*.

What is meant is that the motion of an object in such systems is defined by its inertia, its persistence to remain motionless or to keep moving at a uniform velocity in the same direction as long as the object is not influenced by any physical force. We will assume for the following discussion that we are inside one of these inertial frames. My desk is an appropriate approximation for this purpose. A colleague sitting in a train moving past me at a constant speed is then also inside an inertial frame.

H. Günther, *Elementary Approach to Special Relativity*,
https://doi.org/10.1007/978-981-15-3168-2_4

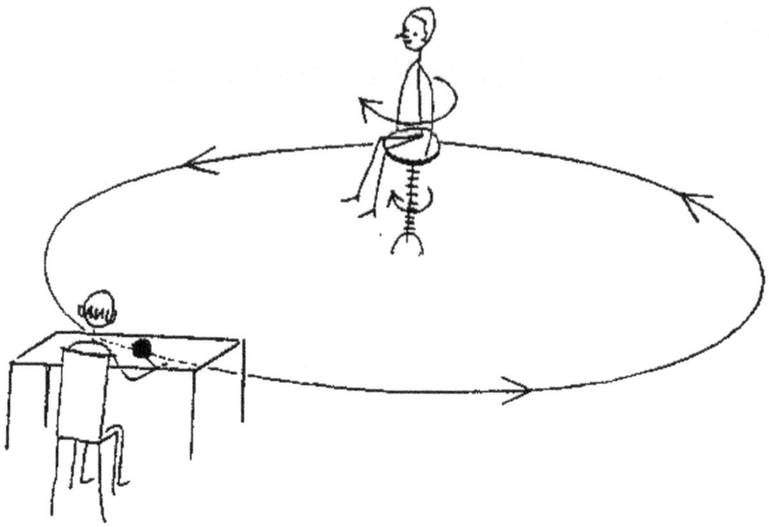

Fig. 4.1 Observer at a desk and an observer sitting on a swivelling chair. The latter observes that the mass resting on the desk is orbiting him

This changes if the train slows down or goes through a curve—this could cause us to feel dizzy. Any accelerated motion causes us to leave our inertial frame.[1]

Inertial systems are especially characterised by their mutual relative motion with a constant velocity and their motion relative to all other systems of reference. They underlie the GALILEIan *principle of relativity*: 'Observers in two different inertial systems notice that the same laws of physics apply'. The total sum of all possible physical processes is identical to all inertial systems. Basically, the GALILEIan principle of relativity was only intended for the laws of mechanics and is thus called *The First* NEWTON*ian Axiom*: It is impossible to determine our special inertial frame just from the motion of an object.[2]

[1]Here one could raise an objection. My well-being includes the fact that the earth has its usual gravitational effect on me. A colleague, who is far away from the earth in a rocket travelling with a constant velocity relative to my desk, also finds himself in an inertial frame. However, he is not affected by the earth's gravitation. If his rocket is accelerated by $9,81\ \mathrm{m/s^2}$, he experiences the gravitation he is used to. The difference is, however, that he is no longer in an inertial frame, whereas I am. Using this simple example, we can see that as soon as gravitational effects (the universal attraction between masses) are included, everything becomes far more complicated. We will strictly ignore all effects of gravitation.

[2]One can however experimentally determine if I am in an inertial frame or not. The organs of balance capable of measuring my state of motion are found in my ear. If I do not find myself in an inertial frame, the organs detect this and send signals to my brain telling me to become dizzy. The sensitivity of the organs in the ear is insufficient to be able to measure the daily rotation of the earth along its own axis. We can however measure this rotation using a FOUCAULTian pendulum, see e.g. H. GOLDSTEIN [29].

We assume, due to our experience, that inertial frames actually exist. The postulation of relativity for all physical processes already introduced in the GALILEIan principle was expressed in the view of mechanics, because it was thought that principally all physical processes and phenomena were of mechanical origin. EINSTEIN's analysis introduced the new idea that relativity—now independent from the field of mechanics—could also be applied for electromagnetic phenomena. The fundamental physical consequences were summarised in the previous chapter.[3]

We are now interested in the general law for the motion of a mass m in an arbitrary inertial system under the action of a force. This law was formulated in 1687 by I. NEWTON and is called *The Second* NEWTON*ian Axiom*. Here, for the momentum $\mathbf{p} = m \cdot \mathbf{v}$ of a mass m moving at the velocity $\mathbf{v}$ under the influence of a force F, the following is valid,

$$\frac{d}{dt}\mathbf{p} = \mathbf{F}. \tag{11}$$

It is as simple as that. The time derivative of the momentum $\mathbf{p}$ is equal to the acting force $\mathbf{F}$. If we mark the time derivative of a physical quantity with a dot, we show a simplified version of (11)

$$\dot{\mathbf{p}} = \mathbf{F}. \tag{11a}$$

If several forces are involved, e.g. $\mathbf{F}_a$, $a = 1, 2, \ldots, n$, the acting force $\mathbf{F}$ in (11) is the resulting vectorial sum,

$$\mathbf{F} = \sum_{\mathbf{a}=1}^{\mathbf{n}} \mathbf{F_a}. \tag{12}$$

Both equations are valid for all forces, independent of their origin. It is of no difference if the forces involved are elegant electric and magnetic forces, or simple forces applied when expanding a spring, pure muscle force or the frictional force. According to Eq. (12), all forces are added up equally by the process of vector addition that then results in the total force (11).

The fundamental mathematical property of arbitrary forces and vectorial addition was stressed by NEWTON as a *corollarium* in his mechanics. Occasionally, this is also called *The Fourth* NEWTON*ian Axiom*; see Fig. 4.2. The sometimes very different properties of forces play no role for the validity of the law of motion (11). NEWTON's mechanics makes no statement about the physical mode of action of forces, with one important exception. There is one general valid mathematical interrelationship for all so-called interaction forces. These interaction forces are forces with which two arbitrary masses m_1 and m_2 can interact. This property is defined as *The Third*

[3] In a purely mathematical view, the EINSTEINian relativity as we will see in Chaps. 13–15 generates the so-called LORENTZ transformation. A limiting case of the LORENTZ transformation is the GALILEI transformation in NEWTONian mechanics, cf. also Chap. 14. These transformations compare the data of space and time measurements of an observed event for two different inertial systems.

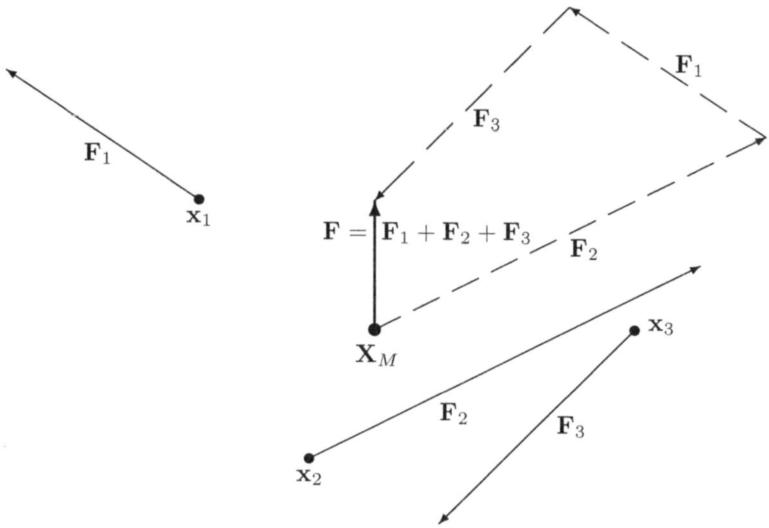

Fig. 4.2 Forces $\mathbf{F}_1$, $\mathbf{F}_2$ and $\mathbf{F}_3$ acting on the masses m_1 at $\mathbf{x}_1$, m_2 at $\mathbf{x}_2$ and m_3 at $\mathbf{x}_3$ respectively can be added up vectorially to a total force $\mathbf{F}$, which then acts at the centre of mass $\mathbf{X}_M$. This centre of mass then moves according to Eq. (11), $\frac{d}{dt}(M\frac{d\mathbf{X}_M}{dt}) = \mathbf{F}$, where $M = m_1 + m_2 + m_3$ and $M\,\mathbf{X}_M = m_1\,\mathbf{x}_1 + m_2\,\mathbf{x}_2 + m_3\,\mathbf{x}_3$

NEWTON*ian Axiom* and has also become known under the term 'actio = reactio'. To be more precise, if the mass m_1 acts on the mass m_2 with the force $\mathbf{F}$, then the mass m_2 acts on the mass m_1 with the force $-\mathbf{F}$. The interaction forces that influence each other are of the same size but have the opposite direction. We will concern ourselves with this axiom later on.

We will now consider a mass m that can only move along the x-axis where it possesses a position of equilibrium x_o and where no total resulting force acts on it. A small deflection $s = x - x_o$ out of the position of equilibrium x_o should result in a force $F = -D\,s$ with the so-called force constant $D > 0$, which drives the mass back into the position x_o. We also take as granted that the mass m stays constant throughout the whole state of motion, $m =$ const. With the velocity $v = \frac{d}{dt}x = \frac{d}{dt}(x_o + s) = \dot{s}$, $p = \frac{d}{dt}(m\,x) = m\,\dot{s}$ is then valid for the momentum p. According to NEWTON's law (11), the motion of the mass m along the x-axis follows the law of harmonic oscillation,

$$m\,\ddot{s} = -D\,s. \tag{13}$$

Out of the general solution of the equation of oscillation (13),

$$s = s_o\,\cos(\omega\,t + \phi), \tag{14}$$

with both constants of integration s_o and ϕ we receive for $\phi = -1/2\pi$ the special solution $s = s_o \sin(\omega t)$. Because of $\dot{s} = s_o\omega \cos(\omega t)$, the mass m reaches maximum velocity $v_o = s_o \omega$ at time $t = 0$. For $\phi = 0$, so that $s = s_o \cos(\omega t)$ it possesses at $t = 0$ the maximum deflection out of the position of equilibrium x_o. The mass oscillates harmonically with the angular velocity $\omega = 2\pi\nu$ around the position of equilibrium (oscillation period $T = 1/\nu$). The function (14) is a solution for (13) if and only if

$$\omega = \sqrt{\frac{D}{m}}. \tag{15}$$

The total energy U of the oscillation is equal to the maximum value of the kinetic energy $E_{kin} = \frac{1}{2}m\,\dot{s}^2$,

$$U = \max(E_{kin}) = \frac{m}{2}\,v_o^2 = \frac{m}{2}\,s_o^2\,\omega^2 = \frac{1}{2}\,s_o^2\,D. \tag{16}$$

We will now consider the harmonic oscillation (14). We count the number ν of times the oscillation passes through the position of equilibrium per second, i.e. determine the frequency ν. We can also measure the kinetic energy which is equal to the total energy U when passing through the position of equilibrium. If for example the oscillation is brought to a standstill at its position of equilibrium by friction, we receive U out of the heat that was set free during the process.

We now observe a remarkable property. As long as we can only measure the angular velocity $\omega = 2\pi\nu = \sqrt{D/m}$ and the total energy $U = \frac{1}{2}\,m\,s_o^2\,D$, it is impossible to determine separately the oscillating mass m and the force constant D. We cannot decide if a mass of one tonne $m = 1000$ kg under the influence of the force constant $D = 10^7$ Nm^{-1} (NEWTON per metre), or if a tiny milligram with the mass $m = 10^{-6}$ kg oscillates under the influence of the force constant $D = 10^{-2}$ Nm^{-1}. In both cases, we get the same ω,

$$\omega = \sqrt{\frac{D}{m}} = \sqrt{\frac{10^7\ \text{N}}{10^3}\frac{1}{\text{m kg}}} = \sqrt{\frac{10^{-2}\ \text{N}}{10^{-6}}\frac{1}{\text{m kg}}} = 10^2\sqrt{\frac{\text{kg m}}{\text{m s}^2\ \text{kg}}} = 10^2\ \text{s}^{-1},$$

in other words,

$$\nu = \frac{\omega}{2\pi} = 15,9\ \text{Hz}.$$

Even the measurement of the total energy cannot change this, because U also contains the amplitude s_o of an oscillation as a constant of integration. We can therefore say:

> Both characteristics of a harmonic oscillation, its angular velocity ω and its total energy U, do not allow us to draw a conclusion about the mechanical state of the oscillating system.

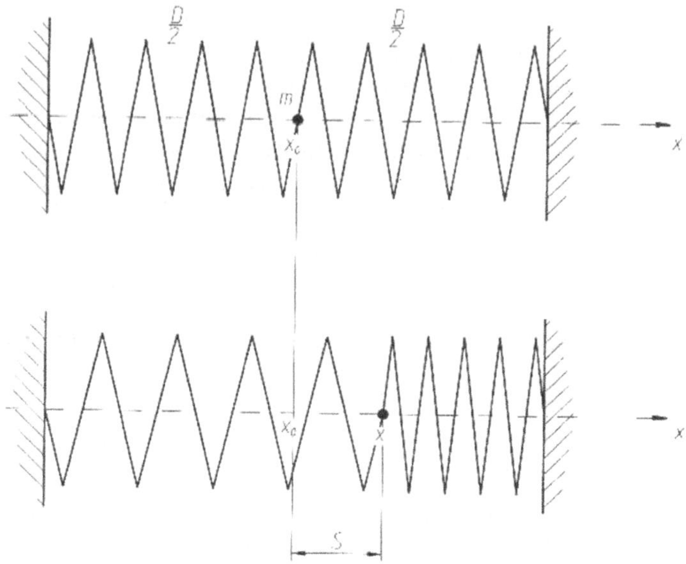

Fig. 4.3 Model of a harmonic oscillating mass m

We can use a single mass m suspended between two walls at a position of rest x_o by two springs, which possess the force constants $D/2$ as a model for harmonic oscillation; see Fig. 4.3. We once again denote the deflection out of the position of equilibrium as $s = x - x_o$. Therefore, the left spring pulls the mass m with the same force $f_l = -\frac{1}{2} D\, s$ to the left as the right spring pushes the mass to the right with the force $F_r = -\frac{1}{2} D\, s$. The total resulting force acting on the mass m is $F = f_l + f_r = -D\, s$, and the equation for an oscillation (13) is valid.

We now consider an oscillating system made up out of two equal masses of the size $m/2$ coupled together with the force constant $(3/4)D$ and attached to the wall with the force constant $D/2$; see Fig. 4.4.

The instantaneous positions of the first and second mass are x_1 and x_2. Let x_{o1} and x_{o2} be their positions of equilibrium, so that $x_1 = x_{o1} + s_1$ and $x_2 = x_{o2} + s_2$ with the deflections s_1 and s_2 out of their positions of rest. Their velocities are therefore $v_1 = \dot{x}_1 = \dot{s}_1$, $v_2 = \dot{x}_2 = \dot{s}_2$ and the forces F_1 or F_2 act on both masses according to

$$F_1 = -\frac{1}{2} D\,(x_1 - x_{o1}) + \frac{3}{4} D\,[(x_2 - x_1) - (x_{o2} - x_{o1})] = -\frac{5}{4} D\, s_1 + \frac{3}{4} D\, s_2$$

and

$$F_2 = -\frac{5}{4} D\, s_2 + \frac{3}{4} D\, s_1\,.$$

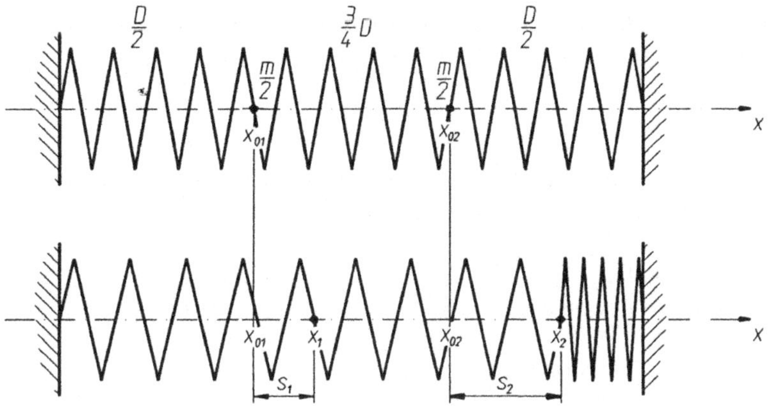

Fig. 4.4 Two oscillating masses $m/2$

We therefore receive the equations for the motion of both masses,

$$\left.\begin{aligned}\frac{m}{2}\,\ddot{s}_1 &= -\frac{5}{4}D\,s_1 + \frac{3}{4}D\,s_2\,,\\ \frac{m}{2}\,\ddot{s}_2 &= -\frac{5}{4}D\,s_2 + \frac{3}{4}D\,s_1\,.\end{aligned}\right\} \tag{17}$$

The motions s_1 and s_2 of both masses $m/2$ can be reduced to two elementary oscillations, the additive displacement a and the relative displacement r of the masses in this system according to

$$\left.\begin{aligned}&a = \frac{1}{2}(s_2 + s_1)\,,\ r = \frac{1}{2}(s_2 - s_1)\\ &\text{and}\\ &s_1 = a - r\,, \qquad s_2 = a + r\,.\end{aligned}\right\} \tag{18}$$

Using addition or subtraction of Eq. (17), we find for a and r

$$\left.\begin{aligned}m\,\ddot{a} &= -D\,a\,,\\ m\,\ddot{r} &= -4D\,r\,.\end{aligned}\right\} \tag{19}$$

The system of Eq. (17) is thus separated. In other words, we have for our system containing two masses, two independent harmonic equations of oscillation, each equation having the form of a harmonic equation of oscillation of a single mass (13). Our system can therefore oscillate harmonically in phase if alone the solution of the first Eq. (19) differs from zero,

$$\left.\begin{aligned}a(t) &= a_o\cos(\omega t + \phi)\,,\\ r(t) &= 0\,,\end{aligned}\right\} \quad \longleftrightarrow \quad \left.\begin{aligned}s_1 &= a_o\cos(\omega t + \phi)\,,\\ s_2 &= a_o\cos(\omega t + \phi)\,,\end{aligned}\right\} \tag{20}$$

here in our example with the same frequency ω, as observed in the oscillation for single masses (13),

$$\omega = \sqrt{\frac{D}{m}}. \tag{15}$$

If alone the solution of the second Eq. (19) is different from zero, do both masses oscillate harmonically against each other, so that

$$\left.\begin{array}{l} a(t) = 0\,, \\ r(t) = r_o \cos(\bar{\omega}t + \bar{\phi})\,, \end{array}\right\} \longleftrightarrow \left.\begin{array}{l} s_1 = -r_o \cos(\bar{\omega}t + \bar{\phi})\,, \\ s_2 = +r_o \cos(\bar{\omega}t + \bar{\phi})\,, \end{array}\right\} \tag{21}$$

with a frequency $\bar{\omega}$ according to

$$\bar{\omega} = \sqrt{\frac{4D}{m}} = 2\,\omega. \tag{22}$$

Here, we purposely chose the force constant so that $\bar{\omega}$ is a multiple of ω. The general solutions of the problem of motion of both masses are

$$\left.\begin{array}{l} a = a_o \cos(\omega t + \phi)\,, \\ r = r_o \cos(\bar{\omega}t + \bar{\phi})\,. \end{array}\right\} \tag{23}$$

We can now show that this simple mechanical oscillating system possesses all those properties that are characteristic of the phenomena of wave propagation.

To do this, we consider the transportation of energy. Using a wave, energy is transported with a characteristic velocity through a medium (or through space) in such a manner that the medium (or space) returns to its original state after the energy was transported through it.

Here, the medium of our oscillating system is the two masses coupled by the springs. At the first glance, this may seem to be strange. Here, however, the key can be found for further wide-ranging generalisations, e.g. when we move from coupled oscillation systems made up of more and more masses, to the atomic lattice of a crystal.

The special values $\phi = \bar{\phi} = -\pi/2$ applied to (23) lead to the solution

$$\left.\begin{array}{l} a(t) = a_o \cos(\omega t + \phi)\,, \\ r(t) = 0\,, \end{array}\right\} \longleftrightarrow \left.\begin{array}{l} s_1 = a_o \cos(\omega t + \phi)\,, \\ s_2 = a_o \cos(\omega t + \phi)\,. \end{array}\right\} \tag{24}$$

With the help of the constant v_o, we now have both free parameters of motion a_o and r_o at our disposal, according to $2\omega\, a_o = v_o,\ 4\omega\, r_o = -v_o$. For the motion of the single masses, we receive from (18) and (24)

$$\left.\begin{aligned}
s_1(t) &= \frac{v_o}{2\omega}\,(\sin\omega t + \frac{1}{2}\sin 2\omega t)\,,\\
s_2(t) &= \frac{v_o}{2\omega}\,(\sin\omega t - \frac{1}{2}\sin 2\omega t)\\
&\text{with the velocities}\\
\dot{s_1}(t) &= \frac{1}{2}\,v_o\,(\cos\omega t + \cos 2\omega t)\,,\\
\dot{s_2}(t) &= \frac{1}{2}\,v_o\,(\cos\omega t - \cos 2\omega t)\,.
\end{aligned}\right\} \tag{25}$$

We can read out of Eq. (25) the stated wave-like energy propagation in our mechanical oscillation system for both masses.

Both masses have the distance $(x_{o2} - x_{o1})$ in their position of equilibrium. At time $t = 0$, only the first mass moves. The second mass is at rest,

$$s_1(0) = 0\,,\quad \dot{s_1}(0) = v_o\,,\quad s_2(0) = 0\,,\quad \dot{s_2}(0) = 0\,. \tag{26}$$

According to (26), the first mass possesses (e.g. through an elastic blow) a starting velocity v_o and has at time $t = 0$ the total system energy of $U = \frac{1}{2}mv_o^2$ in the form of kinetic energy. After one half of an oscillation, after time $t = T/2 = \pi/\omega$, the first mass completely stops moving.

Now, only the second mass moves precisely in the opposite direction,

$$s_1(\frac{\pi}{\omega}) = 0\,,\quad \dot{s_1}(\frac{\pi}{\omega}) = 0\,,\quad s_2(\frac{\pi}{\omega}) = 0\,,\quad \dot{s_2}(\frac{\pi}{\omega}) = -v_o. \tag{27}$$

The total energy $U = \frac{1}{2}mv_o^2$ is at time $t = T/2 = \pi/\omega$ completely at the second mass in the form of kinetic energy. Due to system limitations, i.e. the walls, the energy cannot pass on. It flows back to the first mass and is there at time $t = T = 2\pi/\omega$. The process then restarts from the beginning.

The system made up out of two masses and three springs produces a transportation of energy from the first mass to the second and back again. This energy transport occurs with the velocity $c = covered\ distance/needed\ time$, so with a distance L between both walls,

$$c = \frac{2L}{T} = L\,\frac{\omega}{\pi} = L\,\frac{1}{\pi}\sqrt{\frac{D}{m}}. \tag{28}$$

Here, the distance L is just as much a material parameter of our oscillation device, as is the frequency $\omega = \sqrt{D/m}$. The parameter c is the characteristic velocity for the propagation of a wave front through our medium, which is represented here by the two masses coupled together by two elastic springs; c is thus the *critical signal velocity*. The energy is transported with this velocity through our medium. This result deserves to be especially brought forward:

The transportation of energy, as we understand it from the fields of electromagnetic or acoustic waves, is already realised as an elementary property of a mechanical system made up of only two elastic coupled masses. The signal velocity c is a system parameter that we can understand on the basis of NEWTONian mechanics, from its linear dimension L and the quotient of the force constant D of the elastic springs and the inertia m of both masses.

We will formulate this differently: The motion behaviour of elastic coupled masses is determined only by NEWTON's equations. Wave phenomena occur with it. These can be verified and proven using only our system of two elastic coupled masses. These wave phenomena become characteristic for systems made up of many elastic coupled masses. In the limiting case of infinitely many masses, these wave phenomena occur in their pure form. This means that the NEWTONian motion of such a system is described by the wave equation. This we will prove in Chap. 5.

Next, we will illustrate using the model of a so-called linear chain, how we arrive at the well-known elastic oscillations of a medium from the oscillations of single coupled elastic masses.

We distribute in equal distances along a length L a total mass m as equal point masses $\Delta m = m/N$ in such a manner that the Nth mass is at the end of the length L. Then at the beginning of L, there is no mass. We also consider a spring of the length L with the force constant D. This spring is then segmented into exactly the same number of partial springs as there are N masses to be coupled, i.e. N partial springs of the length L/N with N masses. We then assure ourselves that the force constant D_n, belonging to the spring length L/N, has the value $D_n = N\,D$. The force constant D is the relationship of a force F to the absolute deflection s of the spring, $D = F/s$. If one hangs the same weight on two identical springs suspended above each other, then the deflection s is doubled. This can easily be verified by anyone using two springs. The directional force D is halved when the length of the spring is doubled, and the force is doubled when the length of the spring is halved. Up to now, we have represented this in Fig. 4.5.

Now the two end points of the length L are merged together to form one point. One can imagine this by joining the length L at both of its ends together and forming a circle. Of course, this would only make sense if $N \geq 3$. This is, however, no restriction for us, because we are interested in a very large value for N. We get Fig. 4.6. The second representation of this arrangement can be shown by lining up identical copies of the length L with its N masses in both directions. As a result, there is now also a mass at the beginning of the length L which is identical to the mass at the end of the neighbouring length, i.e. their motions and forces are identical. This enables us to have an infinite number of masses, however the $(k + N)$th mass is identical to the kth mass. Such a periodical arrangement of coupled elastic masses

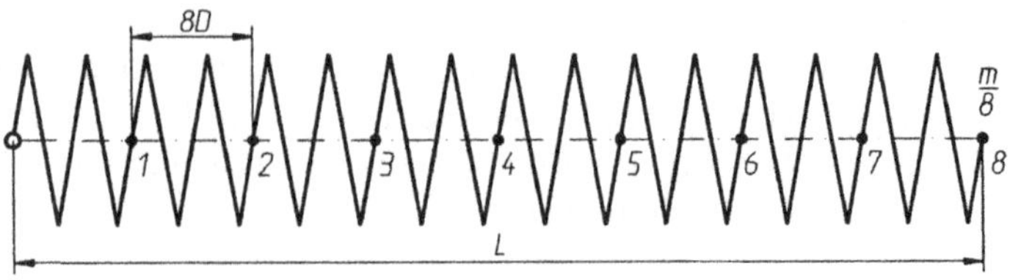

Fig. 4.5 Subdivision of a spring of the length L into eight identical parts

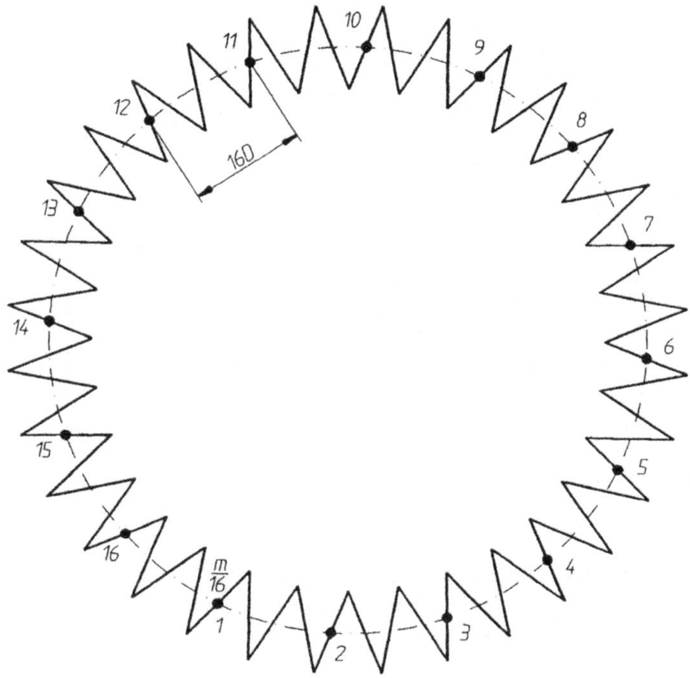

Fig. 4.6 Closed linear chain of the length L = circumference of the circle for $N = 16$

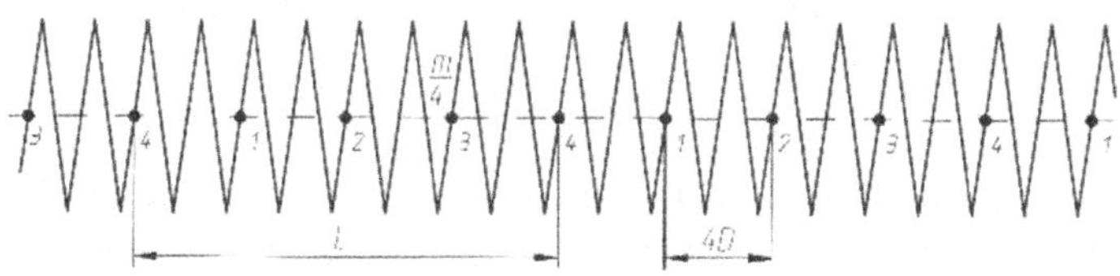

Fig. 4.7 Linear chain of the length L for $N = 4$

is known as a linear chain; see Fig. 4.7. A force F acting on the end of a length L results in a deflection s according to $F = D\,s$. As in the theory of elasticity, we are interested in the relative elongation and write $F = E_{eff}\,s/L$ with an effective modulus of elasticity $E_{eff} = D\,L$. We also introduce an effective specific mass density ρ_{eff},

$$\left.\begin{array}{ll} \rho_{eff} = \dfrac{m}{L}, & \text{Effective mass density of the linear chain} \\ E_{eff} = D\,L. & \text{Effective modulus of elasticity of the linear chain} \end{array}\right\} \quad (29)$$

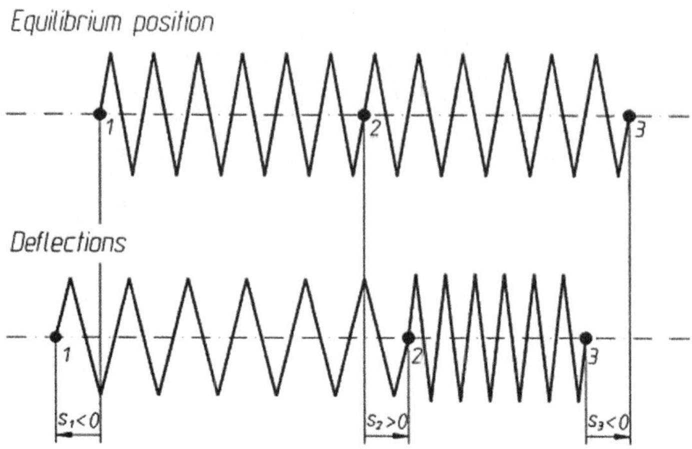

Fig. 4.8 Deviation of the equation of motion for a linear chain

We will now introduce the equation of motion for the N masses $\Delta m = m/N$ of a linear chain. All springs have the identical restoring force D_n, and the ith mass can be found at the position x_i,

$$\left.\begin{aligned} x_i &= \frac{L}{N}i, \quad i = 1, \cdots N\,, \\ D_N &= N\,D, \\ \Delta m &= \frac{m}{N}. \end{aligned}\right\} \qquad \text{Linear chain} \tag{30}$$

The deflection of the ith mass out of its position of equilibrium is s_i. The mass experiences an accelerating force from its neighbouring masses, whose deflections have the values s_{i-l} and s_{i+l}. We consider for example the second mass; see Fig. 4.8. The spring between the first and the second mass is stretched by $(s_2 - s_1)$ and pulls the second mass back into its starting position. The spring between the second and third mass is stretched by $(s_3 - s_2)$ and pulls the second mass away from its position of equilibrium. The equation of motion for the mass Δm with a deflection s_2 is therefore $\Delta m\,\ddot{s}_2 = -D_N(s_2 - s_1) + D_N(s_3 - s_2)$, i.e.

$$\Delta m \cdot \ddot{s}_2 = D_N s_1 - 2D_N s_2 + D_N s_3\,.$$

Hence, the equations of motion for all N masses of a linear chain are

$$\left.\begin{aligned}
\Delta m \cdot \ddot{s}_1 &= D_N s_N - 2D_N s_1 + D_N s_2\,,\\
\Delta m \cdot \ddot{s}_2 &= D_N s_1 - 2D_N s_2 + D_N s_3\,,\\
&\vdots\\
\Delta m \cdot \ddot{s}_i &= D_N s_{i-1} - 2D_N s_i + D_N s_{i+1}\,,\\
&\vdots\\
\Delta m \cdot \ddot{s}_N &= D_N s_{N-1} - 2D_N s_N + D_N s_{N+1}\,.
\end{aligned}\right\} \tag{31}$$

(Here, $s_{N+1} \equiv s_1$ as explained above).

Equation (31) is a coupled system of N differential equations for the motions of the N masses. The general solution of these equations is searched for as $(N-1)$ harmonic oscillations of the single masses (a trivial solution is always the rigid motion of the whole chain). The nth oscillation is written for the ith mass as $s_i^{(n)}(t)$. To determine the eigen frequencies ω_n of a linear chain, we will try the following ansatz,

$$s_i^{(n)}(t) = \cos(\omega_n\, t)\, \cos 2(\frac{i}{N}\, n\, 2\pi)\,, \quad n = 1, 2, \cdots, N-1. \tag{32}$$

This ansatz fulfils the conditions of periodicity: The $(i+N)$th mass oscillates according to (32) just as the ith mass oscillates. The position of the ith mass is according to (30) at $x_i = \frac{L}{N}\, i$. Equation (32) shows us the largest 'wavelength' λ_1 with its corresponding smallest wave number $k_1 = 2\pi/\lambda_1$ according to,

$$\lambda_1 = L, \qquad k_1 = \frac{2\pi}{L} \tag{33}$$

and for the nth harmonic vibration the following is valid,

$$\lambda_n = \frac{L}{n}, \qquad k_n = \frac{2\pi n}{L}, \qquad n = 1, 2, \cdots, N-1. \tag{34}$$

We insert Eq. (32) into Eq. (31) and discover, whilst shortening the common factor $\cos(\omega_n t)$,

$$\begin{aligned}
\Delta m\, \omega_n^2 \cos(i\frac{n}{N}2\pi) &= D_N[\cos(i\frac{n}{N}2\pi - \frac{n}{N}2\pi) + \cos(i\frac{n}{N}2\pi + \frac{n}{N}2\pi)\\
&\quad - 2\cos(i\frac{n}{N}2\pi)]\\
&= D_N[\cos(i\frac{n}{N}2\pi)\cos(\frac{n}{N}2\pi) + \sin(i\frac{n}{N}2\pi)\sin(\frac{n}{N}2\pi)\\
&\quad + \cos(i\frac{n}{N}2\pi)\cos(\frac{n}{N}2\pi) - \sin(\frac{n}{N}2\pi)\sin(\frac{n}{N}2\pi)\\
&\quad - 2\cos(i\frac{n}{N}2\pi)]\,,
\end{aligned}$$

hence

$$\Delta m\, \omega_n^2 \cos(i\frac{n}{N}2\pi) = 2D_N[\cos(i\frac{n}{N}2\pi)\cos(\frac{n}{N}2\pi) - \cos(i\frac{n}{N}2\pi)]\,,$$

and

$$\Delta m\, \omega_n^2 = 2D_N[1 - \cos(\frac{n}{N}2\pi)]\,,$$

so with (30)

$$\frac{m}{N}\omega_n^2 = 2N\, D\,[1 - \cos(\frac{n}{N}2\pi)]$$

and finally

$$\omega_n = N\sqrt{\frac{2D}{m}\big[1 - \cos(\frac{n}{N}2\pi)\big]}, \quad n = 1,\, 2,\, \ldots,\, N-1. \tag{35}$$

The starting solution (32) therefore leads us to a solution for the whole system of differential Eq. (31) if the angular velocity ω_n has the value (35).

We will now turn our attention to a very large N. We will take for example a (one-dimensional) one-metre-long metal rod, whose atoms have a lattice parameter of 10^{-8} cm. This has as a result that at every oscillation of the rod, $N = 10^{10}$ atoms are involved. We will now take only those oscillations into consideration that cause many atoms to oscillate in one sine period. In other words, the following condition has to be fulfilled:

$$n = 1,\, 2,\, 3,\, \ldots \;\; ; \quad n \ll N. \tag{36}$$

We will explain later on why we can in fact principally, because of the Third Newtonian Axiom, let N go to infinity. Hence, condition (36) can principally be fulfilled with arbitrary exactness. For $n/N \ll 1$ and $2\pi n/N \ll 1$, we can replace the cosine term in Eq. (35) with the formula $\cos\alpha = 1 - \alpha^2/2$, so that

$$\omega_n = N\sqrt{\frac{2D}{m}[1 - (1 - \frac{1}{2}(\frac{n}{N}2\pi)^2)]} = N\sqrt{\frac{2D}{m}\frac{1}{2}(\frac{n}{N}2\pi)^2}\,,$$

and

$$\omega_n = n\,2\pi\sqrt{\frac{D}{m}} \quad \text{for} \quad n \ll N. \tag{37}$$

Hence, we get the result that the possible angular velocities ω_n of our elastic coupled linear chain under the condition (36) are an integral multiple of a fundamental angular velocity ω_1,

$$\omega_1 = 2\pi\sqrt{\frac{D}{m}}, \quad \omega_n = n\,\omega_1 \quad \text{for} \quad n \ll N. \tag{38}$$

Taking (36) into consideration the relation between the angular velocity ω_n and the wave number k_n according to (34) and (38), it is

$$\omega_n = c_K\, k_n = c_K\, \frac{2\pi n}{L} \quad \text{with} \quad c_K = L\sqrt{\frac{D}{m}} \quad \text{for} \quad n \ll N. \tag{39}$$

In this case, we receive for the oscillations (32) of the linear chain

$$s_i^{(n)}(t) = \cos(\tfrac{2\pi n}{N} i)\cos(\tfrac{2\pi n}{L} c_K t) \text{ for } n \ll N. \qquad \text{Oscillations of a linear chain} \tag{40}$$

The new introduced parameter c_K is a characteristic velocity of a linear chain. c_K is the velocity with which a perturbation in a linear chain can propagate. Even though we have not explicitly shown this here, it could be easily shown as we have already done in a system made up out of two coupled masses. In the following chapter, we will show the transition from the system of Eq. (31) to the wave equation and will thus spare ourselves this calculation. The exact definition of the velocity c_K is

$$c_K^{(n)} = \frac{\partial \omega_n}{\partial k_n} \tag{41}$$

with a velocity $c_K^{(n)}$ generally dependent on the order n. For $n \ll N$ this dependency disappears according to (39), in other words, $c_K^{(n)}$ becomes independent from the order n of waves. Such a medium is said to be free of dispersion:
for $n \ll N$, the linear chain is free of dispersion.

We can write for c_K using the notation introduced in (29)

$$c_K^{(n)} = c_K = \sqrt{\frac{E_{eff}}{\rho_{eff}}} \quad \text{for} \quad n \ll N. \tag{42}$$

This relationship reminds us of the theory of the elasticity of a rod to which we will soon turn our attention. If we drop the requirement $n \ll N$ we then allow certain oscillations of a linear chain, where only a few masses are found on a wavelength λ_n. This results in the loss of the dispersion free property. We introduce the term k_n into the conditions of frequency (35) and find

$$\omega_n = \omega(k_n) = N\sqrt{\frac{2D}{m}\Big[1 - \cos\frac{L\, k_n}{N}\Big]}\,. \tag{43}$$

We now get a velocity $c_K^{(n)}$ that is dependent on the order n,

$$c_K^{(n)} = \frac{\partial \omega_n}{\partial k_n} = N \frac{2D}{m} \frac{L}{N} \frac{\sin \frac{L\,k_n}{N}}{2\sqrt{\frac{2D}{m}[1 - \cos \frac{L\,k_n}{N}]}},$$

hence

$$c_K^{(n)} = \frac{\partial \omega_n}{\partial k_n} = L\sqrt{\frac{D}{m}} \frac{\sin \frac{L\,k_n}{N}}{\sqrt{2[1 - \cos \frac{L\,k_n}{N}]}}$$

and with (39)

$$c_K^{(n)} = c_K = \frac{\sin \frac{L\,k_n}{N}}{\sqrt{2[1 - \cos \frac{L\,k_n}{N}]}} = \frac{\sin \frac{2\pi n}{N}}{\sqrt{2[1 - \cos \frac{2\pi n}{N}]}}. \tag{44}$$

Equations (43) and (44) describe the strict dispersion relation of a linear chain. An approximation formula for a not too large $x = L\,k_n/N = 2\pi n/N$ can be found using the TAYLOR series expansion according to $\sin x = x - \frac{1}{6}x^3$, $\cos x = 1 - \frac{1}{2}x^2 + \frac{1}{24}x^4$, so that

$$\frac{\sin x}{\sqrt{2 - 2\cos x}} = \frac{x - \frac{1}{6}x^3}{\sqrt{x^2 - \frac{1}{12}x^4}} = \frac{1 - \frac{1}{6}x^2}{\sqrt{1 - \frac{1}{12}x^2}} = (1 - \frac{1}{6}x^2)(1 + \frac{1}{24}x^2) = 1 - \frac{1}{8}x^2,$$

hence

$$c_K^{(n)} = c_K\big(1 - \frac{\pi^2}{2}\frac{n^2}{N^2}\big). \tag{44a}$$

Due to the quadratic dependency of n/N, Eq. (39) gives for $n \ll N$ a very good approximation for Eq. (44a). The real dispersion in every linear chain of single coupled masses is only detectable when the wavelength becomes small enough. Using the aforementioned rod with $N = 10^{10}$ atoms in one metre, we receive for an $n = 10^{-5}N = 10^5$ for the harmonic oscillation of the nth order the value $c_K^{10^5} = c_K(1 - 5 \times 10^{-10})$. For a wavelength value of $\lambda_{10^5} = L/10^5 = 10^{-5}$m, the deviation from the propagation velocity c_K is only 5×10^{-8} %. This deviation stays under the value 5×10^{-8}% for all wavelengths with $\lambda \geq 10^{-5}$m. In other words, if the wavelength $\lambda \geq 10^5$a is valid for a lattice parameter a, then the dispersion can only be measured by measuring $c_K^{(n)}$ if the relative inaccuracy of the process remains smaller than 5×10^{-8}%.

We will now compare the oscillations of a linear chain with the longitudinal natural oscillations of an elastic rod, as we had considered in Chap. 2. We take a homogenous rod with a constant mass density ρ and the modulus of elasticity E. For an elastic deflection s out of the position of equilibrium x at time t the D'ALEMBERTian equation applies

$$\frac{\partial^2}{\partial x^2}s(x,t) - \frac{1}{c^2}\frac{\partial^2}{\partial t^2}s(x,t) = 0 \tag{1}$$

with the signal velocity c,

$$c = \sqrt{\frac{E}{\rho}}. \tag{2}$$

The derivation of these equations that we have taken from the field of acoustics will be shown in the following chapter.

We now look for those solutions of Eq. (1) that correspond to the periodical boundary conditions of our linear chain. In order to do this, we will examine a rod of the length L that can oscillate freely at both of its ends. Both ends of the rod should have the same state of motion, and they oscillate in the same phase as one says. One can then line up identical copies of this rod (in exactly the same way as we did it with the linear chain), so that the states of oscillation are continuously repeated. We then have to choose those from the general solution (6) of the wave Eq. (1) that fulfils the boundary conditions at the ends of the rod. All the solutions of (1) are strictly free of dispersion. This is a characteristic of the theory of linear elasticity. Therefore, (5) is valid,

$$\omega = c\,k = c\,\frac{2\pi}{\lambda}. \tag{5}$$

We get the solutions of the rod oscillating freely at both of its ends, by superimposing a wave running to the left and a wave running to the right, whose angular velocity ω and amplitude B are identical according to

$$\begin{aligned} s(x,t) &= B\,[\cos(kx - \omega t) + \cos(kx + \omega t)] \\ &= B\,[\cos(kx)\,\cos(\omega t) + \sin(kx)\,\sin(\omega t) + \cos(kx)\,\cos(\omega t) \\ &\quad - \sin(kx)\,\sin(\omega t)], \end{aligned}$$

so that we have to choose from the general solution (6) those waves that

$$s(x,t) = 2B\,\cos(kx)\cos(\omega t). \qquad \text{Free oscillating rod with periodical boundary conditions} \tag{45}$$

We now have to take the length of the rod L into consideration. Because the ends of the rod should oscillate freely, L is the largest wavelength $\lambda_1 = L$.[4] We can generally incorporate n wavelengths of the length λ_n in the rod. For a 'real' continuum n is boundless. If we take (5) into consideration, the following condition has to be fulfilled:

---

[4]Notice that for $\lambda = L/2$ both ends of the rod do not oscillate in phase.

$$\lambda_n = \frac{L}{n} \quad , \quad \omega_n = c\,k_n = \frac{2\pi n}{L}\,c, \quad n = 1,\, 2,\, \cdots,\, \infty. \tag{46}$$

We insert (46) into (45) and discover the possible states of oscillation $s_n(x, t)$ of an elastic rod of the length L, because we also have the amplitude with $B = 1/2$ at our disposal,

$$s_n(x,t) = \cos(\frac{2\pi n}{L}x)\,\cos(\frac{2\pi n}{L}ct), \quad n = 1,\, 2,\, \cdots,\, \infty. \qquad \text{Free oscillating rod of the length } L \tag{47}$$

We consider the oscillations (47) at the positions $x_i = \frac{L}{N}i$ at those places where we positioned the single masses $\Delta m = m/L$ in the linear chain and find

$$s_n(x,t) = \cos(\tfrac{i}{N}n\,2\pi)\,\cos(\tfrac{2\pi n}{L}c\,t)\,, \quad n = 1,\, 2,\, \cdots,\, \infty\,. \qquad \text{Free oscillating rod} \tag{47a}$$

These motions are however identical to the motions (40) of the masses Δm of the linear chain for $n \ll N$ if we just dimension them so that their velocity c_K corresponds to the sound velocity c. If the mass m and the length L of the linear chain and the rod are identical, then $\rho_{eff} = \rho$ is valid. The N single masses Δm of the linear chain are then the centres of inertia of the rod of the N pieces at $x_i = \frac{L}{N}i$. We now only need to dimension the force constant D of our spring at the length L, so that $D \cdot L = E_{eff} = E$ and thus $c_K = c$ is fulfilled. The oscillations of the linear chain are then not distinguishable from the oscillations of an elastic rod. We see the following:

> The oscillations of a linear chain consisting of N elastic coupled single masses, where one wavelength is built up out of n particles, are indistinguishable from the oscillations of a homogenous elastic rod under the condition that $n \ll N$.

Is it possible to make a distinction between a linear chain made up out of single masses and an elastic rod only by means of measurement? Is the continuum, here the homogenous rod maybe just a mathematical model, with which we can easier calculate? In the regions of sufficiently high frequencies ω_n, the condition (37) for the linear chain is not fulfilled and the correct formula (43) applies. The linear chain shows a dispersion. The signal velocity $c_K^{(n)}$ is dependent according to (44) on the wavelength λ_n (respectively on the wave number $k_n = 2\pi/\lambda_n$), the same way we discussed it above with the help of the formula of approximation (44a). This dispersion can in fact be measured in the abovementioned example using a metal rod. We can determine with high-precision experiments that the rod, that behaves like a continuum when just normally observed, is in fact only a linear chain of single masses. However, as long we do not have the device to measure this, we cannot distinguish between them.

> Maybe every so-called continuum turns out to be a discontinuous structure if sufficiently accurate modes of measurements could be applied.

Should we include our physical spacetime continuum into these problems? To delve deeper into this problem would inevitably lead us to extend our thoughts to the quantum theory and the General Theory of Relativity. We do not want to do this here. In Chap. 15, we will however compare the observations made on the material crystal lattice and some very old arguments for a discrete background of our physical spacetime.

Where does the wave equation come from? Up to now, we have shown the following. When an elastic rod is divided up into equal parts, the masses of the parts oscillate according to the wave equation exactly the way single elastic coupled masses oscillate according to the NEWTONian equations, as long as we maintain condition (36) whilst dividing the rod into sufficient many parts. Is it possible to deduce the wave equation directly from NEWTONian mechanics? This question will be answered in the next chapter as follows: The wave equation can be completely understood from the postulates of NEWTON's point mechanics. We will generally show how NEWTONian mechanics can in fact imply a continuum such that a mechanical continuum can be described with arbitrary accuracy using point mechanics.
Notice, independent of that, it is a well-known fact that the wave equation for electromagnetic phenomena derives from MAXWELL's field theory.

Chapter 5
The Wave Equation and the Third Axiom

The NEWTONian equations have a remarkable property. They are strictly a point mechanics, but do not have atomistics as a condition. This is simply because NEWTONian equations do not observe the physical positions of particles, but the equations observe the geometrical centres of inertia. This we can clarify with an example.

Two particles with the masses m_1 and m_2 move along a straight path, e.g. along the x-axis towards each other with the velocity v_1 and v_2 until they 'collide'. We assume the conservation of the kinetic energy, hence we observe an elastic collision and denote the resulting velocities of the masses after the collision as v_1' and v_2', so that the conservation laws for momentum and energy are then

$$\left.\begin{aligned} m_1 v_1 + m_2 v_2 &= m_1 v_1' + m_2 v_2' \ , \\ \tfrac{1}{2} m_1 v_1^2 + \tfrac{1}{2} m_2 v_2^2 &= \tfrac{1}{2} m_1 v_1'^2 + \tfrac{1}{2} m_2 v_2'^2 \ . \end{aligned}\right\} \tag{48}$$

Equation (48) after a simple calculation can be also be formulated as

$$\left.\begin{aligned} m_1 (v_1' - v_1) &= m_2 (v_2' - v_2) \ , \\ m_1 (v_1' + v_1)(v_1' - v_1) &= m_2 (v_2' + v_2)(v_2' - v_2) \ . \end{aligned}\right\} \tag{49}$$

At first, Eq. (49) leads us to

$$v_1 + v_1' = v_2 + v_2' \tag{50}$$

and we then find after a bit of calculation the result that can be found in every physics textbook. The velocities after a collision are

H. Günther, *Elementary Approach to Special Relativity*,
https://doi.org/10.1007/978-981-15-3168-2_5

$$\left.\begin{aligned} v_1' &= \frac{(m_1 - m_2)v_1 + 2m_2 v_2}{m_1 + m_2} \ , \\ v_2' &= \frac{2m_1 v_1 + (m_2 - m_1)v_2}{m_1 + m_2} \ . \end{aligned}\right\} \tag{51}$$

That these solutions are correct can be easily proven experimentally , e.g. letting two steel ball bearings collide. In the extreme case where one of the masses has an infinitely large mass, $m_2 \longrightarrow \infty$, an elastic reflection against a wall is the result, $v_1' = -v_1$ and $v_2' = v_2 = 0$.

However, Eq. (51) are not the entire truth, because they result out of the solutions (50). Equation (50) however was a result of Eq. (49), where we divided by either $(v_1' - v_1)$ or $(v_2' - v_2)$. We had to presume that these quantities were not zero, which lead us to lose a complete class of solutions. As one can immediately see out of (49), further solutions differing from (51) are

$$\left.\begin{aligned} v_1' &= v_1 \ , \\ v_2' &= v_2 \ . \end{aligned}\right\} \tag{52}$$

According to (52), the particles m_1 and m_2 maintain their velocities without having influenced each other. We could also state that they pass through each other, e.g. the particle passes through the wall instead of reflecting from it. This second class of solutions can also be experimentally proven and understood using NEWTONian mechanics. For example, it is not necessary that the physical particles belonging to the two masses m_1 and m_2 on the x-axis are in fact positioned on the x-axis. Only their geometrical centres of inertia have to be positioned there. These centres of inertia are pure mathematical quantities. Take for example a mass m_1, a cube where eight (neutral) particles with the mass $m_1/8$ are positioned at its corners. A second mass m_2 with only one mass particle is positioned on the x-axis. The following 'elastic collision' between the cube m_1 and the mass m_2 can be described for all, but finitely many positions of the cube by the solution (52) and not by (51), as shown in Fig. 5.1.

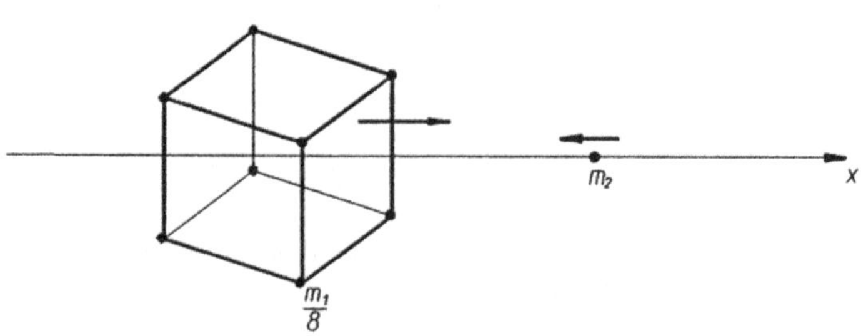

Fig. 5.1 Collision of two particles with the masses m_1 and m_2. In the illustration, it is presumed that the total mass m_1 is divided up equally into eight particles and distributed at the corners of the cube. The concentrated mass m_2 therefore experiences no interactions for virtually all positions of the cube with its $m_1/8$ masses, when a 'collision' takes place

We will now illustrate how to arrive at the mechanics for a continuum from NEWTONian point mechanics. Our special attention will be turned to the derivation of the wave equation (1), which in Chap. 2 we took as granted from the field of acoustics, from the NEWTONian system of point mechanics.

We will presume that interaction forces, i.e. internal forces $\mathbf{F}_{BA}$ (= force of mass M_B on mass M_A) and further external forces $\mathbf{F}_A$ (= external forces on M_A) for a system of N masses M_A, $A = 1, 2, \ldots, N$ are present. For the positions $\mathbf{X}_A$ of the masses M_A in an inertial system (which we always presume), the following NEWTONian equations are valid:

$$\left.\begin{aligned} \frac{d}{dt}\left(M_A \frac{d}{dt}\mathbf{X}_A\right) &= \sum_{B \neq A}^{N} \mathbf{F}_{BA} + \mathbf{F}_A \,, \\ \mathbf{F}_{AB} &= -\mathbf{F}_{BA} \,. \end{aligned}\right\} \tag{53}$$

($B \neq A$ means that the force from M_A does not influence itself). The second equation of (53) is NEWTON's reaction axiom, the Third Axiom. It can however be (and in most cases usually is) that $\mathbf{X}_A$ do not represent the positions of physical particles, but represent the centres of inertia of n_A masses m_a at the position $\mathbf{x}_a$, $a = 1, 2, \ldots, n_A$, so that

$$\left.\begin{aligned} M_A &= \sum_{a=1}^{n_A} m_a \,, \\ \mathbf{X}_A &= \frac{\sum_{a=1}^{n_A} m_a \mathbf{x}_a}{M_A} \,, \end{aligned}\right\} \tag{54}$$

see Fig. 5.2. However, because of the validity of NEWTON's mechanics between all n masses m_a ($n = \sum_{A=1}^{N} n_A$), internal forces $\mathbf{f}_{ba}$ that suffice the Third Axiom as well as other external forces $\mathbf{f}_a$ apply, so that once again the NEWTONian equations are applicable,

$$\left.\begin{aligned} \frac{d}{dt}\left(m_a \frac{d}{dt}\mathbf{x}_a\right) &= \sum_{b \neq a}^{n} \mathbf{f}_{ba} + \mathbf{f}_a \,, \\ \mathbf{f}_{ab} &= -\mathbf{f}_{ba} \,. \end{aligned}\right\} \tag{55}$$

The supposed forces $\mathbf{F}_{BA}$ and $\mathbf{F}_A$ are defined by the forces $\mathbf{f}_{ba}$ and $\mathbf{f}_a$ as

$$\left.\begin{aligned} \mathbf{F}_{BA} &= \sum_{b=1}^{n_B} \sum_{a=1}^{n_A} \mathbf{f}_{ba} \,, \\ \mathbf{F}_A &= \sum_{a=1}^{n_A} \mathbf{f}_a \,. \end{aligned}\right\} \tag{56}$$

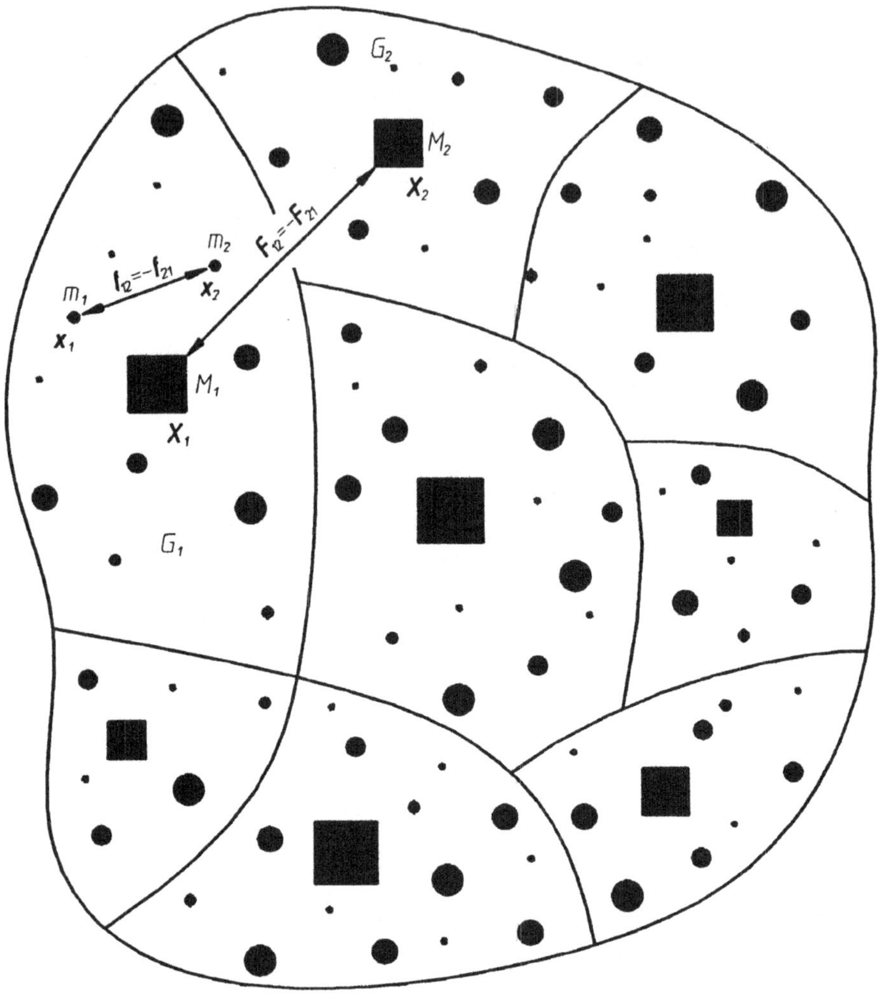

Fig. 5.2 Third Axiom's mode of operation. Only the geometrical centres of inertia are accounted for in the NEWTONian equations. These are not identical to physical particle locations. It is therefore impossible to state exactly how many physical particles are involved in motion. The NEWTONian equations are equally valid for the illustrated small, round particles (small letters) and for the illustrated large, quadratic objects (capital letters). The total of the 'small particles' has to have the same mass and the same centre of inertia as the 'large particles' in the marked areas $G_1, G_2, \ldots$

The earlier postulated Eq. (53) now have to be a result of Eq. (56). The reaction axiom for the forces $\mathbf{F}_{BA}$, the second of Eq. (53), is now a direct result of the reaction axiom for the forces $\mathbf{f}_{ba}$. Summing up Eq. (55), one finds under repeated application of the Third NEWTONian Axiom that the above postulated NEWTONian equation (53) for the motion of the centres of inertia $\mathbf{X}_A$ are valid. Even though this fact is well known, it is still a remarkable result. We will explicitly show this at the end of this chapter. We can therefore state

NEWTONian point mechanics, the number of particles involved in motion remains unknown.

The $\mathbf{x}_a$ can now themselves become the centres of inertia of many small particles, etc. Without further knowledge of the systems physical constitution, we can make no comment about the number of physical particles in our NEWTONian axiomatic system, up to an infinitely large number of particles, or a continuous mass distribution:

Only an experiment can decide if we are dealing with a continuum or if we are dealing with a certain number of particles involved in a motion.

In regard to a later application, we will once again write down the equations of motion for a system of particles whose coordinates q_a have here not been closely defined,

$$\left.\begin{aligned} &\frac{d}{dt}\left(m_a \frac{d}{dt} q_a\right) = \sum_{b \neq a} F_{ba} + F_a \,, \\ &F_{ab} = -F_{ba} \,. \end{aligned}\right\} \tag{57}$$

In the light of this, we now understand the connection between our N discrete oscillating single masses and the oscillations of an elastic rod. The positions of the equally spaced N masses according to (30), $x_i = \frac{L}{N} i$, can be seen as the coordinates of the centres of inertia of an improved equally spaced distribution of mass. Springs work between the masses. If we halve the single masses and keep the equal distribution, we have to halve the distance. This way the working single springs are halved. We explained earlier that the halved spring has the doubled force constant.

The quantity s_i is in Eq. (31) the displacement at the position x_i. We can therefore write $s_i(t) = s(x_i, t)$ and according to (31) following is then valid,

$$\Delta m \frac{\partial^2}{\partial t^2} s(x_i, t) = D_N[s(x_{i-1}, t) - 2s(x_i, t) + s(x_{i+1}, t)] \,.$$

Here, we insert $x_i = \frac{L}{N} i$ and after a simple rearrangement we write for the distances between the particles $\Delta x = L/N$, hence

$$\Delta m \frac{\partial^2}{\partial t^2} s\left(\frac{L}{N} i, t\right) = D_N \left[s\left(\frac{L}{N} i + \Delta x, t\right) - s\left(\frac{L}{N} i, t\right) - \left(s\left(\frac{L}{N} i, t\right) - s\left(\frac{L}{N} i - \Delta x, t\right)\right)\right] .$$

According to our explanations above, we can in the light of the NEWTONian postulates make N as large as we wish. The value $\frac{L}{N} i$ approaches every point on the length of the rod, in other words we can instead take a continuous variable x with optional precision. We therefore accordingly substitute $\frac{L}{N} i \longrightarrow x$ and receive

$$\Delta m \frac{\partial^2}{\partial t^2} s(x, t) = D_N\big[\big(s(x + \Delta x, t) - s(x, t)\big) - \big(s(x, t) - s(x - \Delta x, t)\big)\big] \,.$$

For a very large N, Δx is very small. We therefore write according to the TAYLOR formula

$$s(x+\Delta x,t)-s(x,t)=\Delta x\,\frac{\partial}{\partial x}s(x,t)\,,$$
$$s(x,t)-s(x-\Delta x,t)=\Delta x\,\frac{\partial}{\partial x}s(x-\Delta x,t)$$

and thus

$$\Delta m\frac{\partial^2}{\partial t^2}s(x,t)=D_N\,\Delta x\left[\frac{\partial}{\partial x}s(x,t)-\frac{\partial}{\partial x}s(x-\Delta x,t)\right]$$

and after another application of the TAYLOR formula

$$\frac{\partial}{\partial x}s(x-\Delta x,t)=\frac{\partial}{\partial x}s(x,t)-\Delta x\,\frac{\partial^2}{\partial x^2}s(x,t)\,,$$

so that

$$\frac{\Delta m}{\Delta x}\frac{\partial^2}{\partial t^2}s(x,t)=D_N\,\Delta x\,\frac{\partial^2}{\partial x^2}s(x,t)\,. \tag{58}$$

With $\Delta x = L/N$ and according to (30) $\Delta m = m/N$ we get $\Delta m/\Delta x = m/L$, which is according to our instructions of a continuous equal division nothing else than the mass density ρ of a homogenous rod,

$$\rho=\lim_{\Delta x\to 0}\frac{\Delta m}{\Delta x}=\lim_{N\to\infty}\frac{m/N}{L/N}=\frac{m}{L}\,. \tag{59}$$

The force constant $D_N = N \cdot D$ (where D belongs to the spring with the length L, see (30)) is the ratio of the force $F = D_N \cdot \Delta s$ to the absolute elongation Δs on the length L/N. Therefore, D_N becomes $D_N = N \cdot D$ with an infinitely increasing N, when the length of the spring approaches zero. In the passing to the limit to the continuum, the force is therefore referred to the relative strain $\Delta s/\Delta x$,[1] and we write for the limit $F = (D_N\,\Delta x)\cdot \partial s/\partial x$. The value $D_N \cdot \Delta x$ is called the modulus of elasticity E.[2] E has the dimension of a force, because the strain $\partial s/\partial x$ is without dimension. We see that

$$E=\lim_{\Delta x\to 0}(D_N\cdot\Delta x)=\lim_{N\to\infty}\left(N\,D\,\frac{L}{N}\right)=D\,L\,. \tag{60}$$

---

[1]The relative strain ε of a homogeneous rod with the original length L elongated by ΔL is $\varepsilon = \Delta L/L$.

[2]To avoid confusing this quantity E with energy, we will, when concerning ourselves with transversal oscillations belonging to dislocations, substitute the modulus of elasticity with the equivalent line tension σ.

We insert (59) and (60) in (58) and receive, based alone on NEWTONian mechanics, starting out from our equation for oscillations (31), for a linear chain with the periodical boundary conditions in the passing to the limit $N \longrightarrow \infty$ the D'ALEMBERTian wave equation (1) and also the explanation for the sound velocity c from mechanical parameters,

$$\left.\begin{aligned} \frac{\partial^2}{\partial x^2} s(x,t) - \frac{1}{c^2}\frac{\partial^2}{\partial t^2} s(x,t) = 0\,, \\ c = \sqrt{\frac{E}{\rho}}\,. \end{aligned}\right\} \tag{61}$$

With (59) and (60) for ρ and E, we find our formula (39) for the sound velocity which was introduced in the linear chain

$$c = \sqrt{\frac{D\,L}{m/L}} = L\sqrt{\frac{D}{m}}\,. \tag{62}$$

We can therefore state.

> The wave equation can be fully understood alone out of the postulates of NEWTON's point mechanics.

We still have not achieved much with this knowledge. Sound waves can be comprehended without the help of Special Relativity. We, of course, wish for more and think here of the vision of E. MACH, cf. THIELE [93], 'Light somewhat like sound. Sound somewhat like light', wherein a sense yet to be determined, the dynamics of light, as well as sound are supposedly equivalent.[3] In spite of everything, the LORENTZ transformation, the centrepiece of the Special Theory of Relativity was discovered in 1887 by W. VOIGT [96] long before the Special Theory of Relativity, from the 'differential equations for the oscillations of an incompressible medium', thus from the wave equation, without however having received much attention. In his paper, on the DOPPLER effect (cf. also Chap. 13 and the fourth footnote on Chap. 18) VOIGT primarily had electromagnetic waves in mind based on an elastic ether model. He also expressively had acoustic waves in mind, a 'ringing bell', see VOIGT [96], so that we in fact first come across the LORENTZ transformation in the field of acoustics. (Here VOIGT's equations differ by a common coefficient from the 'correct' Eq. (151) on for LORENTZ transformation, which however is of no great importance). In 1908 VOIGT [97] verified, ' . . . already then [in 1887] some results were formed which latter were obtained from the electromagnetic theory'.

[3]A mechanical theory of light is in this view as absurd as an electromagnetic theory of sound. A Special Theory of Relativity for the elastic disturbances on a lattice, for sound, really cannot exist as we will see in the following chapter. There is however a second possibility of motion on a lattice that of plastic displacement; see Chap. 7. The analysis of this will inevitably lead us to relativistic concepts including all details as far as to the secondary relativistic effects (e.g. pair creation which we will bring to attention in Chap. 26).

We now turn to the accepted conclusion of the equivalence of Eq. (53) with Eq. (55) and will prove this using the definitions (54) and (56). In doing this, we accept the validity of Eq. (55). The complete area of space occupied by masses is divided up into the areas $G_K, k = 1, 2, \ldots, N$; see Eq. (16). Thus,

$$\frac{d}{dt}\left(m_a \frac{d}{dt}\mathbf{x}_a\right) = \sum_{b \neq a}^{n} \mathbf{f}_{ba} + \mathbf{f}_a$$

and without loss of generality $\mathbf{x}_a \in G_1$. Addition of all 'particles' in the area G_1 results in

$$\sum_{a \in G_1} \frac{d}{dt}\left(m_a \frac{d}{dt}\mathbf{x}_a\right) = \sum_{b \neq a}^{n} \sum_{a \in G_1} \mathbf{f}_{ba} + \sum_{a \in G_1} \mathbf{f}_a \,,$$

so including the definitions (54) and (56)

$$\begin{aligned}\frac{d}{dt}\left(M_1 \frac{d}{dt}\mathbf{X}_1\right) &= \left(\sum_{(b \neq a) \in G_1} + \sum_{b \in G_2} + \cdots + \sum_{b \in G_N}\right) \sum_{a \in G_1} \mathbf{f}_{ba} + \mathbf{F}_1 \\ &= \sum_{b \in G_1} \sum_{a \in G_1 (a \neq b)} \mathbf{f}_{ba} + \sum_{b \in G_2} \sum_{a \in G_1} \mathbf{f}_{ba} + \cdots + \sum_{b \in G_N} \sum_{a \in G_1} \mathbf{f}_{ba} + \mathbf{F}_1\end{aligned}$$

and once again using the definitions (54) and (56),

$$\frac{d}{dt}\left(M_1 \frac{d}{dt}\mathbf{X}_1\right) = \mathbf{F}_{11} + \mathbf{F}_{21} + \cdots + \mathbf{F}_{N1} + \mathbf{F}_1 \,.$$

Here is

$$\mathbf{F}_{11} = \sum_{b \in G_1} \sum_{a \in G_1} \mathbf{f}_{ba} = 0$$

because of the Third Axiom for $\mathbf{f}_{ab}$ according to (55). Because G_1 was an arbitrary area, we receive the NEWTONian equation (53) for the arbitrarily considered mass M_A

$$\frac{d}{dt}\left(M_A \frac{d}{dt}\mathbf{X}_A\right) = \sum_{B \neq A}^{N} \mathbf{F}_{BA} + \mathbf{F}_A \,,$$

and once again (under observation of the reaction axiom for forces according to (55), $\mathbf{f}_{ab} = -\mathbf{f}_{ba}$) the reaction axiom is also valid without loss of generality for the forces $\mathbf{F}_{AB}$, for example for $\mathbf{F}_{21}$,

$$\mathbf{F}_{21} = \sum_{b \in G_2} \sum_{a \in G_1} \mathbf{f}_{ba} = \sum_{b \in G_2} \sum_{a \in G_1} (-\mathbf{f}_{ab}) = -\sum_{a \in G_1} \sum_{b \in G_2} \mathbf{f}_{ab} = -\mathbf{F}_{12} \,.$$

Therefore, we have proven that (53) is derived from (55).

Chapter 6
Lattice and the Continuum

In the last chapter,we conducted a step-by-step transition from a point lattice to a continuum. We showed starting from a linear chain of elastic interacting single masses how to arrive at an oscillating rod. All we have to do to achieve this is to continuously halve the masses and the connecting elastic springs. The mathematical result at the end of this process, from a coupled system of ordinary differential equations for single masses (31), is the wave equation for an elastic rod (61). One says of such a constructed continuum, that its internal forces are based on *contact interaction* or on the *continuous action*. In this connection, we talk about the hypothesis of continuous action. We would say in reference to single masses that every mass of the N-particle system only interacts with its direct neighbours. This factual situation should be maintained in the passing to the limit $N \longrightarrow \infty$. We will now show how the transition from normal differential equations for single masses Δm_i to the partial differential equations for a continuum of the mass density ρ can easily be described due to the continuous action hypothesis. We will once again consider the one-dimensional case, where we distinguish two situations, for which we however, at the end, receive one and the same result.

(a) The strict one-dimensional problem, in other words longitudinal oscillations: Up to now, we have only taken this case into consideration and observed the longitudinal oscillations of a rod. The displacements $s = s(x, t)$ were made in the direction of the rod, which we positioned on the x-axis. The Newtonian equations (57) are valid for the single masses Δm_i. For reasons of simplicity, we ignored the external forces F_a (in the passing to the limit of a continual mass distribution the external forces act as a force field $F(x)$). The total sum of the interaction forces refers only to its direct neighbours due to the hypothesis of continuous action, so that

$$\left.\begin{aligned} &\frac{d}{dt}\left(\Delta m_i \frac{d}{dt} q_i\right) = \sum_{k\neq i} F_{ki} = F_{i-1\,i} + F_{i+1\,i}\,,\\ &F_{ik} = -F_{ki}\,. \end{aligned}\right\} \tag{63}$$

H. Günther, *Elementary Approach to Special Relativity*,
https://doi.org/10.1007/978-981-15-3168-2_6

For every point of time t in our one-dimensional chain, the single force $F_{i+1\,i}$ that acts on the mass x_i to the right, is per definition equal to the *tension* $\tau(x_i, t)$. All of the forces are directed at the particle's line of junction, in the x-direction, and can therefore simply be added like numbers (their vector properties do not apply here). Therefore, $F_{i-1\,i} = -F_{i\,i-1} = -\tau(x_{i-1}, t)$ and the following are valid,

$$F_{i-1\,i} + F_{i+1\,i} = -F_{i\,i-1} + F_{i+1\,i} = -\tau(x_{i-1}, t) + \tau(x_i, t) \; .$$

(Only in the one-dimensional case tension is also a force; in a three dimensional continuum, tension is a force/area, see Chap. 23). The equally distributed positions of masses $x_i = \frac{L}{N} i$ are, in the passing to the limit assimilated into the continuous variable x, so that

$$x_i = \frac{L}{N} i \longrightarrow x \; , \quad x_{i+1} \longrightarrow x + \Delta x$$

and

$$F_{i-1\,i} + F_{i+1\,i} \longrightarrow \tau(x - \Delta x, t) + \tau(x, t) \; .$$

Using the TAYLOR series expansion for τ with a sufficiently small Δx, this becomes

$$F_{i-1\,i} + F_{i+1\,i} \longrightarrow \left(\tau(x, t) - \frac{\partial \tau(x, t)}{\partial x} \Delta x \right) + \tau(x, t) = \frac{\partial \tau(x, t)}{\partial x} \Delta x \; . \tag{64}$$

We replace the single masses with a continuous mass density ρ and the time-dependent displacements $q_i(t)$ of the single masses with the function $s(x, t)$,

$$\Delta m_i \longrightarrow \rho \Delta x \; , \quad q_i(t) \longrightarrow s(x, t) \; . \tag{65}$$

For the passing to the limit from the NEWTONian equations for single masses to a continuum, we place

$$\left. \begin{aligned} \frac{d}{dt} \left(\Delta m_i \frac{d}{dt} q_i \right) &\longrightarrow \Delta x \, \rho \frac{\partial}{\partial t} \frac{\partial s(x, t)}{\partial t} \; , \\ \sum_{k \neq i} F_{ki} &\longrightarrow \frac{\partial \tau(x, t)}{\partial x} \Delta x \; . \end{aligned} \right\} \tag{66}$$

The first passing to the limit (66) contains an important approximation that is somewhat complicated. The NEWTONian inertia term $dP/dt = \frac{d}{dt}(\Delta m \, v)$ describes the change with respect to time of a momentum carried by a certain moving mass Δm. The volume ΔV_m (in the one-dimensional case Δx_m) occupied by this mass also changes. During the transition to a continuum, whose motion we will describe in the scope of a field theory, we are interested in a completely other physical quantity. This quantity is the change with respect to time of momentum contained in a *fixed spatial element* of volume ΔV (in the one-dimensional case Δx), which we observe in a predetermined position x, a constant position of our reference system. We therefore have to calculate which statement of the NEWTONian inertia term $dP/dt = \frac{d}{dt}(\Delta m\, v)$ results in a fixed static volume ΔV. These calculations will be shown thoroughly in the appendix, in Chap. 25, where we will generally concern ourselves with this problem. The passing to the limit (66) is a special case of the derived passing to the limit (324). Here, we will have to be satisfied with the following statement. This limit is to be understood within the meaning of the elasticity theory, which alone interests us in the following. This means, in the so-called total time derivative for the velocity of matter v, $dv/dt = \partial v/\partial t + (\partial v/\partial x)\,(dx/dt)$, the second, non-linear term is simply dropped, and in the same way, for the quantity $v = ds/dt = \partial s/\partial t + (\partial s/\partial x)\,(dx/dt)$ only the partial derivative with respect to time for the displacement s should be considered. This leads to the approximation in (66). For the mathematics, we refer to Chap. 25.

Here, we want to point out that the application of the linear elasticity theory is not always justified. For example, using linear elasticity, we cannot calculate the elastic deformations that cause a so-called dislocation (see Chap. 7) in its immediate neighbourhood. Nevertheless, the continuation of our considerations will not be affected by this limitation. With the exception of Chaps. 25–27, where we will be dealing explicitly with linear problems of the theory of elasticity, we will be occupied only with plastic deformations (cf. Chap. 7), especially with the deformations of the dislocation line under the influence of the non-linear lattice potential, for example the creation and motion of so-called kinks in the dislocation line, see Chaps. 9 and 10.

We insert (66) in (63) and find by cancelling Δx,

$$\rho \frac{\partial}{\partial t}\frac{\partial s(x,t)}{\partial t} = \frac{\partial \tau(x,t)}{\partial x}\,. \tag{67}$$

(However, in a three-dimensional lattice we get the Eq. (325) from Chap. 25.)

We will now formulate the three fundamental assumptions of the linearised theory of elasticity:

1. In the NEWTONian inertia term for the total time derivative for the momentum of moving masses, the partial time derivatives of the matter velocity v or rather of the displacement s are simply to be set.
2. We presume, that the changes in mass density ρ that inevitably occur during an oscillation can be ignored, so that the density ρ is a constant.
3. HOOKE's law is valid; in other words, for every point of time t, the stress τ at position x is directly proportional to the relative strain ε,

$$\left.\begin{aligned} &\frac{d}{dt}(\Delta m\, v) = \Delta x\, \rho \frac{\partial}{\partial t} v = \Delta x\, \rho \frac{\partial^2 s}{\partial t^2}\,, \\ &\rho = \text{const.}\,, \\ &\tau(x,t) = E\,\varepsilon(x,t) = E\,\frac{\partial s(x,t)}{\partial s}\,. \end{aligned}\right\} \text{Linearised theory of elasticity} \tag{68}$$

Within the linearised theory of elasticity, we also presume that the factor of proportionality E of HOOKE's law as well as the mass density ρ is a constant in space and time. Because we have limited our considerations to monocrystalline solids, whose structures will be dealt with in the next chapter, we do not have to consider viscose material properties, where stress is also dependent upon the strain rate.

When we apply the linearised theory of elasticity (68) on Eq. (67), due to the constancy of E and ρ, we immediately discover our wave equation (61),

$$\left.\begin{aligned} &\frac{\partial^2}{\partial x^2}s(x,t) - \frac{1}{c^2}\frac{\partial^2}{\partial t^2}s(x,t) = 0\,, \\ &c = \sqrt{\frac{E}{\rho}}\,. \end{aligned}\right\} \tag{61}$$

Due to the fact that we have considered longitudinal displacements, the quantity c in (61) is the longitudinal sound velocity. See also the second footnote in Chap. 5.

(b) Transversal oscillations:

In this case, we maintain the continuum, or rather the linear chain in its one dimensionality, but we now presume that the displacements $q_i(t)$ or rather $s = s(x,t)$ take place perpendicular to the line direction. In other words, we consider the transversal oscillations of a one- dimensional system. Instead of a rod, we will use as an example an oscillating string. We immediately notice that compared to the longitudinal oscillations of a rod, we need a supplementary condition. If we take the string, hold it at both ends and put it on a table, it cannot perform a small harmonic oscillation. We would have to produce a very large transversal deflection of the string in order to be able to produce an oscillation. Luckily, we are not interested in such oscillations with very large transversal amplitudes s, because they are mathematically harder and therefore more complicated to understand. This is however different if the string is stretched, as for example in the case of a violin or a guitar. Then, every small excitation leading to a starting deflection leads to harmonic oscillations that we can hear as tones. We now want to show that all the equations in this chapter remain valid for these transversal oscillations, as long as we presume that a tension σ is already present in the direction of the line when the string is in a state of rest; in other words, the string is stretched with the force σ. Figure 6.1 shows what this looks like in a linear chain. The forces of the amount σ from the left and the right act equally on every particle at rest. This is the tension of the chain.

Figure 6.2 is the result of a transversal deflection. We have to take the vector properties of the forces, i.e. their directions between the particles into consideration.

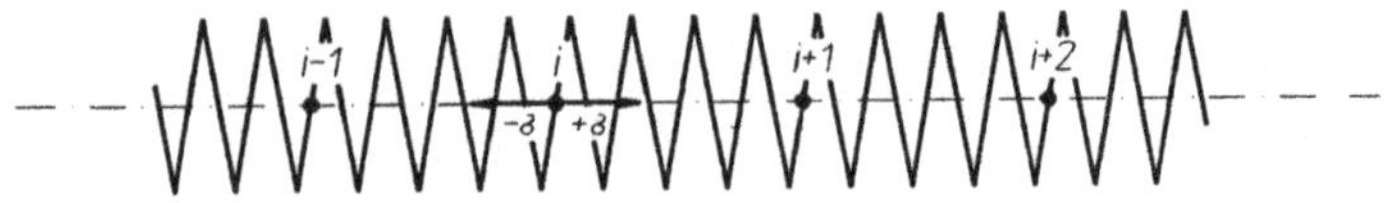

Fig. 6.1 Equilibrium in the linear chain stretched by the line tension σ

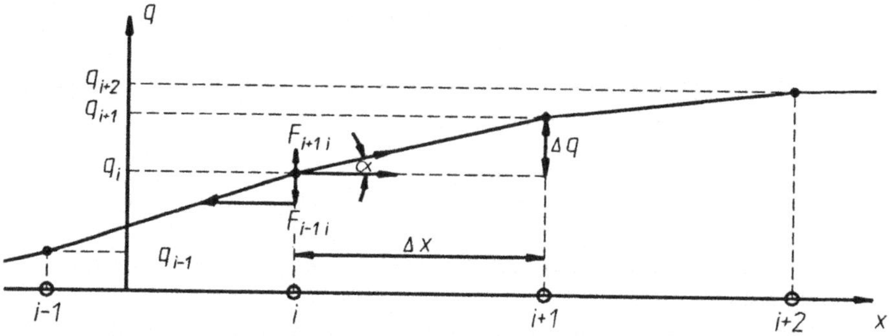

Fig. 6.2 Transversal deflection q of the masses of a linear chain in dependence on its position on the x-axis. Because the horizontal forces cancel each other, according to our approximation (69), the interaction forces F_{ki} can be seen as purely transversal (see explanations in the text)

Here, we make a principle assumption of approximation. An angle α is defined according to $\tan\alpha = \Delta q_i/\Delta x$, with the difference Δq_i of the deflections of two neighbouring particles, $\Delta q_i = q_{i+1} - q_i$ and their distance $\Delta x = x_{i+1} - x_i$. We can assume of all of these angles α that they are 'small'; in other words,,

$$\left.\begin{aligned} &\alpha \approx \sin\alpha \approx \tan\alpha \approx \frac{\Delta q}{\Delta x} \ , \\ &\cos\alpha \approx 1 \ . \end{aligned}\right\} \tag{69}$$

Now, we can immediately see out of Fig. 6.2 that for all tangential forces $F_x = \sigma \cos\alpha$ is valid. Because the forces acting on every particle of its direct neighbours have opposite directions, the tangential forces cancel each other due to the approximation (69). We therefore do not need to observe these for a motion. Therefore, all the forces in Eq. (57), where we dropped the external forces F_a, are purely transversal and lead to pure transversal motions. In case **(b)**, for the transversal oscillations, we read in Eqs. (63) all the deflections q_i and all the forces F_{ki} as pure transversal quantities. With this additional explanation, we can literally practically use all the details of the case **(a)** for longitudinal oscillations.

The single transversal force $F_{i+1\,i}$ that acts on a single mass x from the right is per definition equal to the (transversal) tension $\tau(x_i, t)$, and for the sum of both left and right transversal forces acting on a mass x_i, the following is once again valid,

$$F_{i-1\,i} + F_{i+1\,i} = -F_{i\,i-1} + F_{i+1\,i} = -\tau(x_{i-1}, t) + \tau(x_i, t) \ ,$$

with the result (64) when Δx is sufficiently small. We also arrive at Eqs. (65)–(67) in exactly the same way as shown above. Hooke's law (68) can be directly seen in Fig. 6.2. For the transversal force, for example on a particle at x_i, the following is valid:

$$F_{i+1\,i} \equiv \tau(x_i, t) = \sigma \; \sin\alpha = \sigma \; \tan\alpha \;\; ,$$

hence taking the approximation (69) into account, $\tan\alpha = \Delta q_i(t)/\Delta x$, in the passing to the limit of continuous quantities,

$$\tau(x,t) = \sigma \, \frac{\partial s(x,t)}{\partial x} \;\; , \quad \rho = \text{const.} \tag{70}$$

This is however nothing else than one of the basic assumptions (68) of the linearised theory of elasticity.

For transversal oscillations, we only get an additional statement telling us, that the modulus of elasticity is identical to the base tension σ of the line. Finally, we arrive once again at our wave equation (61). In case **(b)** of transversal oscillations, c stands for the transversal sound velocity, and the following is valid,

$$\text{Modulus of elasticity E} = \text{line tension}\sigma \;\; . \tag{71}$$

We have finally found a very simple procedure, starting from the NEWTONian equations (57) (or rather (63)), moving to the system of ordinary differential equations (31) of a linear chain and then to the wave equation (61) of the one-dimensional continuum, for which we had to do a great deal of calculation in the last chapter.

Transition from single masses to a continuum:

With the passing to the limit (66), we move over from NEWTONian point mechanics (63) to the continuum and assume the condition (68) of a linearised theory of elasticity.

Such a simplified transition from single masses to an approximation of our solid by a continuum, which we have seen to be valid for both longitudinal and transversal displacements, will come to save us a lot of work. This allows us to include all the phenomena on a lattice elegantly in a mathematical form, which could otherwise not be so simply shown. A typical example would be the transportation of energy by a wave motion in a lattice. This wave motion is very easily described in the linearised continuum theory, as shown in the appendix, in Chap. 25. Every mechanical process in the continuum picture has an equivalent on the lattice, where it however has to be mathematically described with a lot of effort as seen in Chap. 4 when we dealt with the linear chain.

Equation (61) is a wave equation for the displacement s of continuous distributed masses. To be able to measure (61), we have to determine the momentary positions and the equilibrium positions of the masses involved. Equation (61) has a background of moveable single masses as a condition, bound to equilibrium positions. (We do however know that these are only the centres of mass coordinates). We can free ourselves from this fixed background if we ask for the equation of the relative strain $\varepsilon = \partial s/\partial x$. Then, partial differentiation of (61) with respect to x results in

$$\frac{\partial^2}{\partial x^2}\varepsilon(x,t) - \frac{1}{c^2}\frac{\partial^2}{\partial t^2}\varepsilon(x,t) = 0 \;\; . \tag{72}$$

The quantity ε is a field in one-dimensional space as defined by the continuum (the rod) and satisfies D'ALEMBERT's wave equation. In the view of this, ε is in fact comparable to other fields in physics, for example electromagnetic fields. However, the space in which electromagnetic waves propagate is three dimensional, whereas the one-dimensional lattice, which we idealised to a one- dimensional continuum (an infinitely long rod), is the space in which the field ε propagates. The ε-waves are nothing more than sound waves. We could also say that the rod is the ether for sound, just as we could talk about the propagation of light through the ether, as long as we only mean the propagation through our three-dimensional space. Ether and space will be used synonymously. The lattice (or the equivalent continuum) as a carrier, the space for sound waves is the aspect that we will delve deeper into in the course of our consideration. Here, we will not however just deal with and investigate sound waves, but we will concentrate on other physical states of the lattice, just as we can consider other physical states in three-dimensional space, not only electromagnetic waves, but also elementary particles for example.

The relative elastic strain ε is the physical quantity that represents elastic deformations in far more complicated lattices (or rather continua) that can be measured directly or indirectly for every state of deformation. The relative strain of a rod can be described using one function alone, $\varepsilon = \varepsilon(x, t)$, whereas the elastic strain of two- and three- dimensional atomic lattices have to be represented by a more complicated mathematical quantity. Strain is a tensor ε with three, or rather six independent functions for the two dimensional- or the three-dimensional case. The same applies for the elastic stress caused by strain. These relations are not necessary for the representation of our problem. In Chaps. 25–27, we will show the mathematically interested reader more about the three-dimensional lattice and the continua. We will, however, only make use of the shown results in supplementary explanations. With regard to the physical meaning of strain ε it must be said that when dealing with complicated, multidimensional atomic lattices, we have no choice but to work with this relative strain ε. An elastic displacement $\mathbf{s}$ out of a position of equilibrium cannot be defined for complicated atomic lattices that deviate from the ideal structure (and this applies to most of them). All crystalline solids found in nature show characteristic deviations from the ideal structure, which are only found in small local areas of a lattice. These deviations are produced by the so-called dislocations, which are the key to understanding plastic deformations of crystals. They are of utmost importance for practically all mechanical properties of crystalline solids. We will have to examine the properties of dislocations more closely.

For the strict one-dimensional lattice or continuum, there is only one longitudinal sound velocity. A two- or three- dimensional lattice, however, additionally allows for transversal motions which produce transversal sound velocities. The longitudinal sound velocity always differs from the transversal sound velocity. This is a problem of the two- and three-dimensional lattices. As we know from the theory of elasticity and what we will explicitly show in Chap. 25, there are always two sound velocities which are different from each other even in the mathematically simplest, multi-dimensional lattice. In the view of this, our lattice, the ‘space for the sound waves’ as we wanted to say differs substantially from our space for the light with its single critical vacuum speed of light c_L. Those who know the Special Theory of Relativity, immediately know that two critical velocities for Special Relativity are impossible. There is no LORENTZ transformation with its characteristic consequences of time dilatation and LORENTZ contractions for two critical velocities. In the light of this, sound is *not* somewhat like light, at least not inside of an atomic lattice. One could at best think of liquid or gaseous substances that also only possess one critical velocity. These however, do not find our interests, because they have no internal mechanical structures. These mediums remain empty for further questions concerning Special Relativity.

In our case, it will not be the theory of elasticity, but plasticity on which the emphasis lies, and this is determined by the properties of one-dimensional dislocations. As we will see in the following chapter, the dislocations inside of a crystal actually, in a sense, define their own one-dimensional lattice, from which we consider the transversal deflections, which then leads to one single defined critical velocity.

Of course, there is a relation between dislocations and elastic deformations, but these we can neglect in the first approximation. In Chaps. 26–27, we will also discuss the relations between the dislocations and the elasticity of three- dimensional lattices, including those that have more than two critical velocities. There we will prove that we can separate a well-defined part, the so-called structural eigen stresses, from the elastic deformations caused by dislocations. These structural eigen stresses are, in comparison to all elastic deformations, dependent only upon one single critical velocity, which has the surprising consequence that in the area of plasticity including the 'structural elasticity', LORENTZ symmetry of a solid is valid. In other words, expressed in the language of physics: In a solid, we discover a symmetry group which is isomorphic with the symmetry group of our physical spacetime. Here, we wish to express that the terms 'LORENTZ transformations', 'LORENTZ symmetry', 'LORENTZ factor', etc. are in fact reserved in literature for corresponding terms that possess the parameter of the speed of light, and we should therefore use a terminus 'proper MINKOWSKI rotation', compare to J. A. SCHOUTEN [85], Chap. I. H. F. GOENNER [27], Chap. 1, has pointed to the isomorphism of these correlations with arbitrary changes in the finite critical velocity. We will keep the terms 'LORENTZ transformation', 'LORENTZ symmetry', etc. even for critical velocities occurring inside a solid, because here confusion cannot occur, (cf. also Chap. 13) and thus use A. SEEGER's terminology, see for example SEEGER [86]. For the relations between Special Relativity and classical field theory see H. F. GOENNER [28].

The various sound velocities for the complete theory of elasticity, however, lead once again to a breaking of the LORENTZ symmetry for the complete system of dislocations and elastic deformations. The valid LORENTZ symmetry of a solid will thus not always be easy to recognise. It is thus of interest to pick out such a class of physical phenomena for solids, for which this LORENTZ symmetry applies in its complete clearness. To this end, we will next examine crystals and their dislocations more closely in the following chapter.

Chapter 7
The Crystalline Solid—Dislocations

Since time out of mind, the continuous re-emerging forms of crystals have occupied the thoughts of mankind—be it that their constantly new, self-generating, unchangeable structural order in the domain of the microscopes gives them the aura of eternity, or that their valuable mechanical properties lay hidden in these regularities. In 1912, M. v. Laue made a decisive contribution in the decoding of the basic principles of this order with his famous diffraction experiments. The atoms (or groups of atoms) of an ideal crystal are thus arranged in a periodical, infinite, three-dimensional space lattice in such a manner, that one can always distinguish a group of atoms whose continuous translation in three different independent directions could lead to a continuously expanding crystal, see Fig. 7.1. Here, we examine only monocrystalline structures, not polycrystals made up of many single crystals. Independent of this, every real crystal that can thus only have a finite expansion must principally differ from the ideal symmetry alone by its boundaries.

If one sends X-rays through such a crystal whose wavelength is about as large as the distance between two lattice planes, the well-known Laue diagrams are produced. With the help of these diagrams, we can calculate the structure of the lattice, see Fig. 7.2.

Crystallography deals with the mathematically possible details and arrangements of lattice structures. For our purposes, it will be sufficient to use a simple cubic lattice. Even the properties of this lattice will be made of restricted use in the passing to the limit to the continuum mechanics. In fact, here we only need a modelled extension of our observed linear chain in three-dimensional space.

A very important fact when dealing with lattices is that we are dealing with two completely different possible mechanical kinds of motion not including the third possible kind of motion, the motion of the complete lattice as a rigid solid, which does not interest us.

H. Günther, *Elementary Approach to Special Relativity*,
https://doi.org/10.1007/978-981-15-3168-2_7

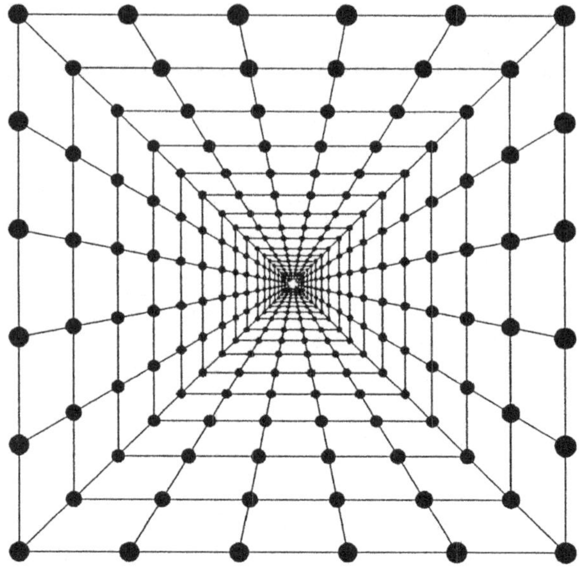

Fig. 7.1 Continuous re-emerging structures in a crystal lattice

1. The first mechanical form of motion has already been discussed in detail. Elastic waves that transport mechanical energy through a solid (for example, the energy of a hammer blow after I hit the end of a rod). Here, we can also say that signals are transmitted through a solid with the sound velocity. It is of no importance that we have only examined a one-dimensional lattice, the rod. The mathematical situation for three-dimensional lattices is more complicated (in fact generally far more complicated) and results in a complicated system of directional and polarisational dependent sound velocities.[1] However, nothing changes our fundamental physical statement concerning the signal transmission with one of the sound velocities. In the appendix, Chap. 25, we will discuss the simplest cases of a three- dimensional lattice and continuum. Physically, this mechanical kind of motion of the lattice is generally defined by the term *elastic deformation*. An atom takes a position inside of the elastically deformed crystal, from which it then moves back to its original lattice position after the reason for the elastic deformation has been removed, see Figs. 7.4 and 7.5. It is characteristic for an elastic deformation to completely disappear when the cause for the deformation is removed. There are static and dynamic elastic deformations.

2. We have also made day-to-day experiences with the second mechanical kind of motion. We bend a wire, roll a metal sheet, make a dent in a car, forge iron, stretch, hammer, engrave, chisel, etc. The physical superimposed concept for this mechanical kind of motion of atoms in a solid is *plastic deformation*. We can thus note:

[1] We understand the direction of the oscillating vector **s** of an elastic displacement in comparison to the direction of propagation **k** of the wave by polarisation. One speaks of longitudinal waves, when **s** and **k** oscillate parallel to each other and one speaks of transversal waves when **s** and **k** oscillate orthogonal to each other.

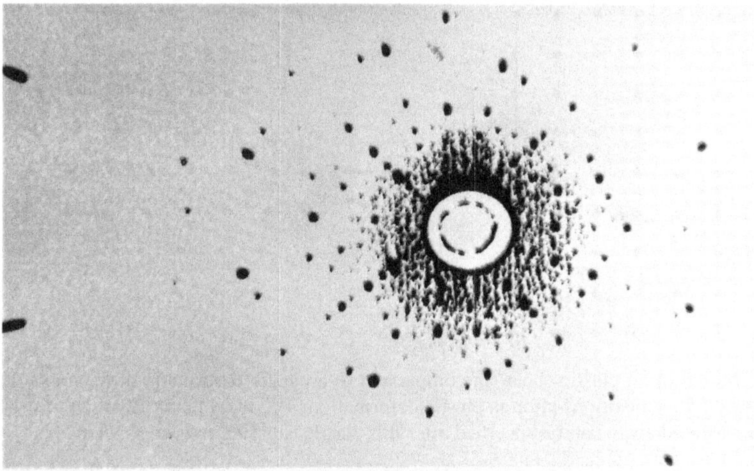

Fig. 7.2 LAUE pattern of a silicon crystal according to J. WASHBURN's photograph in C. KITTEL [46], 1

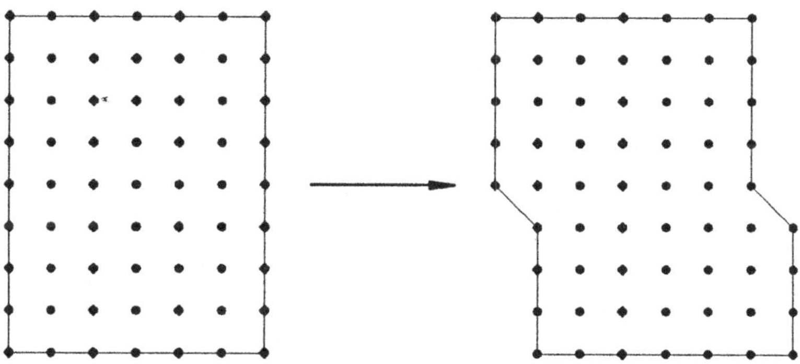

Fig. 7.3 Model of a 'minimal' plastic deformation in a cubic lattice. The change in form when compared with the initial state is visible. After the plastic deformation has taken place, all the atoms once again occupy ideal lattice positions. It is therefore not possible to determine after the deformation has taken place where the atoms in the ideal lattice were originally situated. The same plastic deformation could also have been produced by a different suitable displacement of atoms. In contrast to elastic deformations compare with Fig. 7.4

There are two characteristic completely different possible mechanical kinds of lattice motion: elastic and the plastic deformations.

Simplified one can say that elastic deformations only occur when the involved forces are small enough. If the mechanical influences on the solid (our hammering) increase, then plastic deformations also occur. Plastic deformations lead to a permanent change of a solid, elastic deformations do not.

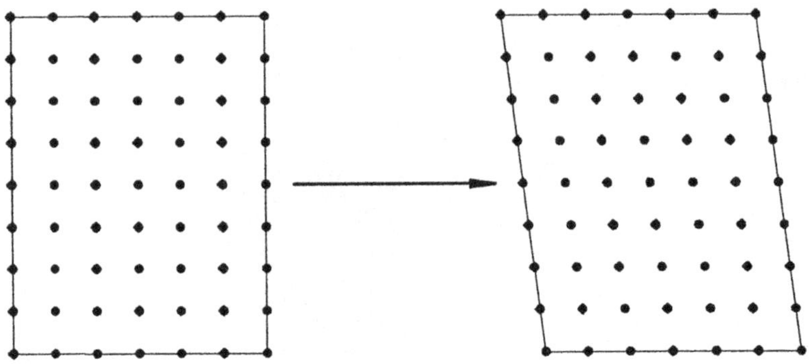

Fig. 7.4 Model of an elastic shear. In comparison to a plastic deformation, the atoms do return to their ideal lattice positions after an elastic deformation. The ideal lattice position of every atom elastically sheared away can be specified after this elastic shearing took place. This is also shown in Fig. 7.5

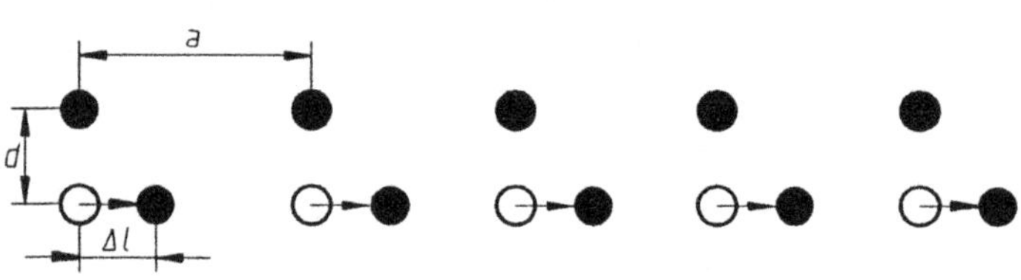

Fig. 7.5 Illustration of the process of elastic shear

There are *two* characteristic differences to an elastic deformation. The *first* difference is that after a plastic deformation some of the atoms occupy different positions, namely neighbouring lattice positions.

Now to our *second*, for plastic deformations characteristic difference. First, we will form the hypothesis that the smallest plastic deformation is the displacement of the upper part of the crystal to its lower part by one lattice parameter. We will later see that this idea will have to be corrected twice. We have shown such a minimal plastic deformation in a cubic lattice before and after from a head on view in Fig. 7.3. To test this hypothesis of a minimal plastic deformation, we will calculate an approximated value for the amount of the force needed to produce such a plastic deformation. This estimation goes back to J. Frenkel [23]. Using sufficiently small forces, the deformation stays elastic and we receive Fig. 7.4. We have shown a section of this in Fig. 7.5.

We can also, as with Hooke's law, cf. (68), for one-dimensional elongation $\tau = E\,\varepsilon$, start with a linear relationship between the elastic shear $\Delta l/d$ (with the displacement Δl of the lower plane and a distance d between the lattice planes above it) and the force F acting on the lower area A, the shear stress $\sigma = F/A$. Using a constant of proportionality, the modulus of shear G, we can write

$$\sigma = G\,\frac{\Delta l}{d}\,. \tag{73}$$

However, it is easily to see that the stress has to increase to $\Delta l/d$ slower then proportional with increasing strain. The force has to completely disappear after the masses of the lower plane have moved to the intermediate position of the higher plane, because the atoms could just as well come from the other side with the opposite force. If we traverse across the neutral position, we have to change the sign of the force. We presume a distance a of the atoms in the planes (this distance is generally somewhat different from the plane distance d). At $\Delta l = a/2$, the force is zero just as it is zero at $\Delta l = a$. This dependence of the force on the displacement Δl can be reproduced in an approximation using a sine function. In place of (73), we can write for arbitrary Δl,

$$\sigma = G\,\frac{a}{2\pi d}\,\sin\left(\frac{2\pi}{a}\Delta l\right)\,. \tag{74}$$

Because of $\sin x \approx x$ for $x \ll 1$ Eq. (74) leads to Eq. (73) for small Δl. Equation (74) also fulfils the demand that the force for $\Delta l = n\,a/2$ has the value zero. The largest force $F = A\,\sigma$ that, according to this model still manages to cause elastic displacements follows from the maximum value of s when $\Delta l = a/2$, so that

$$\sigma_{max} = G\,\frac{a}{2\pi d}\,. \tag{75}$$

Only then, when this value for stress is exceeded does the lower plane glide one atomic position according to Fig. 7.5; in other words, the lattice is plastically displaced or deformed. The term used for the stress value that leads to a plastic deformation is called the critical shear stress σ_c. We calculate with $a \approx d$, so that $\frac{a}{2\pi d} \approx 1/6$. Hence, this critical value σ_c for a plastic change in a crystal should, according to our model and Eq. (75), lay around a sixth of the value for the modules of shear G. However, values around 100–1000 times smaller are in fact observed for the critical shear stress σ_c. Our model for the process of plastic deformations thus can not be correct.

We will have to look for the solution to this riddle in the fact that the atoms of a lattice do really also take other positions then those attributed by ideal space lattice symmetry rules. Such non-ideal lattice places are occupied in astoundingly large numbers when a crystal grows naturally resulting in ideal lattice positions being left empty. Here, we distinguish the *real crystal* from the *ideal crystal* where all ideal lattice positions and only these positions are occupied. All real crystals have small areas with an ideal lattice structure (neglecting surface effects). The unoccupied ideal lattice positions in a crystal are called *imperfections* of a crystal (*structural defects, crystal defects*). In this area where such defects occur, atoms do not occupy their ideal lattice positions, but other positions where they do not sit as stabile as they would in their ideal positions:

Imperfections of a lattice are areas of higher mechanical mobility inside of a crystal.

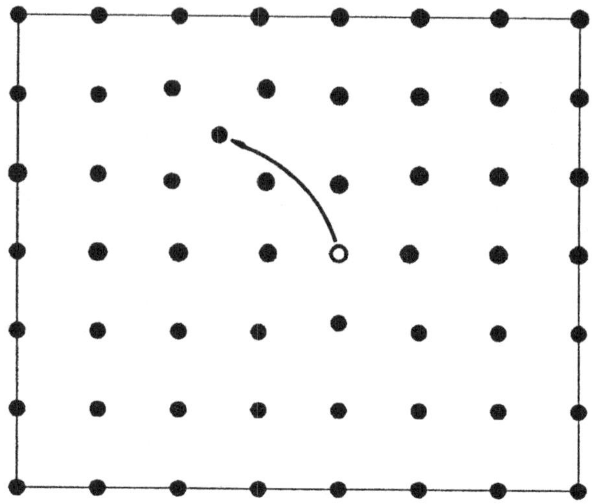

Fig. 7.6 Point defects do not cause plastic deformations. The lattice atoms in the region of the point defect (vacancy or interstitial atom) only experience elastic displacements

We will now search for a relationship between the motion of lattice imperfections and plastic deformations. There are structural defects that are restricted to one lattice position, for example interstitial atoms, *vacancies*. One can easily see that those displacements starting from such *point defects* do not cause a plastic deformation, see Fig. 7.6. A different class of defects leaves the ideal lattice positions along a complete finite curve in a lattice unoccupied. Such defects are called *dislocations*. The curve is the dislocation line[2]. Dislocation lines can never end inside of a crystal; in other words, they are either closed or end on the crystal's surface. This property will, in a moment, be clarified by a description of dislocations. For a detailed description of the situation concerning dislocations in crystals, we refer to E. KRÖNER's [47] book published in 1958 that in its clarity and preciseness gives us a very good orientation of this field, cf. also the English version of this subject in E. KRÖNER's [48]. For our questions, we will have to be content with the following short description of the most important properties. In Chaps. 26 and 27, we will show for the mathematically interested reader a somewhat more complex approach to dislocations. For further developments, see F. HEHL [40] and E. KÖNER.

To elucidate the characteristics of a dislocation it is helpful to illustrate a property of an ideal crystal containing no dislocations, see Fig. 7.7. All atoms are positioned

[2]It can occur that vacancies align themselves along a line, which however cannot lead to plastic deformations. We will see in Chap. 8 that an important property of deformations of a dislocation line originates from the base tension of this line, the so-called line energy, which remains preserved for all deformations and can principally not be passed on to the neighbouring atoms of the lattice. Founded on this is the decomposition (80) for the forces on a dislocation line into two physically different terms (compare (81) and (86)). Such a property does not apply for lines composed of vacancies. The line energy of a chain of vacancies can in fact be passed on to neighbouring atoms in a lattice leading to the decomposition of the chain which is impossible for a dislocation. Our explanations in Chap. 8 lead to the sine-GORDON equation for a dislocation can therefore principally not be transferred to a line composed of vacancies.

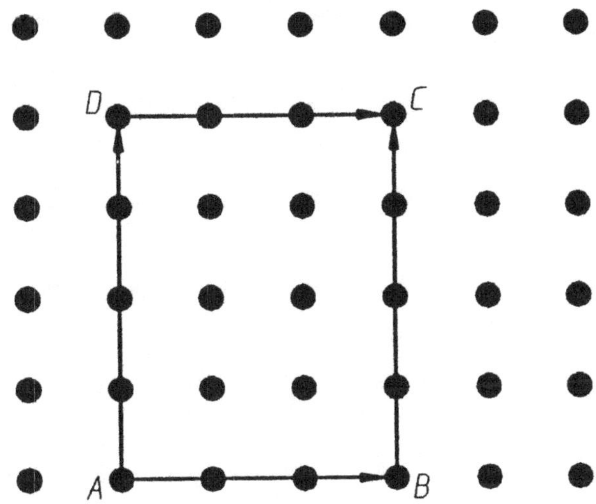

Fig. 7.7 Closed path in an ideal crystal. From atom A, we reach atom B in three steps in the first lattice direction, and from here, we reach atom C in the second direction in four steps. If we start from A and move four steps in the second direction until we reach atom D and then move three steps in the first direction, we should, being inside an ideal crystal, reach the same atom C

in ideal lattice positions. From an arbitrary atom A, we move n lattice parameters in a lattice direction, which we have chosen as the x-axis, until we reach an atom B. From here, we move m lattice parameters in a second, different lattice direction, which we have chosen as the y-axis, until we reach atom C. Now, we go back to atom A and move m lattice parameters in the y-direction and arrive at atom D. From atom D, we move n lattice parameters in the x-direction. If we have covered an area that has an ideal crystalline structure, we should arrive at the same atom C. This is characteristic for a crystal containing no dislocations. Had the covered area has included a dislocation, we would get a different result as seen and explained in Fig. 7.8.

There are two types of dislocations, edge dislocations and screw dislocations. The *edge dislocations* were in fact discovered three times in 1934 independently from each other by E. Orowan [69], M. Polanyi [76] and John G. Taylor [92]. In edge dislocations, one lattice plane ends inside a crystal. The line along this lattice plane is the dislocation line. It is described by the direction vector **t** (see Fig. 7.10). Evidently, this line cannot end inside of a crystal. The vector leading from the lattice plane containing the dislocation line, to the neighbouring plane is called the Burgers vector **b** of a dislocation. In edge dislocations, the Burgers vector is perpendicular to the dislocation line; in other words, the scalar product of both vectors **b** and **t** disappears, $\mathbf{b} \cdot \mathbf{t} = \mid \mathbf{b} \mid \mid \mathbf{t} \mid \cos(\mathbf{b}, \mathbf{t}) = 0$. The Burgers vector is explained by a so-called Burgers circuit that we describe in Fig. 7.8.

In *screw dislocations*, the lattice plane winds itself around the dislocation line in the form of a screw. It is also evident that this process cannot end inside of a crystal, compare with Fig. 7.9. The Burgers vector is now the pitch of the screw; in other words, in a screw dislocation, the dislocation line is directed parallel to the Burgers vector so that $\mathbf{b} \cdot \mathbf{t} = \mid \mathbf{b} \mid \mid \mathbf{t} \mid$. These screw dislocations were discovered in 1939 by J. M. Burgers [7, 8]. In order to describe the process of plastic deformation, we now

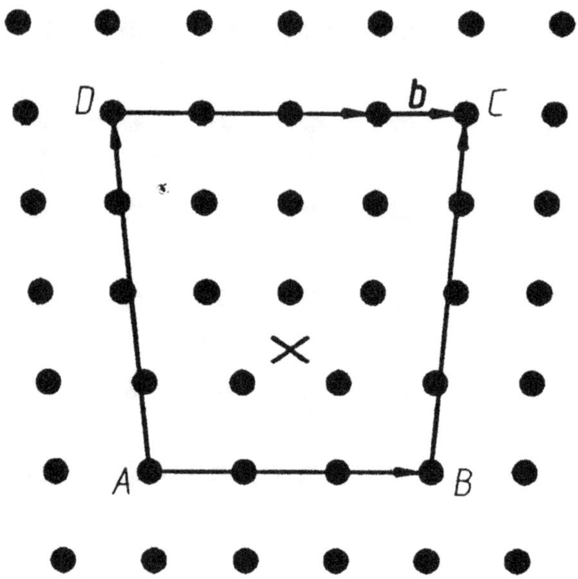

Fig. 7.8 Edge dislocation in the plan view according to E. KRÖNER [47, 48]. The direction vector **t** of the dislocation points vertically out of the image plane. The dislocation is defined by a BURGERS circuit as follows: As in Fig. 7.7, we start from A and move m lattice parameters (here $m = 3$) in the first crystal direction, here the x-axis and arrive at B. We move from B n steps (here $n = 4$) in the second direction, here the y-axis and arrive at C. Afterwards, we take n steps from A in the y-direction and reach point D, and then, we move m steps in the x-direction towards the atom at C. If we miss the lattice point C by the length **b**, we have enclosed a dislocation with the BURGERS vector **b**. For the calculation of **b**, the elastic displacement of the atoms has to be subtracted. Because our considerations are completely restricted to the linear theory of elasticity, we need not differentiate between the true and the local BURGERS vector, see F. C. FRANK [21]

consider for the sake of simplicity a straight edge dislocation in a cubic lattice, see Fig. 7.10. The plastic deformation, as seen and described in Fig. 7.3 in one step, now consists of many small migrational steps of a dislocation through the crystal.

In comparison to our simple Fig. 7.3 where the complete lower crystal block had to be moved as a whole, our new conception of a dislocation motion in one step will only move as much matter as is moved when the dislocation line is moved by one lattice parameter, a BURGERS vector **b**. The relatively small value of the critical shear stress σ_c for the starting of a plastic deformation can be explained by this dislocation motion.

We see: The *second* characteristic of a plastic deformation is that the ideal lattice structure is disordered along whole lines, the so-called dislocation lines. These dislocation lines change their position in a crystal when plastic deformation takes place. The description of these dislocation motions in linear approximation goes back to E. KRÖNER [49] and G. RIEDER, and later was able to be formulated in non-linear terms,

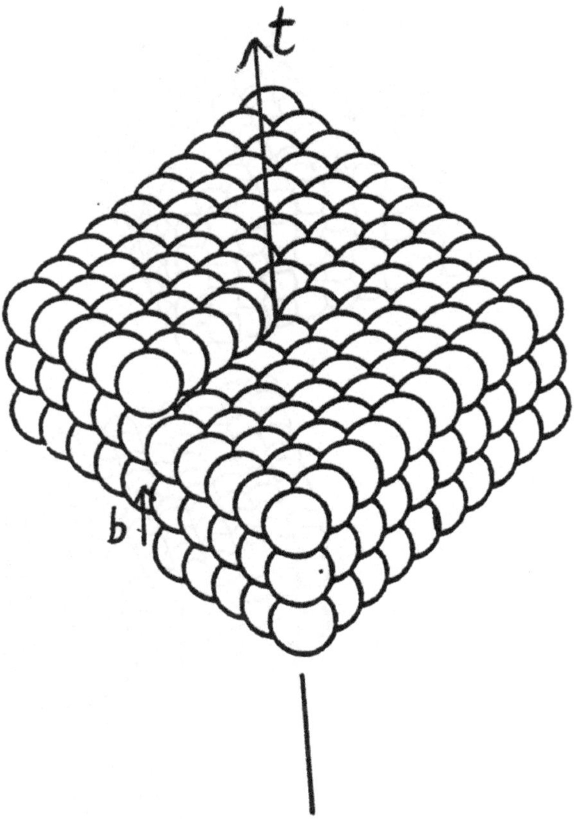

Fig. 7.9 Screw dislocation according to E. KRÖNER [47, 48], visibly ending on the surface of the crystal. The line direction **t** of the dislocation and its BURGERS vector **b** are parallel to each other

compare H. GÜNTHER [31, 32]. In comparison to plastic deformations in the elastic case, the atoms of a lattice return to their old positions, be they ideal or non-ideal lattice positions when the force causing the elastic deformation is removed. This is the first correction of our original description of the process of a plastic deformation. For a quantitative description of the motion of dislocations, we wish to refer to the appendix in Chap. 26.

The second correction starts with our idea that a dislocation line, which can also be found in a bent form, if it is not just a pure screw dislocation moves as a whole in one step whilst either keeping or changing its geometrical form. This type of a dislocation motion takes place inside a crystal when large forces cause a fast and strong plastic lattice deformation. Such global dislocation motions and its relationship to elastic deformations is not our main concern. We will talk about this in the appendix, in Chaps. 26 and 27. Instead, we will investigate the question, which motions are possible for the single sectors relative to each other in a dislocation line. The motion possibilities inside of a single dislocation line will be investigated

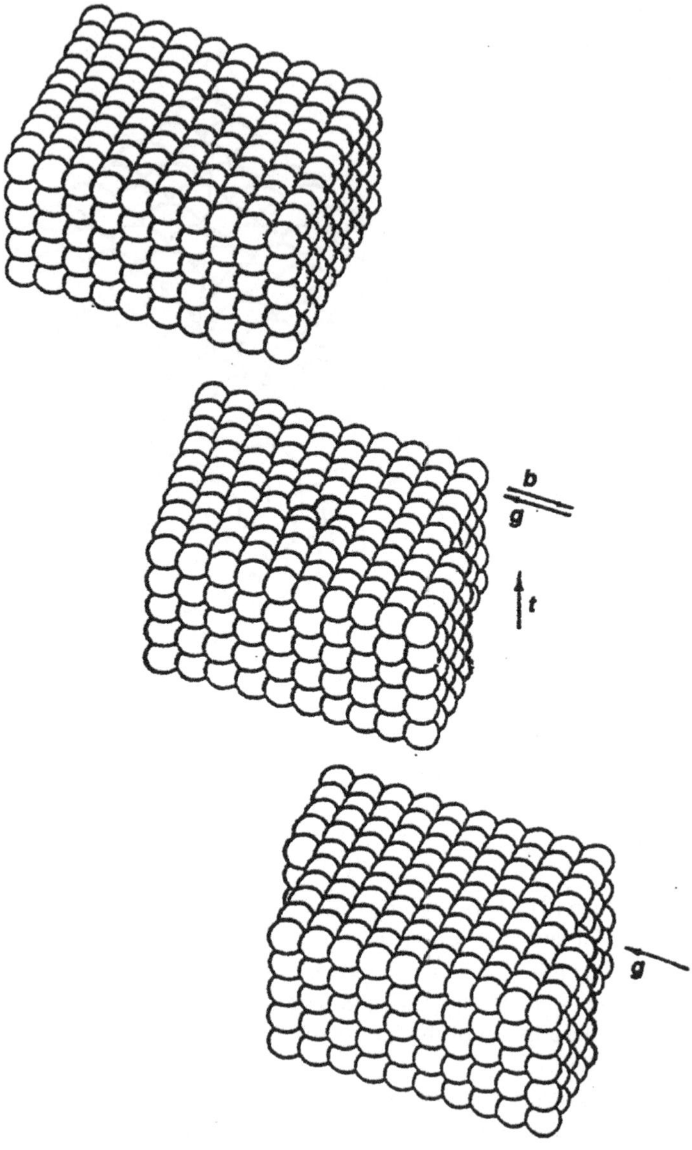

Fig. 7.10 Passage of an edge dislocation with the Burgers vector **b** and the line direction **t** through a crystal block inside a primitive cubic lattice according to E. KRÖNER [47, 48]. The vector $\mathbf{g} = -\mathbf{b}$ gives us the relative displacement of both sides of the slip plane

more closely. This will be the second correction of our original model conception of plasticity. Such internal motions of dislocation lines are termed as microplasticity. They dominate when the involved forces are small.

Chapter 8
The sine-GORDON Equation of a Dislocation

The distances between the points of a lattice make up a characteristic length for the properties of a dislocation. We have already observed this in our elementary considerations of the critical shear stress in the last chapter. This length is about 10^{-8} cm. Hence, we find ourselves, measured in lattice units inside a small crystal (with a linear dimension in the cm-region) infinitely far away from its surface. For the following, we will use as a good approximation for this situation an infinitely extended crystal. The starting point of our considerations will be the geometrically simplest, and out of reasons of symmetry, the physically simplest form of a straight, never-ending infinitely long dislocation line in a crystal. We have seen that the displacement of the straight dislocation line as a whole is a somewhat suitable model for plastic deformations. However, does the dislocation really get displaced as a whole, rigid as a stick, or is the displacement process of a dislocation line by one lattice position, compare with Fig. 7.10, generally more complex? Are there other neighbouring physical stabile geometric forms existing alongside the straight dislocation with its energetically preferred line form, thus allowing them to exist force free inside of a crystal as for an arbitrary time, where these forms are composed partially of straight lines, dents and edges? One must also think of those line forms that can only exist whilst participating in characteristic motions. Such lines are called dynamically stabile.

The mathematical condition for all these sorts of dislocations in the proximity of a straight dislocation line is made up of one equation, the sine-GORDON equation. The sine-GORDON equation has a long history. It describes the energetically possible, physically stabile forms of a dislocation in the neighbourhood of a straight dislocation line. It was a long hard trek until the uniform and mathematically simple condition was discovered for this. After years of research by L. PRANDTL [78], U. DEHLINGER [10] and J. FRENKEL and T. KONTOROVA [25], A. SEEGER [87] in 1949 (cf. also A. SEEGER [88] and P. SCHILLER) and later F. C. FRANK [22] and J. H. VAN DER MERWE managed to formulate the sine-GORDON equation for a dislocation. Ten years later this equation

H. Günther, *Elementary Approach to Special Relativity*,
https://doi.org/10.1007/978-981-15-3168-2_8

was discussed whilst searching for field theoretic models for elementary particles, see T. H. R. SKYRME [84] [91] (who gave the equation its name) and also the survey by C. REBBI [80] and G. SOLIANI. The attractiveness of the sine-GORDON equation in various areas of physics has, up to today not diminished.

The sine-GORDON equation is similar to the wave equation, and it has been shown how we arrived at this equation. We will therefore use the experiences we made with the wave equation in order to help us develop the sine-GORDON equation.

Our physical starting point is the NEWTONian equations (53) and (57), in which we assume those forces are included that can describe the motions of the masses of a crystal lattice. In the simplest case, these masses are positioned in the 'proximity' of ideal lattice points where they oscillate.

An ideal crystal is attributed a mathematically well defined, unbounded space lattice. If all the points in a mathematical lattice are occupied with atoms or groups of atoms, we get an ideal crystal. If, however, the space lattice points of one half plane, for example the plane that ends on the x-axis at the coordinate origin, i.e. $y = 0$, $z < 0$, are not occupied by atoms, we get an ideal crystal with a gap between two half planes. If both half planes are joint together, we receive the projection shown in Fig. 7.8 the lattice plane of a crystal with a straight dislocation line pointing directly out towards the reader, also see KRÖNER [47, 48].

A dislocation is thus a one-dimensional line of disturbance in a crystal. This line of disturbance only exists in relation to the surrounding lattice. If we remove the lattice, we also remove the line of disturbance. The motion of a dislocation line is defined alone by the motion of its surrounding lattice atoms and can therefore be exclusively described as motion relative to a lattice. We will choose the centres of inertia $q_\alpha = q_\alpha(t)$ of the lattice atoms surrounding the dislocation line as the coordinates for the motion of this line. In fact, 'all' lattice atoms have to be considered, but only those in near proximity are in fact affected. In Fig. 7.8, we have marked a dislocation coordinate with a cross. This is the geometrical centre of the dislocation. The index α counts the lattice planes perpendicular to the drawing plane. The q_α do not necessarily have to lie directly in the planes shown in Fig. 7.8. In Chap. 5, we showed that the centres of inertia q_α of NEWTONian mass distributions principally satisfy the equations of NEWTONian type (57). The dislocation coordinates q_α also follow the following equation, which will be explained below in more detail,

$$\frac{d}{dt}\left(\Delta m_\alpha \frac{d}{dt} q_\alpha\right) = \sum_{b \neq \alpha} F_{b\alpha} \quad \text{with} \quad F_{b\alpha} = -F_{\alpha b}. \tag{76}$$

A strict calculation of the coefficients Δm_α and the interaction forces $F_{\alpha b}$ out of the system of the NEWTONian equations (53) or (57) should now be made. We will not try this here, and only at the end of this chapter, we will think about an approximation for the magnitude of these inertial masses Δm_α. Instead of this, we will concern ourselves with the physical interpretation of these quantities and their role in the development of the sine-GORDON equation, and then, we will develop the sine-GORDON equation strictly out of Eq. (76).

The quantity Δm_α describes per definition the inertia during a (plastic) displacement of the dislocation coordinates q_α relative to the lattice. In 1964 KRÖNER [50] showed the necessity of introducing such a resistance of dislocations against an acceleration relative to the lattice, the inertia of dislocations with respect to this lattice, an effective mass of dislocations.

In 1939, J. FRENKEL [25] and T. KONTOROVA using U. DEHLINGER [10] decisive preparations from 1929 wrote down the system of differential equations for the deflection q_i of a one-dimensional row of lattice atoms, which were directly next to the geometrical centre of the dislocation line. For the NEWTONian equations (55) of the deflections q_i of the lattice atoms laying transversal to the dislocation line the authors found

$$\Delta m_i \frac{d^2 q_i}{dt^2} = \sigma\, q_{i-1} - 2\sigma q_i + \sigma\, q_{i+1} - D \sin\left(\frac{2\pi}{a} q_i\right) \quad \text{FRENKEL–KONTOROVA equation} \tag{77}$$

with the mass of the lattice atoms Δm_i and the two quantities σ and D calculated from the parameters of the lattice (see A. SEEGER [86]). Equation (77) is identical with the Eq. (31) of a linear chain of elastic coupled masses, which are additionally under the influence of an external force $-D \sin\left(\frac{2\pi}{a} q_i\right)$. The force on the atom with the number i and the deflection q_i originates from the influence of the surrounding lattice and takes into consideration that when a deflection of the size of a multiple lattice parameter takes place, $q_i = n\,a$, then periodically, the same force is produced. We notice that the sine function is not in the least prescribed, compare also with (74) and can only boast the property of simplicity.[1]

Due to the fact that the q_i are the supposed transversal deflections of the chain, the quantity σ contains the base tension of this atom row as we have seen in Chap. 6. An important result of the investigations by DEHLINGER, FRENKEL and KONTOROVA is that the rows of lattice atoms in near proximity to a dislocation line show a special *base tension*. In the years 1948/49, A. SEEGER [87] was able to derive the sine-GORDON equation out of Eq. (77). This requires the conclusion from the first three terms on the right side of (77) to the term $E\,\Delta x\, d^2 s/ds^2$ in the continuum approximation of the chain with $q_i(t) \longrightarrow s(x,t)$, in the same way as we did it in Chap. 4 when dealing with a linear chain and its longitudinal deflections (see Eqs. (63)–(68)) and also later with the transversal deflections.

The FRENKEL–KONTOROVA Eq. (77) and also SEEGER's sine-GORDON equation (cf. SEEGER [87, 88]) were originally derived and considered only for lattice atoms. Later, it was A. SEEGER who also applied the equation on the dislocation line itself. Here, we wish to remind the reader of the fundamental importance that the conclusion from the lattice atom masses Δm_i in (77) has on the relative inertial masses Δm_α of a dislocation in (76). Only the inertial masses Δm_α of a dislocation can principally

[1]The choice of the (non-linear) sine function moved J. RUBINSTEIN to name the equation the sine-Gordon equation in order to remind people of its linear approximation, the KLEIN-GORDON equation.

move through the whole lattice. The lattice atoms with their masses Δm_i remain with slight displacements at their lattice positions.

We will show in the appendix, Chap. 25, that there are principally two different forms of dislocation motion: Gliding and climbing. Both of these motions lead to plastic deformation of the lattice. However, only gliding, the so-called conservative dislocation motion, is a pure translation of the lattice distorsion made by the dislocation. This allows us to identify the inertia term in (76) with the momentum of the dislocation. In comparison to this, we have climbing, the so-called non-conservative dislocation motion, which plays a more important role with increasing temperatures, generally only occurring in connection with point defects. Here, in this case, the inertia term in (76) describes the coupled properties of dislocations and point defects. For our considerations, we will assume that the dislocations do not climb, which can be physically realised if we stay at sufficiently low temperatures.

Here, we note: Conservative motion means 'conservation of crystal volume' and does not mean the conservation of energy as in point mechanics. Implicitly KRÖNER's idea of a relative, inertial mass of dislocations in the so-called line tension model of dislocations (or string model) was anticipated. This model is explained in great detail in A. SEEGER [88] and P. SCHILLER's paper dated from 1966. Here, the dislocation line is examined as if it were a tense string. This is nothing else than a continuum approximation of a linear chain of elastic coupled inertial masses under the influence of a base tension as seen and dealt with in Chap. 6. The model of line tension is based on the assumption that the above-introduced dislocation coordinates q_α and the inertia's Δm_α move relative to the lattice. In order to register the influence of the surrounding lattice on the 'tense string' the authors determined a periodical potential U, namely

$$U = U(q) = -D\left[\cos\left(\frac{2\pi}{a}q\right) - 1\right], \tag{78}$$

which gives us a force F defined by the lattice according to $-dU/dq = F = = \Delta m \partial^2 q/\partial t^2$. This force gives the dislocation piece along the length Δx a certain acceleration $\partial^2 q/\partial t^2$ relative to the crystal lattice, according to its inertia Δm. Therefore this Δm is KRÖNER's relative inertial mass for the gliding of a dislocation considered by SEEGER and SCHILLER. We therefore arrive at our formulated base equation, according to (76), for the dynamics of dislocations, cf. also H. GÜNTHER [33]. Due to the fact that we want to derive the sine-GORDON equation from this equation, we will formulate axiomatically our starting point:

A dislocation constitutes a linear chain of elastic coupled, inertial masses Δm_α with the positions q_α. The conservative motion of these inertial masses, defined relative to the crystal lattice, is determined by the NEWTONian equations if we replace the inertial system in the NEWTONian postulates with a crystal lattice,

$$\left.\begin{aligned} \frac{d}{dt}\left(\Delta m_\alpha \frac{d}{dt} q_\alpha\right) &= \sum_{b\neq\alpha} F_{b\,\alpha}, \\ F_{b\,\alpha} &= -F_{\alpha\,b}. \end{aligned}\right\} \tag{79}$$

Mathematically formulated, thc condition for the exclusive gliding of a dislocation is that the three vectors, the displacement vector $\delta\mathbf{q}$ of a dislocation, its BURGERS vector $\mathbf{b}$ and its line direction vector $\mathbf{t}$ are not allowed to set a different volume than zero, in other words $\delta\mathbf{q}\cdot(\mathbf{b}\times\mathbf{t})=\mathbf{0}$. We can fulfil this motion restriction from the beginning, as it is done when possible, in mechanics by the choice of our variables q as a transversal deflection, cf. also GÜNTHER [33]. Therefore we do not further need to take this condition into consideration.

Here the assumption of a linear chain means that the sum of the forces $F_{b\alpha}$ is made up of two parts. Firstly the sum of the forces $F_{\beta\alpha}$ which describe the elastic interactions of the dislocation masses Δm_α with its next neighbours and secondly the sum of the forces $F_{g\alpha}$ that the masses m_g of the lattice atoms around the chain exert on m_α. Principally we would have to include all lattice atoms, but we will of course again assume a continuous action and will thus only take the direct neighbours into consideration. Therefore the following splitting of the sum of the forces is valid,

$$\sum_{b\neq\alpha} F_{b\alpha} = \sum_{\beta=\alpha\pm1} F_{\beta\alpha} + \sum_{g} F_{g\alpha}. \tag{80}$$

In other words, the effective dislocation masses Δm_α positioned at q_α constitute such a linear chain as was discussed in detail in Chap. 4, with the exception that this chain is now embedded inside of a crystal and is defined exclusively relative to this crystal.

An ideal lattice always has the property as described in Fig. 7.7: On the lattice we move k times in the x-direction to the next lattice point. We then take l 'lattice steps' in the y-direction. After this we then move back in the opposite direction: firstly k lattice steps in the negative x-direction and then l lattice steps in the negative y-direction. If we have an ideal lattice, to be more precise if the lattice inside of the traversed area is ideally structured we should arrive back at our starting location. This is not the case when dealing with dislocations. As we have seen and characterised in Fig. 7.8, a dislocation of the BURGERS vector $\mathbf{b}$, where the dislocation is perpendicular to the traversed area (BURGERS circuit) leads to the fact that we do not arrive at the starting point, we arrive at a point at the distance $\mathbf{b}$ away from the starting point. One could also say, the dislocation is a topological singularity of the crystal lattice (see also the appendix, Chap. 26, Eqs. (367) and (368) and the Figs. 26.1 and 26.2). In the light of this background we can emphasise our general dynamics of dislocations as follows:

> The physical property of the topological singularity ∗dislocation∗ inside of a crystal is that the dislocation possesses, in a NEWTONian sense, an inertial mass with respect to the lattice.

For those who are familiar with the experiments on plastic deformations, where we discovered that the smallest element is the motion of a dislocation, see Chap. 26 for more details, our Eq. (79) will at first cause some irritation. Dislocations do not seem to fulfil the inertia property of an already existing velocity so characteristic of NEWTONian mechanics. Here we wish to state that it is just as difficult to observe the elementary laws of NEWTON's mechanics in the motions of a plough ploughing a field. In Chap. 21 we will discuss a more detailed description about the inertial properties of dislocations. There we will also try to combine this with C. F. v. WEIZSÄCKER's [98] theory of ultimate alternatives ("Uralternatives").

An external experimentalist will declare the inertia of a dislocation in reference to the lattice as an effective mass. He will also declare the forces between such effective masses as so-called configuration forces. For the observer in the centre of the crystal, where we will also position ourselves, such a notation makes no sense. We will thus, in the view of the NEWTONian equations (79) speak of the masses of the dislocations and the forces between them.

In the following we will concern ourselves with the motions q_α perpendicular to the dislocation line. The elastic displacements of atoms in the vicinity of a dislocation, that are always produced as a reaction to the presence of a dislocation will be ignored.[2]

If all q_α move a distance of 10^{-8} cm simultaneously then the dislocation has moved one lattice parameter. This would be the assumed elementary step as shown in the dislocation model of a plastic deformation in the previous Chap. 7. As in the discussion of the critical shear modulus, cf. (74), we assume that once again a sine function is responsible for the dependency of the total force of the surrounding lattice to the displacements q_α of the masses Δm_α. For small deflections q_α, the lattice acts like an elastic spring on the dislocation, a spring that forces the dislocation mass Δm_α with the force constant D_m back into its position of equilibrium. For larger q_α this force decreases so that it then disappears in the centre position of the dislocation at $q_\alpha = a/2$ between both positions of equilibrium $q_\alpha = 0$ and $q_\alpha = a$. For the second sum in (80) we can therefore formulate the equation

$$\sum_g F_{g\alpha} = -D_m \sin\left(\frac{2\pi}{a} q_\alpha\right) \tag{81}$$

with a disappearing force of the lattice acting on the dislocation at $q_\alpha = n\,a/2, n = 0, 1, 2 \ldots$. According to the above assumed neglection of the elastic reaction on the lattice we write in (81) the parameter of an ideal, undisturbed lattice.

We will now allow the various positions q_α of the dislocation line to be various functions of time. This is in contrast to the model of rigid motion of a dislocation (with which we have sufficiently been able to describe macroscopic plasticity),

$$q_\alpha = q_\alpha(t). \tag{82}$$

With the additional Eq. (81) that takes the circumstance of the linear chain being embedded inside the crystal into consideration, we can in well-known fashion move across from single masses to a continuous distribution of the centres of inertia in our linear chain (which is nothing more than our dislocation). To be able to do this we only have to abide strictly to the rules found in Chap. 6, see Eqs. (65)–(68). Equations (80) and (81) are firstly inserted into (79) and we find that

[2]The calculation of the elastic deformation caused by the dislocation is the object of the theory of elasticity of dislocations. We will discuss this in the appendix, Chaps. 26 and 27.

$$\left.\begin{aligned} &\frac{d}{dt}\left(\Delta m_\alpha \frac{d}{dt} q_\alpha\right) = \sum_{\beta=\alpha\pm1} F_{\beta\alpha} - D_m \sin\left(\frac{2\pi}{a} q_\alpha\right), \\ &F_{\beta\alpha} = -F_{\alpha\beta}. \end{aligned}\right\} \tag{83}$$

We have to take into consideration here that the repelling force (81) of the surrounding lattice refers to the single masses m_α, thus therefore to the corresponding dislocation section Δx. The passing to the limit (65) also has to be supplemented by $D_m \longrightarrow D\,\Delta x$ and we then arrive at the continuous distribution according to

$$\left.\begin{aligned} \Delta m_\alpha &\longrightarrow \rho_o \Delta x, \\ q_\alpha(t) &\longrightarrow q(x,t), \\ D_m &\longrightarrow D\Delta x. \end{aligned}\right\} \tag{84}$$

We have thus introduced an effective mass density ρ_o along the dislocation line, and for the deflection of this dislocation line transversal to its distinguished position of a straight line along the x-axis, we have written, in place of the discrete values $q_\alpha(t)$ for the single masses m_α, the continuous function $q(x,t)$. In Eq. (83) we once again make the three base postulates (68) of the linear theory of elasticity. The inertia of a dislocation line with respect to the lattice is taken into account by the constant mass density ρ_o. For the modulus of elasticity E we write the line tension σ and receive

$$\left.\begin{aligned} &\frac{d}{dt}\left(\Delta m_\alpha \frac{d}{dt} q_\alpha\right) \longrightarrow \rho_o\,\Delta x \frac{\partial}{\partial t}\frac{\partial q(x,t)}{\partial t}, \\ &\rho_o = \text{const.}, \\ &\tau(x,t) = \sigma\,\frac{\partial q(x,t)}{\partial x}. \end{aligned}\right\} \tag{85}$$

with the modulus of elasticity σ of a linear dislocation chain. The relative elongation $\partial q/\partial x$ of this chain, the chain that was extended transversally, has to described by the change of the deflection q of the dislocation transversally to its position of equilibrium. For the right-hand side of (83), we can note in the passing to the limit to a continuum

$$\left.\begin{aligned} &\sum_{\beta=\alpha\pm1} F_{\beta\alpha} \longrightarrow \frac{\partial\tau(x,t)}{\partial x}\Delta x, \\ &D_m \sin\left(\frac{2\pi}{a} q_\alpha\right) \longrightarrow D \sin\left(\frac{2\pi}{a} q(x,t)\right)\Delta x. \end{aligned}\right\} \tag{86}$$

We introduce (85) and (86) into (83) and get for the moment by crossing out Δx that

$$\rho_o \frac{\partial}{\partial t}\frac{\partial q(x,t)}{\partial t} = \frac{\partial\tau(x,t)}{\partial x} - D \sin\left(\frac{2\pi}{a} q(x,t)\right). \tag{87}$$

The second term on the right-hand side of this equation is the force resulting from out fore mentioned lattice potential (78). For $\partial\tau(x,t)/\partial x$ we once again use our assumption of linear elasticity (85) and finally arrive at

$$\left.\begin{aligned} &\frac{\partial^2}{\partial x^2}q(x,t) - \frac{1}{c_o^2}\frac{\partial^2}{\partial t^2}q(x,t) = \frac{D}{\sigma}\sin\left(\frac{2\pi}{a}q(x,t)\right), \\ &c_o = \sqrt{\frac{\sigma}{\rho_o}}. \end{aligned}\right\} \text{sine–GORDON equation} \tag{88}$$

This is already the famous sine-GORDON equation of a dislocation.

A particular property of this equation is its non-linearity based on the sine function on the right hand side. Didn't we just make the base assumption (68) of the linear theory of elasticity? Well yes, but only for the one dimensional theory of elasticity of a linear chain. This linear chain finds itself under the influence of the non-linear potential U of the surrounding lattice according to (78). This produces the non-linearity in (88).

Because the relative deflection $\partial q/\partial x$ is without dimension, the modulus σ has according to (85) the dimension of tension τ, it is therefore a force, an energy per unit of length. The modulus of elasticity must be equated to the tension of equilibrium of the chain when the deflection is perpendicular to the chain as seen in Chap. 6. Hence σ is nothing else than the line tension in the string model of a dislocation, see SEEGER [88] and P. SCHILLER. A dislocation however cannot in comparison to a string end inside of a crystal. Another difference is that a dislocation cannot release its energy by just cutting through it, whereas a string can. From this point of view, the string model is not sufficient for the dislocation. Furthermore, the quantity $L\,\rho_o = m$ is in the case of a dislocation not just simply the mass of the lattice atoms on the length L—because there are no atoms on this length!—it is in fact the effective mass that should be calculated by inertia measurements of dislocations as a whole with respect to the lattice.

Our considerations are not sufficient to be able to numerically calculate the inertia of a dislocation. The determination of the inertia of a dislocation is according to the second equation (88) dependent on the line tension σ of a dislocation and the velocity c_o. The quantity c_o is the critical signal velocity of the sine-GORDON equation. Its relationship to the sound velocity will be discussed later (see Chap. 12). Furthermore, there are elementary lines of thought that lead to an explanation of the quantity of the line tension σ of a dislocation. We will delve further into this in Chap. 23, and there we will make an estimation that the mass m_a of a dislocation of the length a of a lattice parameter in one of our considered examples will be around 3% of the mass of the surrounding lattice atoms, see Eqs. (295) and (299).

Such a statement can easily be misunderstood. The calculated mass m_a of the dislocation section is its inertia with respect to the crystal lattice; in other words, m_a is the ratio of the acting force (the configuration force to be exact) to the produced acceleration of this dislocation section relative to the lattice. The significant difference of this dislocation mass m_a to the mass of the atoms of the lattice becomes clear as soon as we take a look at the energy mass equivalence. The energy of a dislocation, that is closely tied to the line tension, can be determined, for example, (at least in an ideal experiment) by the mutual destruction of two dislocations with opposite signs. The released energy can be found in the oscillations of the lattice and after its dissipation in heat motion of the lattice and can therefore principally be determined through calorimetric measurement. Such a process of mutual destruction of two opposite dislocations is comparable to the conversion of the energy $2m_o c_L^2$ of an electron–positron pair during its annihilation into radiation energy of electromagnetic waves. However, the released energy from the mutual destruction of two dislocations is not allowed to be calculated using the speed of light in a vacuum c_L; it has to be calculated out of the relativistic energy mass equivalence using the signal velocity c_o many times smaller then c_L. We will deal with this in more detail in Chap. 23. An estimation of the determination of the magnitude of the mass of a dislocation will be shown at the end of this chapter.

We will now take a closer look at Eq. (88) and will try to make it easier to work with. In order to do this, we firstly put

$$\mathrm{q} = \frac{2\pi}{a} q. \tag{89}$$

Due to the fact that the deflection q of a dislocation as well as the lattice parameter a have the dimension of a length, we receive q according to (89) as a dimensionless unit for the deflection of a dislocation out of its position of equilibrium at $\mathrm{q} = 0$, such that for $\mathrm{q} = 2\pi$ the neighbouring position of equilibrium, the actual displacement $q = a$ is occupied. We multiply (88) with σ/D and receive

$$\left.\begin{aligned} &\frac{\frac{a}{2\pi}\frac{\sigma}{D}\partial^2}{\partial x^2}\,\mathrm{q}(x,t) - \frac{\frac{a}{2\pi}\frac{\sigma}{D}\partial^2}{c_o^2\partial t^2}\,\mathrm{q}(x,t) = \sin\big(\mathrm{q}(x,t)\big),\\ &c_o = \sqrt{\frac{\sigma}{\rho_o}}. \end{aligned}\right\} \tag{90}$$

Here, $\sqrt{(a\sigma)/(2\pi D)}$ is a characteristic length λ_o of the lattice that we will call one gm ('Gittermeter'). Under the influence of the constants a, σ and D, the quantity λ_o lies in the region of a few Å (ÅNGSTRÖM, 1Å = 10^{-8} cm). In a similar way, $\tau_o = \lambda_o/c_o = \frac{1}{c_o}\sqrt{(a\sigma)/(2\pi D)} = \sqrt{(a\rho_o)/(2\pi D)}$ is a characteristic time τ_o of the lattice that we will call one gs ('Gittersecond'). Under the influence of a, ρ_o and D, it lies in the region of 10^{-12} s (here, we calculated with a magnitude of c_o corresponding to a sound velocity of c_o = 4000 ms^{-1}, see Chap. 12). We note

$$\left.\begin{aligned} &\sqrt{\frac{a\sigma}{2\pi D}} = \lambda_o = 1\,gm \approx 1\,\text{Å},\\ &\frac{\lambda_o}{c_o} = \sqrt{\frac{a\rho_o}{2\pi D}} = \tau_o = 1\,gs \approx 10^{-12}\,s \end{aligned}\right\} \tag{91}$$

and therefore for (90)

$$\frac{\partial^2}{\partial\left(\frac{x}{\lambda_o}\right)^2}\mathrm{q}(x,t) - \frac{\partial^2}{\partial\left(\frac{t}{\tau_o}\right)^2}\mathrm{q}(x,t) = \sin\left(\mathrm{q}(x,t)\right). \tag{92}$$

Using the dimensionless coefficient of measure x for the determination of space and the dimensionless coefficient t for the determination of time,

$$\left.\begin{aligned} \mathrm{x} &= \frac{x}{\lambda_o}, \\ \mathrm{t} &= \frac{t}{\tau_o}, \end{aligned}\right\} \tag{93}$$

the following is developed,

$$\frac{\partial^2}{\partial x^2}\mathrm{q}\left(\lambda_o\mathrm{x},\,\tau_o\mathrm{t}\right) - \frac{\partial^2}{\partial t^2}\mathrm{q}\left(\lambda_o\mathrm{x},\,\tau_o\mathrm{t}\right) = \sin\left(\mathrm{q}(\lambda_o\mathrm{x},\,\tau_o\mathrm{t})\right). \tag{94}$$

In physics, it is normal to remove uncomfortable parameters in equations in such a fashion that the units of measure, in which the quantities found in the equations are defined, are suitably assimilated. Only the product out of the coefficient of measure and the unit of measure can be determined. It is of no difference if we either use 1 cm, or 10^8 Å, if we use either 3600 s, or if we use 1 h, 1 cm = 10^8 Å, 3600 s = 1 h etc. For a further simplification of Eq. (94), we agree to measure all distance in multiples of λ_o; in other words, we measure in gm (the distance λ_o then has the coefficient of measure 1) and all time intervals in multiples of τ_o; in other words, we measure in gs (the time τ_o then has the coefficient of measure 1). Using this agreement, we get a 'simplified' version of the sine-GORDON equation for a function or, as one says in physics, for a (dimensionless) field $\mathrm{q} = \mathrm{q}(\mathrm{x}, \mathrm{t})$ in dependence of the dimensionless variables x and t according to

$$\left.\begin{aligned} &\frac{\partial^2\mathrm{q}}{\partial\mathrm{x}^2} - \frac{\partial^2\mathrm{q}}{\partial\mathrm{t}^2} = \sin\mathrm{q}, \\ &\mathrm{q} = \mathrm{q}(\mathrm{x},\mathrm{t}). \end{aligned}\right\} \text{sine-Gordon equation} \tag{95}$$

Equation (95) is also the sine-GORDON equation of a dislocation. It is purely a matter of taste whether we use (88) or (95) for our calculations. Equation (95) is less tedious in writing terms. One does however have to take into consideration that in (95) all units of length are measured in gm and all units of time in gs, respectively, (see Eqs. (91) and (92)) and that q is according to (89) the ratio of the real deflection q of a dislocation to the lattice parameter a, multiplied by 2π. Once one has found a solution $\mathrm{q} = \mathrm{q}(\mathrm{x}, \mathrm{t})$ of Eq. (95), then the corresponding solution $q(x, t)$ of Eq. (88) is

$$q(x,t) = \frac{a}{2\pi}\mathrm{q}\left(\frac{x}{\lambda_o},\,\frac{t}{\tau_o}\right) = \frac{a}{2\pi}\mathrm{q}\left(\frac{x}{\lambda_o},\,\frac{c_o t}{\lambda_o}\right) \tag{96}$$

or, if we fully write out the lattice parameters,

$$q(x, t) = \frac{a}{2\pi}\mathrm{q}\left(\sqrt{\frac{2\pi D}{a\sigma}}x\,,\,\sqrt{\frac{2\pi D}{a\rho_o}}t\right). \tag{96a}$$

We will preferably use Eq. (95) for the following calculations. The solutions of the equation that interest us can thus be easier examined.

Using the sine-GORDON equation, we have discovered an equation, where the solutions give us the physically possible line forms of a dislocation free from external forces. In the region of micro-plasticity, the transitions between these line forms constitute the second correction in our idea of a mechanism of a plastic deformation.

In the light of the importance, the sine-GORDON equation plays in theoretical physics; it is not uninteresting to remind ourselves that we have founded this equation only on the axiomatic system of NEWTON'ian mechanics (79) and a few other well-known approximations of continuum mechanics. The validity of the sine-GORDON equation has, in the sense of this approximation, been sufficiently proven in order to describe a certain class of phenomena in a solid. In Chaps. 15 and 21, we will come to realise inside the frame of this approximation that the crystalline solid is a model for a relativistic spacetime, and furthermore, we will see because of this model character, what becomes of relativity when we observe processes that are not fulfilled for the considered approximations.

We as external observers have found that the sine-GORDON equation can be used to find the physically possible deviations of a once straight dislocation line relative to the surrounding lattice. As long as the internal observers inside of our crystal remain at rest relative to the crystal lattice, they use at most for these measurements different units than an outside observer would use. Everything else remains the same. Both parties will not make any other observations that the other party makes. A problem arises when the internal observers move relative to the crystal lattice. We will concern ourselves with this problem in Chaps. 10–13.

A class of trivial solutions q^{o} of the sine-GORDON equation describes the original state of the crystal in the sense that the dislocation possess its minimal level of energy inside of a crystal. Dislocations then occur as infinite, straight and stabile lines being at rest. We will name these solutions with A. SEEGER [89] vacuum solutions.[3] One can easily confirm by examining Eq. (95) that these solutions are described by

$$\mathrm{q}^{\mathrm{o}} = \pm 2n\pi, \quad n = 0, 1, 2, \ldots. \qquad \text{Vacuum solutions} \tag{97}$$

We also notice that the constant functions

$$\overline{\mathrm{q}^{\mathrm{o}}} = \pm(2n+1)\pi, \quad n = 0, 1, 2, \ldots. \qquad \text{Unstable solutions} \tag{97a}$$

[3] SEEGER uses the term ENNEPER equation for the sine-GORDON equation. In fact, this equation was actually examined in 1870 in the frame of differential geometry by A. ENNEPER [18], so that a row of mathematical facts about the sine-GORDON equation dates back to ENNEPER.

satisfy the sine-GORDON equation. However the straight dislocation is in an unstable position of equilibrium when q $= \pm(2n+1)\pi$. A slight deviation from these positions leads to a lattice force (see also (81)) forcing the dislocation away from its position, instead of forcing it back into its position.

We will in the following chapters concern ourselves with some of the numerous excitation states, in other words with states of a real lattice containing dislocations, lying energetically above a vacuum (97). We will in other words examine the solutions of the sine-GORDON Eq. (88) or (95) as possible line forms of dislocations in a crystal. We will not concern ourselves with the change of present line structures as a result of influencing forces. In order to solve this problem, we would have to incorporate the external forces $\mathbf{F}_A$ in the general NEWTONian equation (53) into our starting Eq. (79). This way, we could take for example load tensions into consideration that leads to a deformation and motion of dislocations by an additive tension term on the right-hand side of the sine-GORDON equation.

We now come once again to the inertial mass m_α of a dislocation. In order to do this, we will consider a cubic crystal block with the edge lengths L made up of N lattice atoms of the mass Δm. The lattice parameter is a.

This object of the total mass $M = N\,\Delta m$ with $N = (L/\alpha)^3$ atoms possesses $3N$ degrees of freedom of motion. One would choose, in the case of small elastic oscillations, for example the $3N$ CARTESian coordinates of the atoms as the variables. In the sense of LAGRANGEian mechanics, we can also use any other generalised coordinates that are capable of describing the well-defined position of the system. The art of using the LAGRANGEian mechanics is to determine such generalised coordinates that were adapted for the physical system and thus allow an especially simple and clear description. We made it clear at the beginning of this chapter that there are two distinct different forms of motion for a crystal lattice: elastic and plastic deformations. The CARTESian coordinates of mass points do not make a difference between these forms of motion and are therefore unsuitable for the determination of plastic deformations in a lattice. LAGRANGEian coordinates of plastic displacements are, for example, the above-introduced coordinates of dislocation q_α. The plastic displacements of a lattice are always coupled to elastic displacements. One has to realise that according to (75), below the critical shear stress, the whole motion is purely elastic. Only the motion of overcoming the potential barrier is classified as plastical. The distinction of this part of the motion of a lattice as plastic is a collective phenomenon. This makes the calculation of that part of the kinetic energy belonging to the coordinates q_α far more difficult. We can, however, present a simple estimation for this:

A dislocation of the length L moves through a crystal block, as shown in Fig. 7.10. Here, the bottom part of the cubic crystal for example glides away. If the dislocation moves with the velocity dq/dt across the distance L, it covers this length after the time T,

$$T = \frac{L}{dq/dt}.$$

During this period of time, the crystal block in our figure has plastically moved one lattice position. The rigid motion of the crystal block of the mass M can be attributed to an average velocity V,

$$V = \frac{a}{T} = \frac{a}{L}\frac{dq}{dt},$$

with a kinetic energy E according to

$$E = \frac{M}{2} = \frac{1}{2}M\left(\frac{a}{L}\right)^2\left(\frac{dq}{dt}\right)^2 = \frac{1}{2}\Delta m\left(\frac{L}{a}\right)^3\left(\frac{a}{L}\right)^2\left(\frac{dq}{dt}\right)^2 = \frac{1}{2}\Delta m\frac{L}{a}\left(\frac{dq}{dt}\right)^2.$$

The kinetic energy of the crystal block is assigned to the kinetic energy E_L of the dislocation. In fact, the complete crystal block of the mass M, however, does not glide rigidly so that

$$E_L < \frac{1}{2}\Delta m\,\frac{L}{a}\left(\frac{dq}{dt}\right)^2.$$

Hence, for the kinetic energy E_a of this dislocation of the length a of a lattice parameter, we find that

$$E_a < \frac{1}{2}\Delta m\left(\frac{dq}{dt}\right)^2.$$

The kinetic energy of an inertial mass Δm stands on the right side of this inequality. This mass moves through the crystal with the velocity of the dislocation dq/dt ; Δm is the mass of the surrounding lattice atoms. Therefore, the inertial mass m_α of a dislocation of the length a of a lattice parameter has to be smaller, i.e.

$$m_\alpha < \Delta m$$

in conformity with the estimation that we will make for a dislocation mass in Chap. 23 on the basis of the determination of line tension.

Chapter 9
Natural Measuring-Rods and Clocks

On the one hand, it is correct that we can choose our units of measure arbitrarily. A yard as a unit of length and a pile of sand running through an egg timer as a unit of time are completely sufficient to formulate our physical laws. However, we will not find this too pleasurable, because randomly chosen units or units specifically chosen for a certain problem can lead to tedious parameters. The largest problem is however its constancy and reproducibility. If we wish our observations to become more detailed and accurate, then our measuring devices also have to become more precise. It is also most unsatisfactory if we have to invent random units of measure when concerning ourselves with such fundamental conceptions as space and time. One thus began to search for natural units of measure, units of measure given to us by the physical objects for which we made our laws, for example, the earth's revolution period of the sun and the circumference of the earth. Modern examples would be the caesium atomic clock and the wavelength of the yellow sodium line.

Here, we wish to do the same when observing our mechanical objects and their motions in the crystal (replaced by a continuum). Due to the fact that we are dealing purely with mechanical processes, we will try to define our standard lengths and clocks with the help of those mechanical 'objects' that can be found inside of our crystal. Here, the sine-GORDON equation comes to our help. As we know, the solutions of this equation describe certain mechanical states such as one-dimensional imperfections distinguished by the crystal. It is thus near at hand to take these easily reproducible and identical lines found inside of our crystal and examine them for simple and characteristic forms suitable to be used as a standard length—a measuring-rod as it will be called hereafter—and a well-defined unit for an oscillation period.

Here, we once again wish to bring your attention to a characteristic of physical quantities: The correct description of any physical quantity contains two details, the unit of measure that gives us a quantity to compare something with and the coefficient of measure that tells us how often we have to take our measure of comparison in

H. Günther, *Elementary Approach to Special Relativity*, https://doi.org/10.1007/978-981-15-3168-2_9

order to be able to examine the physical quantity under consideration. The different qualities of the units of measure are defined as dimensions. If the temperature of an object is determined to be 277 K, this means the defined unit of one KELVIN had to be taken 277 times in order to reach this temperature. Thus, the temperature here has dimension K (KELVIN). In the same manner, the statement that an object has the mass of 43, 5 kg means that I have to place my unit of one kg 43, 5 times on to a scale to achieve the equivalent mass. Here, the mass has the dimension kg (kilogram). Physical measurement is thus a process of comparing and counting. We will now turn to the measurement of length and time in our crystal. Here, we search for those solutions of the sine-GORDON equation that serves our purposes as suitable units of measurement. We will start with the unit of length.

A simple solution of the sine-GORDON equation is the so-called static kink solution $\mathrm{q}_{\mathrm{o}}^{\mathrm{I}}$, which is dependent alone on the (here dimensionless) coordinate x; see A. SEEGER [86, 89],

$$\mathrm{q}_{\mathrm{o}}^{\mathrm{I}} = \mathrm{q}_{\mathrm{o}}^{\mathrm{I}}(\mathrm{x}) = 4 \arctan \mathrm{e}^{\mathrm{x}} \ . \tag{98}$$

The actual displacement q_o^I of a dislocation as a solution of the sine-GORDON Equation (88) is according to (96) and (96a) $q_o^I(x) = \dfrac{a}{2\pi}\mathrm{q}_{\mathrm{o}}^{\mathrm{I}}\Big(\dfrac{x}{\lambda_o}\Big)$, hence

$$q_o^I(x) = \frac{2a}{\pi} \arctan \mathrm{e}^{x/\lambda_o} \tag{98a}$$

or, if we insert the lattice parameters,

$$q_o^I(x) = \frac{2a}{\pi} \arctan \mathrm{e}^{\sqrt{\frac{2\pi D}{a\sigma}}\, x} \ . \tag{98b}$$

For the exponential function in the case of more complicated exponents, we will use the notation $\mathrm{e}^x := \exp[x]$ in the same way as compositions with, for example, $\arctan \mathrm{e}^x = \arctan \exp[x]$. We can thus also write for Eq. (98b)

$$q_o^I(x) = \frac{2a}{\pi} \arctan \exp\Big[\sqrt{\frac{2\pi D}{a\sigma}}\, x\Big] \ . \tag{98b}$$

We check that the function (98) fulfils Eq. (95). It is

$$\sin(4\alpha) = \sin(2 \cdot 2\alpha) = 2\sin(2\alpha)\cos(2\alpha) = 4\sin\alpha\cos\alpha(\cos^2\alpha - \sin^2\alpha)$$

$$= 4\sin\alpha\cos\alpha(1 - 2\sin^2\alpha)\ ,$$

$$\sin(4\alpha) = \cos\alpha(4\sin\alpha - 8\sin^3\alpha)\ .$$

The following is also valid, verifiable with $\tan\alpha = \sin\alpha/\cos\alpha$,

$$\sin\alpha = \frac{\tan\alpha}{\sqrt{1+\tan^2\alpha}}\,, \quad \cos\alpha = \frac{1}{\sqrt{1+\tan^2\alpha}}\,, \text{ hence}$$

$$\sin 4\alpha = \frac{1}{\sqrt{1+\tan^2\alpha}}\left(4\frac{\tan\alpha}{\sqrt{1+\tan^2\alpha}} - 8\frac{\tan^3\alpha}{(\sqrt{1+\tan^2\alpha})^3}\right)$$

$$= 4\frac{\tan\alpha}{1+\tan^2\alpha}\left(1 - 2\frac{\tan^2\alpha}{1+\tan^2\alpha}\right) = 4\frac{\tan\alpha}{1+\tan^2\alpha}\,\frac{1+\tan^2\alpha - 2\tan^2\alpha}{1+\tan^2\alpha}\,,$$

$$\sin 4\alpha = 4\tan\alpha\,\frac{1-\tan^2\alpha}{(1+\tan^2\alpha)^2}$$

and thus with $\alpha = \arctan\,\mathrm{e}^{\mathrm{x}}$

$$\sin(\mathrm{q}_{\mathrm{o}}^{\mathrm{I}}) = \sin(4\arctan\,\mathrm{e}^{\mathrm{x}}) = 4e^{\mathrm{x}}\frac{1-e^{2\mathrm{x}}}{(1+e^{2\mathrm{x}})^2}\,.$$

On the other hand, a repeated differentiation gives us

$$\frac{\partial}{\partial \mathrm{x}}(4\arctan\,\mathrm{e}^{\mathrm{x}}) = 4\frac{e^{\mathrm{x}}}{1+e^{2\mathrm{x}}}\,,$$

$$\frac{\partial^2}{\partial \mathrm{x}^2}(4\arctan\,\mathrm{e}^{\mathrm{x}}) = 4\frac{\partial}{\partial \mathrm{x}}\frac{e^{\mathrm{x}}}{1+e^{2\mathrm{x}}} = 4\frac{e^{\mathrm{x}}(1+e^{2\mathrm{x}}) - e^{\mathrm{x}}2e^{2\mathrm{x}}}{(1+e^{2\mathrm{x}})^2}$$

and thus

$$\frac{\partial^2}{\partial \mathrm{x}^2}\mathrm{q}_{\mathrm{o}}^{\mathrm{I}} = \frac{\partial^2}{\partial \mathrm{x}^2}(4\arctan\,\mathrm{e}^{\mathrm{x}}) = 4e^{\mathrm{x}}\frac{1-e^{2\mathrm{x}}}{(1+e^{2\mathrm{x}})^2}\,.$$

Hence, the time-independent field $\mathrm{q}_{\mathrm{o}}^{\mathrm{I}} = \mathrm{q}_{\mathrm{o}}^{\mathrm{I}}(\mathrm{x})$ does in fact fulfil Eq. (95),

$$\frac{\partial^2 \mathrm{q}_{\mathrm{o}}^{\mathrm{I}}}{\partial \mathrm{x}^2} = \sin \mathrm{q}_{\mathrm{o}}^{\mathrm{I}}\,.$$

The function (98a) describes a line form immediately observable in the continuum. What does this line look geometrically? As one can see from (98) the dislocation line passes for $x \longrightarrow -\infty$ into the position of equilibrium $q = 0$ and changes in the vicinity of $x = 0$ into the neighbouring position of equilibrium $q = a$, into which it passes for $x \longrightarrow +\infty$. Here, the changing into the neighbouring position of equilibrium in the vicinity of $x = 0$ takes place in a relatively small range of the order λ_o; see Fig. 9.1. Those types of kinks can be found in large numbers inside of

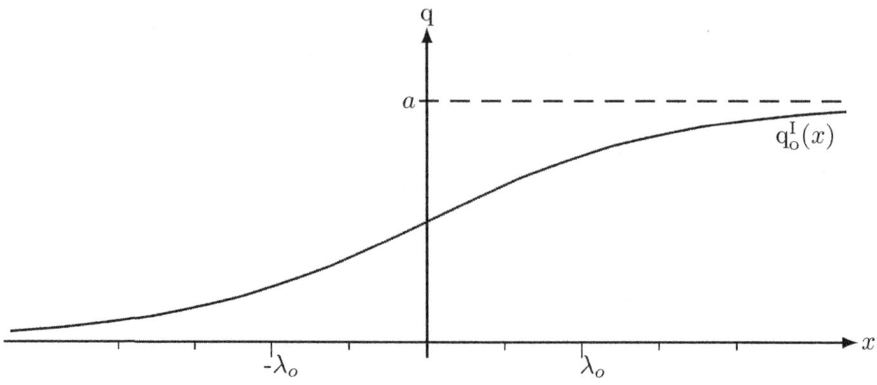

Fig. 9.1 Kink solution (98a), $q_o^I(x) = \frac{2a}{\pi} \arctan \exp[x/\lambda_o]$

a crystal. The internal geometry of these lines is just perfect to be used to introduce a natural unit of measure for length, a standard length, our measuring-rod. Such a measuring-rod would be for example the distance between the x-coordinates for both points with the largest line curvature. It is easier to determine the distance L_o of the x-coordinates for both points of intersection of the tangent through the point of inflection with both asymptotes, cf. H. GÜNTHER [34], see Fig. 9.2. We find

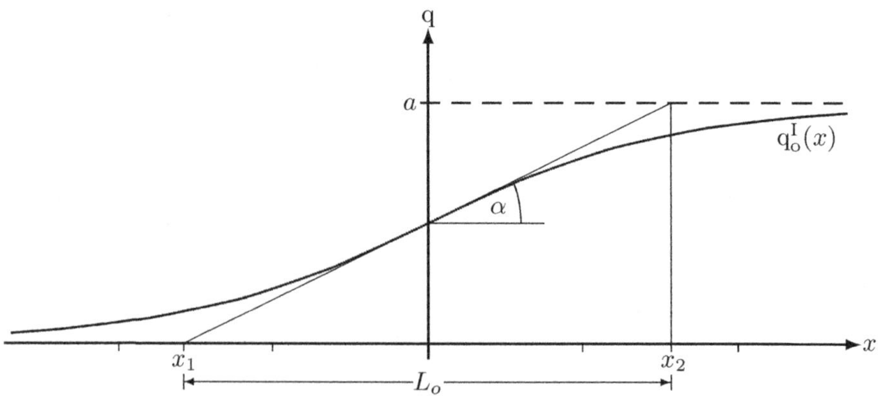

Fig. 9.2 Definition of the measuring-rod L_o out of the kink solution $q_o^I(x)$ (see (98a), Fig. 9.1). The function $y = q_o^I(x)$ has its point of inflection at $x = 0,\ y = q_o^I(0) = a/2$. Here, the tangent has the gradient $\tan\alpha = \frac{\partial q_o^I}{\partial x}\big|_{x=0} = \frac{2a}{\pi}\frac{1}{\lambda_o}\frac{e^0}{1+e^{2\cdot 0}} = \frac{a}{\pi\lambda_o}$ and can thus be described by $y = \frac{a}{2} + \frac{a}{\pi\lambda_o}x$. This line cuts the x-axis at $x_1 = -\pi\,\lambda_o/2$ and the asymptote $y = a$ at $x_2 = +\pi\,\lambda/2$. The distance of both points on the x-axis leads to the standard length L_o, our measuring-rod according to $L_o = x_2 - x_1 = \pi\,\lambda_o$, see (99)

$$L_o = x_2 - x_1 = 2\,\frac{a/2}{\left.\dfrac{\partial q_o^I}{\partial x}\right|_{x=0}} = \frac{a}{\dfrac{2a}{\pi}\dfrac{1}{\lambda_o}\dfrac{e^0}{1+e^0}}\ ,$$

$$L_o = \pi\,\lambda_o\ . \tag{99}$$

Using this, we have defined a natural unit of measure L_o out of the internal geometry of the kink line. We can use this as a measuring-rod in our continuum.

This unit of measure L_o has, according to Eq. (91), the length of a few ÅNGSTROMS which is equivalent to a few lattice parameters. Referring to the lattice would however be a step backwards which is no longer necessary. The unit of length L_o is our natural 'ruler' which can be reproduced at any time by the internal geometry of the kink lines of our continuum.

The decisive point is as follows: Using L_o we can measure any other length. Here, L_o compares less with the Parisian standard metre, where careful preparations have to be taken to ensure its consistency, but it compares more with the wavelength of the yellow sodium line, for example, which is identically reproduced by nature.

Let us recapitulate: The length $\mathbf{X}$ of an object can be determined by seeing how often our measuring-rod L_o fits on to the object, for example $\mathbf{X} = 27L_o$. It is obvious the coefficient of measure x for the length $\mathbf{X}$ increases if the scale of the measuring-rod used as our unit decreases.

Having introduced a standard of length, our measuring-rod, we can now do geometry, in other words we can describe the positions and distances of 'objects' from each other. This fact will play an important role later on in Chap. 12. We however want more. We also want to describe the processes of motion of these objects. We do not just want to know where an object is, and we also want to know when the object is there. We want to describe 'events' as one would say. An event is characterised by the location x where it takes place and by the time t when it happened. Therefore, we still need a clock. Once again we wish to measure the time in our crystal without the help of *outside* instruments. In order to achieve this, we need to refer to the *internal* structures that the crystal itself possesses. It is the sine-GORDON equation that once again delivers us a suitable period of time, a unit to measure time with. To find this, we have to observe the so-called localised breather solution q_o^{III} which represents a time and space-dependent solution of the sine-GORDON equation (95); see A. SEEGER [86] cf. also H. GÜNTHER [34],

$$\mathrm{q_o^{III}} = \mathrm{q_o^{III}(x,t)} = 4\arctan\frac{\sin\frac{t}{\sqrt{2}}}{\cosh\frac{x}{\sqrt{2}}}\ . \tag{100}$$

The actual displacement q_o^{III} of the dislocation as a solution of (88) is according to (96) once again $q_o^{III}(x,t) = \frac{a}{2\pi}q_o^{III}\left(\frac{x}{\lambda_o}, \frac{c_o t}{\lambda_o}\right)$, thus

$$q_o^{III}(x,t) = \frac{2a}{\pi} \arctan \frac{\sin \frac{c_o t}{\lambda_o \sqrt{2}}}{\cosh \frac{x}{\lambda_o \sqrt{2}}} = \frac{2a}{\pi} \arctan \frac{\sin(\Omega_o t)}{\cosh \frac{x}{\lambda_o \sqrt{2}}} \;, \tag{100a}$$

or, if we insert all the parameters,

$$q_o^{III}(x,t) = \frac{2a}{\pi} \arctan \frac{\sin\left(\sqrt{\frac{\pi D}{a\rho}}\, t\right)}{\cosh\left(\sqrt{\frac{\pi D}{a\sigma}}\, x\right)} \;. \tag{100b}$$

We leave the reader with the option of checking that the breather solution (100) really does fulfil the sine-GORDON equation (95).

With $q_o^{III}(x,t)$, we have received a line form that can only exist when in permanent motion. What does this line form look like geometrically? Because of the hyperbolic cosine in the denominator of (100), the dislocation line moves at any arbitrary time t not only for $x \longrightarrow +\infty$, but also for $x \longrightarrow -\infty$ in the stabile position of equilibrium $q = 0$; see Fig. 9.3.

For every fixed x the line oscillates with an angular velocity Ω_o,

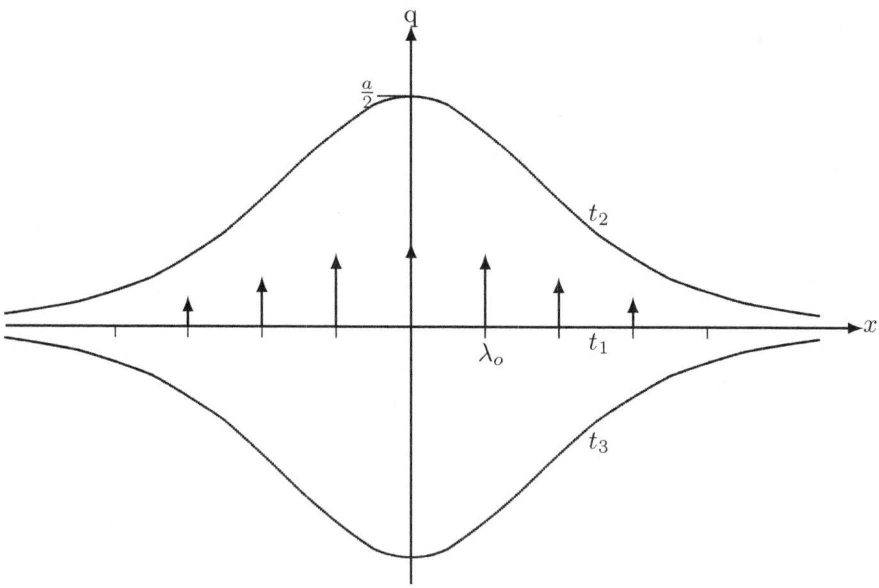

Fig. 9.3 Breather solution (100a), $q_o^{III}(x,t) = \frac{2a}{\pi} \arctan \frac{\sin \frac{c_o t}{\lambda_o \sqrt{2}}}{\cosh \frac{x}{\lambda_o \sqrt{2}}}$, for $t_1 = 0$, $t_2 = \frac{\sqrt{2}\,\pi\lambda_o}{2c_o}$, $t_3 = \frac{3\sqrt{2}\,\pi\lambda_o}{2c_o}$. (The arrows show the direction of motion of the oscillating dislocation line during its passing through the zero position)

$$\Omega_o = \frac{c_o}{\lambda_o\sqrt{2}} = \frac{1}{\tau_o\sqrt{2}} \,, \tag{101}$$

that according to $T_o = 2\pi/\Omega_o$ defines a period of oscillation. Thus we have, out of the internal spacetime geometry of the breather solution $q_o^{III}(x,t)$ introduced a unit of measure for time, the period of oscillation T_o,

$$T_o = 2\pi\sqrt{2}\,\tau_o = 2\pi\sqrt{2}\,\frac{\lambda_o}{c_o} \,, \tag{102}$$

in which τ_o was substituted according to Eq. (91); see Fig. 9.4.

This oscillation reaches the maximum amplitude at $x = 0$. We use these oscillations to define a natural clock, the 'breather clock', see GÜNTHER [34]—in the same way as the oscillations of atoms are used to define a caesium atomic clock. This is the way that clocks have been built. The time designation t on the clock counts the number of oscillations. It is the coefficient of measure. The oscillation period of the clock is the unit of measure. Let us recapitulate: The time **T** for any mechanical process in our crystal can be determined using our breather clock by counting how often an oscillation T_o took place, here for example $\mathbf{T} = 13T_o$. This number 13 can be read from the clock as $t = 13$. Once again the coefficient of measure for the same process increases if the period of the oscillation, used as our unit of measure, decreases.

Together with (99) and (102), we have not only found a natural measuring-rod, but now also a natural standard of time. These standards of course could be calculated mathematically using the lattice parameters according to (91),

$$\left.\begin{aligned} L_o &= \sqrt{\frac{\pi a \sigma}{2D}} \,, \\ T_o &= \sqrt{\frac{4\pi a \rho_o}{D}} \,. \end{aligned}\right\} \tag{103}$$

Once again, this is not necessary. The point is that we can now just forget the lattice and its many parameters and also forget the 'Gittermeter' and 'Gittersecond' we calculated according the (91). We can immediately use our measuring-rods and clocks determined alone out of the internal geometry of those line forms available,

L_o	Natural unit of measure for length
T_o	Natural unit of measure for time

. (104)

The units of measure for length and time measurement of a lattice are realised physically by measuring-rods and clocks of the lattice.

We can therefore determine using our measuring-rods and clocks the location x and the time t of every event occurring inside of our crystal. Can we really do

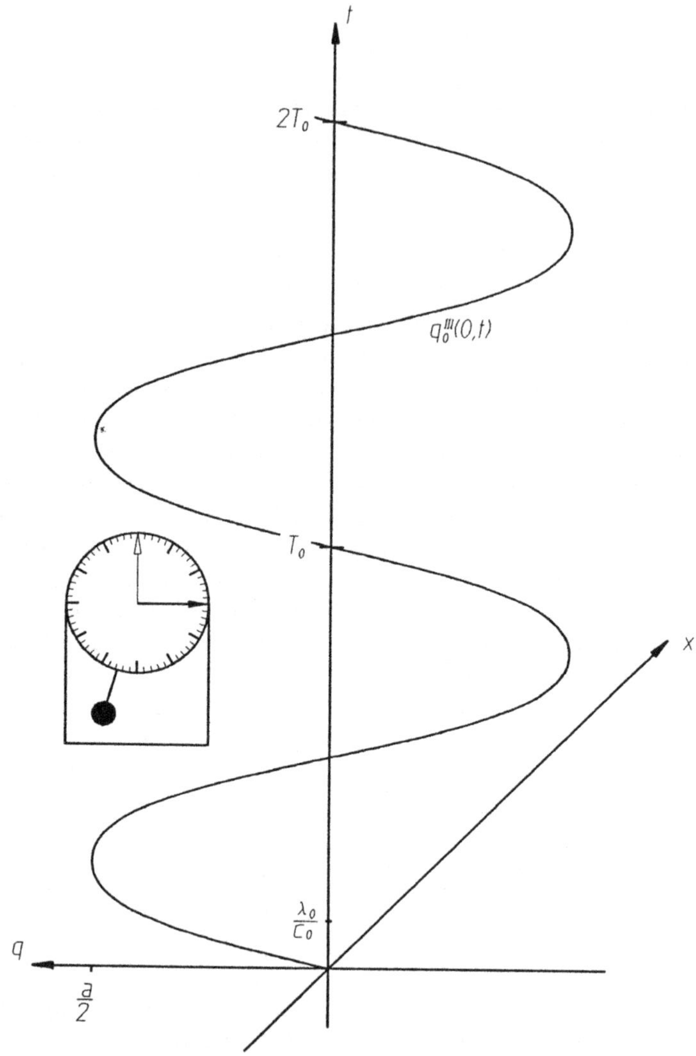

Fig. 9.4 Periodical oscillations of the breather solution (100a) for a fixed x-value, here $x = 0$, thus with $\cosh 0 = 1, q_o^{III}(0, t) = \frac{2a}{\pi} \arctan\left(\sin \frac{c_o t}{\lambda_o \sqrt{2}}\right)$. An instrument that can count these oscillations is a clock. For $t = T_o$, one receives the second passing through the zero position, thus because of the period 2π of the sin-function, $\frac{c_o T_o}{\lambda_o \sqrt{2}} = 2\pi$. Thus, we receive in accordance with (96) for the period of oscillation $T_o = \sqrt{2}\, 2\pi\, \lambda_o / c_o \approx 8,89\, \lambda_o / c_o$

this? To be able to determine the time coordinates of an event, we would have to position a clock at every position and start them all at the same moment in time. We would thus have to synchronise the clocks after having distributed them. How this is achieved has been generally discussed in Chap. 2. Is it possible to spare ourselves the tiresome procedure by starting all of the clocks at the same location and then distributing them? Do the spread clocks still not tick synchronously? This question will be answered in the following chapter.

Let us keep the following in mind:

> The kink, with its unit of length L_o and the breather clock, with its unit of time T_o gives us the primary units of measurement, with which we can measure all distances and time differences and thus the motion processes in our continuum.

Chapter 10
Measuring-Rods and Clocks in Motion

Measuring-rods and clocks have to be able to be moved. One likes to have a ruler and a stopwatch at hand when one rushes off to the position where a motion is to be recorded. However, how does one measure the length of an object that flies by, how do we measure its time of flight from position A to position B? Do we have to jump on to the object and position our ruler and start our stopwatch at position A and stop it at position B? It would be far more comfortable and practical to stay in one position and just let the object fly by. We mark on our ruler the simultaneous positions of both end points of the object. This should represent its length. In order to measure the time of flight between A and B, we would just simply compare the readings of the stopwatches as the object arrives there. Does it make a difference how we arrange and use our instruments of measurement? Is not it just a matter of taste and competitiveness if the observer jumps on to the object or just remains seated and watches it fly by? Well, in both cases, we have to move the instruments of measurement even if only to distribute the stopwatches along the course that we started simultaneously at the starting position, as in the second case.

The process of synchronising clocks positioned at different locations without moving them was described in Chap. 2. In order to do this, we need to have a signal at hand, and we need to know its exact velocity. In our equations for the linear chain (31), we tacitly agreed that at the positions $x_i = \frac{L}{N}\, i$ of all N masses one and the same time t would be measured. It was this fact that allowed us to formulate the solution (32). We thus assumed that we have introduced a synchronised time for the whole linear chain. The question of clock synchronisation was never put forward in NEWTONian mechanics. It was mutually accepted that one could transport an already ticking clock to any position where it could be used to synchronise another clock. It was left to A. EINSTEIN to ask this question: Can we really be certain that moving clocks do not change their pace? How can we be sure that the oscillation period of a moving clock does not change? As strange as this question may seem to the

H. Günther, *Elementary Approach to Special Relativity*,
https://doi.org/10.1007/978-981-15-3168-2_10

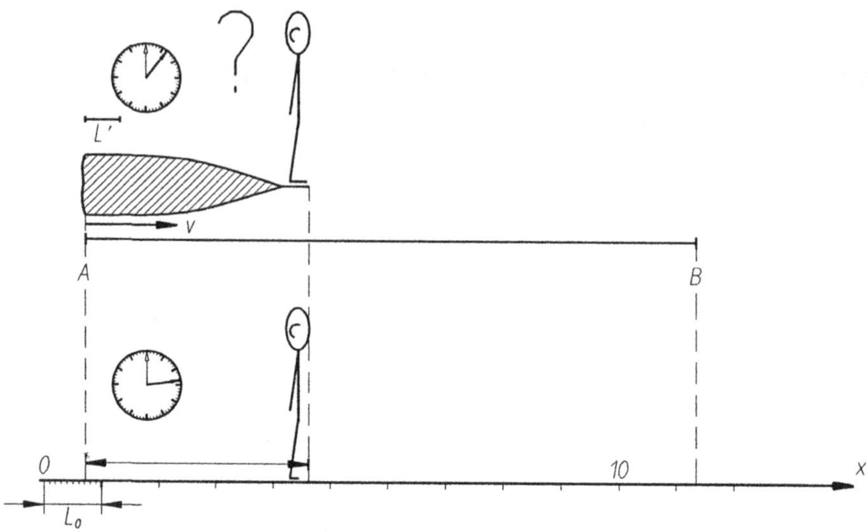

Fig. 10.1 Have the measuring-rods and clocks of the flying observer changed?

unprepared reader, in order to exclude this insecurity, we have to synchronise the clocks with the help of an exactly known velocity *after* we have distributed them.

In order to measure the NEWTONian motions of the linear chain or crystal with the help of the NEWTONian equations, we used that inertial system where the chain's or crystal's centre of inertia is at rest. There we have found the wave equation (61) with the critical signal velocity c and the sine-GORDON equation (88) with its critical velocity c_o. The validity of these equations is therefore secured without any doubt.[1]

The 'flying object' inside of our crystal is any localised structure above an ideal lattice, whose motion inside of a crystal can be attributed a single velocity. In the simplest case, such structures, our 'flying objects' are shapes of dislocation lines moving through a crystal lattice. Here, one can use any structure and the structure can be as complicated as one wishes. In Chap. 23, we will explicitly calculate all the properties of a particle for the simplest of these line forms, the kink which realises our measuring-rod (see (98a) and Fig. 9.1). These properties will of course be defined with respect to the lattice.

We could now use the sound velocity c in order to synchronise two breather clocks, described in the last chapter, in their positions A and B. Using these clocks, we could

[1]Because of our assumption (68) of a linear theory of elasticity (and also because of the enormous difference between the sound velocity and speed of light), we do not take into consideration the corrections that take place during the motions of the masses of a linear chain. These corrections occur due to the dependence of the masses Δm in the NEWTONian equations (63) (from which we formed the wave equation (61)) from their velocity v according to EINSTEIN's Special Theory of Relativity, $\Delta m_v = \Delta m_o\sqrt{1 - v^2/c_L^2}$. We thus exclude the possibility of extremely high frequency oscillations, see also Chap. 21.

then measure the time of flight, as we defined above, of an object between A and B. This is not the problem. As long as we as an observer stay in one and the same inertial system with our crystal, no complications should arise.

However, what happens if we sit on the moving object and observe the crystal, if we then wish to synchronise clocks and use our measuring-rod from our position on the moving object? What happens if we compare the synchronised clocks and the measuring-rods on the moving object with those of the observer at rest with respect to the crystal? See Fig. 33. What velocity does the observer on the flying object use to synchronise his clocks? Does he too use the sound velocity? But the sound velocity changes if the observer moves with respect to the crystal, and it even becomes dependent upon the direction of motion! There are even several sound velocities inside of a crystal. Which of these should we use? We will not found our further considerations on sound velocity, the critical velocity for a propagation of an elastic deformation. In 1905, this was with the discovery of the Special Theory of Relativity by A. EINSTEIN completely different. In those days, the observer independent, untouchable, infallible and unbelievable constancy of the speed of light in a vacuum c_L, which represents the signal velocity of an arbitrary electromagnetic excitation, was made to the infallible principle of the resulting theory, out of which everything else was consequently deduced, cf. EINSTEIN's axioms quoted in Chap. 2, cf. also Chap. 16.

We do not have such a distinguished maximum velocity inside of a crystal—at least not yet. We do however have something else. We know far more about our instruments of measurement, our measuring-rods and clocks. They are solutions of a physical equation, the sine-GORDON equation. For the mysterious question in the Special Theory of Relativity, what happens with the length of the measuring-rod and the oscillation period of a clock if we move the measuring-rod and the clock, we turn to the sine-GORDON equation. If however something happens, if the length and the period of an oscillation during a motion really should change, then we would have to re-think our above naive conception of measurement. In our case, we do not however have to search for the answer to this question as EINSTEIN had to, using brave deductions out of a large theoretical principle. These deductions and the statements won from them would have at that time to be proven by extensive precision measurements.

This is not necessary in our case. We can calculate the answer using the sine-GORDON equation. Here, we wish to express that the physical reliability of our calculations are founded on the NEWTONian equations. These equations also control, according to our Eq. (79), the dynamics of dislocations in crystals.

The starting point of our considerations is the inertial system, we call this the reference system Σ_o, in which we have introduced, with the help of the line form q_o^I a standard of length, a measuring-rod L_o and with the help of an oscillating line q_o^{III} a unit of measure for time, an oscillation period T_o, in other words a clock. We thus write

$$\Sigma_o: \quad \begin{aligned} q_o^I(x) &= \frac{2a}{\pi}\arctan\exp\left(\frac{\pi x}{L_o}\right) \longrightarrow L_o \ , \\ q_o^{III}(x,t) &= \frac{2a}{\pi}\arctan\frac{\sin\frac{2\pi t}{T_o}}{\cosh\frac{\pi x}{L_o\sqrt{2}}} \longrightarrow T_o \ . \end{aligned} \tag{105}$$

A system of coordinates x and t belongs to such a reference system Σ_o with a measuring-rod L_o and an oscillation period T_o. This means the following: We first determine a starting position, the origin of the coordinates. This is an arbitrary 'initial event' at the position 0 at time 0 and thus has the coordinates $x = 0,\ t = 0$. An arbitrary event that takes place at a distance $\mathbf{X} = xL_o$ from the 'initial event' and after a time $\mathbf{T} = t\,T_o$ has the coordinates x and t belonging to the reference system Σ_o. We can also write $\Sigma_o(x,t)$ and can thus say:

> The coordinates x and t are the coefficients of measure for the time and length measurements in the reference system $\Sigma_o(x,t)$ with the units of measure L_o and T_o.

(The shortened versions of time t and space x can be used instead of the coordinates x and t.)

The moving lengths and clocks also have to be solutions of the sine-Gordon equation. Does this even exist? We firstly remind ourselves: If we displace a function $y = f(x)$ along the x-axis by b to the right, then this new function $y = \bar{f}(x)$ has to be described according to (compare with Fig. 2.1)

$$y = \bar{f}(x) = f(x-b) \ . \qquad \text{Displacement of } y = f(x) \text{ to the right by } b \tag{106}$$

A sufficiently large number of measuring-rods L_o and clocks preferred frame Σ_o. A measuring-rod that has its central point lying at $x = b$ is described by the line form

$$q_{ob}^I(x) = \frac{2a}{\pi}\arctan\exp\left(\frac{\pi(x-b)}{L_o}\right) \longrightarrow \text{Measuring-rod } L_o \text{ at } b \text{ in } \Sigma_o \tag{107}$$

and a clock found at the position $x = b$ is described by the line form

$$q_{ob}^{III}(x,t) = \frac{2a}{\pi}\arctan\frac{\sin\frac{2\pi t}{T_o}}{\cosh\frac{\pi(x-b)}{L_o\sqrt{2}}} \longrightarrow \text{Period } T_o \text{ of a clock at } x = b \text{ in } \Sigma_o \ . \tag{108}$$

Because of $\frac{\partial}{\partial x}f(x) = \frac{\partial}{\partial x}f(x-b)$ the functions in (107) and in (108) obviously fulfil the sine-Gordon equation just as the functions in (105) do. We can in fact thus position our measuring-rods and clocks wherever we want.

Now an observer has the constant velocity v against the preferred frame Σ_o. He is then at rest in his own system of reference Σ'. If we see this observer at our time $t = 0$ at the position $x = 0$ as we take it, then we see him at any time t at the position $x = vt$. The measuring-rod and the clock that rest next to the observer are registered by us at

$$x = x(t) = v\,t \ . \qquad \text{Position of measuring-rod and clock of the observer in } \Sigma' \tag{109}$$

Let us first examine the measuring-rod. The measuring-rod moved at the uniform velocity v has to develop out of the displayed function $q_o^I(x)$ in Fig. 9.2 by a displacement of $v \cdot t$ to the right, with the important condition that the sine-GORDON equation maintains its validity. If one simply displaces the function $q_o^I(x) = \frac{2a}{\pi}$ arctan exp$[\pi\, x/L_o]$ by $v\,t$ to the right, we get according to (105) the function

$$\bar{q}_o^I(x) = \frac{2a}{\pi} \arctan \exp\left(\frac{\pi(x - vt)}{L_o}\right) \ . \tag{110}$$

Using this line form, we would receive a moving observer's measuring-rod L_o, and this measuring-rod would be exactly identical to the observer's measuring-rod L_o at $b = v\,t$ in Σ_o according to Eq. (107). In this case, we could say that our anxiety concerning the changing of length of a moving measuring-rod were unfounded.

The decisive point is however that the function (110) is not a possible line form in the lattice. Function (110) does not fulfil the sine-GORDON equation which can easily be checked by mathematical calculation. We thus make an important observation in our crystal:

It is impossible to rigidly displace the line defining our measuring-rod.

In other words, in the reality of our continuum defined by the crystalline background, there are no measuring-rods that stay unchanged during a motion! In Chap. 8, we developed the sine-GORDON equation for this continuum, and this equation now causes a line deformation dependent on the velocity v of the motion. Actually, the line form has to be compressed in order to be moveable in order to be a corresponding solution $q^I = q^I(x, t)$ of the sine-GORDON equation, compare A. SEEGER [86], H. GÜNTHER [34], note the remark on p. 341,

$$\left.\begin{aligned} q^I = q^I(x, t) &= \frac{2a}{\pi} \arctan \exp\left(\frac{\pi(x - vt)}{L_o\gamma}\right) \ , \\ \gamma &= \sqrt{1 - \frac{v^2}{c_o^2}} \ . \end{aligned}\right\} \tag{111}$$

In fact, $q^I(x, t)$ is a line form localised at $x = v\,t$ similar to the function $q_o^I(x)$ that moves with the velocity v to the right along the x-axis, see Fig. 10.2.

We now compare the functions (111) and (107). At time $t = 0$, the moving measuring-rod L' only covers a fraction $\gamma\, L_o$ of the measuring-rod at rest L_o at $b = 0$. The same occurs for all other points in time with $b = v\,t$. The length L' that we measure in Σ_o for the moving measuring-rod is $L' = L_o\,\gamma$. This is already the well-known LORENTZ contraction that the sine-GORDON equation quite simply delivers:

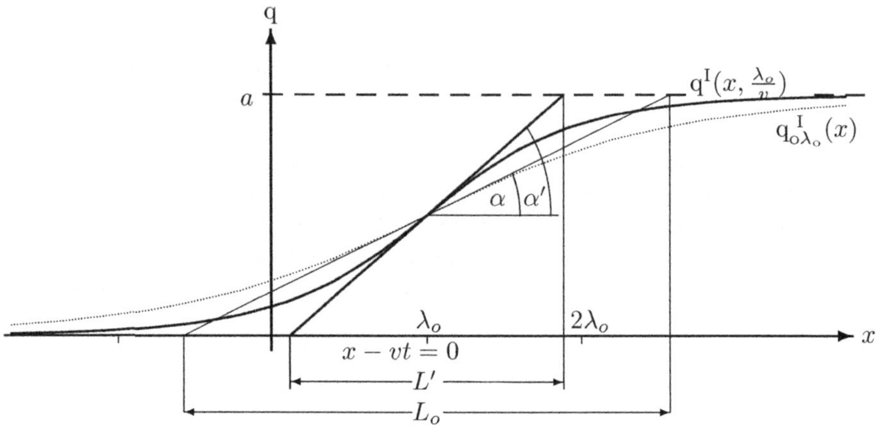

Fig. 10.2 The contraction of a moving kink. We observe the case of a kink (111) moving with the velocity $v = 0,8\,c_o$ at certain point of time t_o, here $t_o = \lambda_o/v$, thus $q^I(x, \frac{\lambda_o}{v}) = \frac{2a}{\pi} \arctan \exp \frac{\pi(x-\lambda_o)}{L_o \gamma}$. In comparison, a static kink at $x = \lambda_o$ according to (107), thus $q^I_{o\lambda_o}(x) = \frac{2a}{\pi} \arctan \exp[\frac{\pi(x-\lambda_o)}{L_o}]$ is illustrated as a thin line. We get in our example, for the length L' of a moving kink according to (112) $L' = \sqrt{1 - 0,64 c_o^2/c_o^2}\; L_o = 0,6\, L_o$

The moving rod is shortened.

$$L' = L_o \sqrt{1 - \frac{v^2}{c_o^2}} \;.$$

LORENTZ contraction
of the moving measuring-rod L' (112)

Here, we explicitly wish to point out that it is the unit of measure for the length, the measuring-rod, that is found in Eq. (112) for the LORENTZ contraction. The shortening of the measuring-rod is that what we can immediately observe.

We now consider in the preferred frame Σ_o the static distance (in Σ_o) $\mathbf{X} = x L_o$ from the origin of the coordinates measured using measuring-rod L_o. The length L' that glides along this distance $\mathbf{X} = x\, L_o = x'\, L'$ fits x' times into this distance. With (112), the following is valid,

$$x' = \frac{x}{\sqrt{1 - v^2/c_o^2}} \;. \tag{113}$$

One condition that has to be fulfilled in the application of the Eq. (113) for LORENTZ contraction, expressed in the coefficients of measure x and x', is that the distance to be measured $\mathbf{X}$ has to be at rest in the reference system Σ_o, see Fig. 10.3. The quantity x' is just the coefficient of measure for the shortened measuring-rod that is used to measure the static length $\mathbf{X}$ in Σ_o. Here, it is $L' = \sqrt{1 - v^2/c_o^2}\, L_o$.

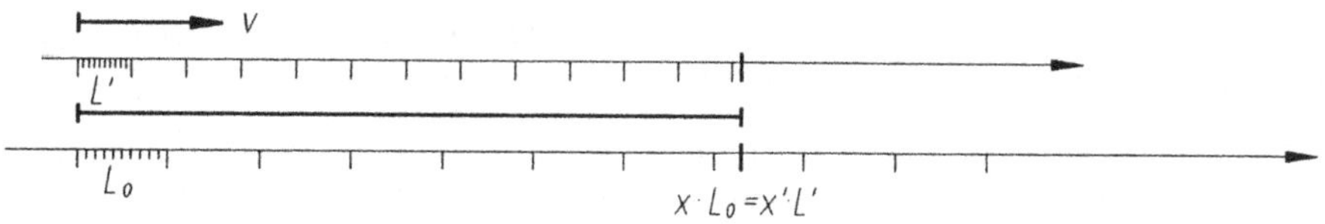

Fig. 10.3 Equation (113) of the Lorentz contraction, $x' = x/\sqrt{1 - v^2/c_o^2}$. Shown is $v = 0,8\,c_o$; thus, $\gamma = \sqrt{1 - v^2/c_o^2} = 0,6$ and therefore $x' = x/0,6$; $L' = 0,6\,L$

In order to avoid irritations, we want to call the reader's attention to the fact that there is another situation which formally leads to the same Eq. (113), but which describes another physical situation. In fact, it is only necessary for the validity of Eq. (113) that both the starting point and the end point of the length **X** are measured at the same point of time t of Σ_o. If both the starting point with its coordinate x_1 and the end point with its coordinate x_2 of a length $\Delta\mathbf{X}$ move in Σ_o with the same constant velocity v, so that $x_1 = x_o + vt$, $x_2 = x_o + \Delta x + vt$, then our measuring-rod L_o should fit, according to the assumptions Δx times in the length $\Delta\mathbf{X}$. However, the moving measuring-rod L' fits $\Delta x'$ times, $\Delta X = \Delta x L_o = \Delta x' L'$, so that we only have to exchange the coordinates with the coordinate differences in (113). Thus, the following is again valid

$$\Delta x' = \frac{\Delta x}{\sqrt{1 - v^2/c_o^2}} \ . \tag{113a}$$

This physically describes the following: It is supposed of a static rod ΔX referred to Σ' which on its part moves with a velocity v with respect to Σ_o, that its coefficient of measure of its length in Σ_o is exactly Δx. The coefficient of measure $\Delta x'$ of its length in Σ' is, due to the Lorentz contraction, determined by Eq. (113). We will return to this case in Chap. 12.

There are certain details in the measuring of lengths that play an important role later on. We have to distinguish these different situations:

1. We determine the spatial distance between two events, let us say the distance between the event O and the event B that we are observing in an *arbitrary* inertial system Σ: In order to measure this distance, we would mark both positions where the events occurred and we would then count how often our measuring-rod, available to us in this reference system, fits between both points. This would give us our coefficient of measure for the distance. We are not confronted with any problems using this method of measurement.

2. The next object to have its length measured is a rod. The objective is to determine the length of this rod in a reference system Σ. Here, two situations have to be differentiated. (a) The rod is stationary in the reference system Σ. I can therefore mark both end points without rushing, at any time and once again count how many times my measuring-rod fits in between. (b) The following situation creates completely new problems. The rod is moving with the uniform velocity v through our reference

system Σ. What is the length of the moving rod in Σ? The length of a moving rod is a new concept that has to be *defined*. This is done as follows:

> Take the coordinates $x_1(t)$ and $x_2(t)$ of both end points of the rod in motion at one and the same moment t in Σ. Then $x_2(t) - x_1(t)$ is the coefficient of measure for the length of the moving rod in Σ.

We see, in order to be able to measure the length of a moving rod, we would have to distribute (identical) clocks throughout our reference system Σ. These clocks would have to be started in such a manner that they run synchronically. Thus, the idea of a length of a moving rod is connected with the idea of the synchronisation of clocks. Our preferred frame Σ_o is the static crystal lattice. In this lattice, we have derived the sine-GORDON equation, with which we can directly calculate the length of a moving measuring-rod. This leads us to Eq. (112). Thus, we have principally solved the problem of measuring lengths in this preferred frame Σ_o. If however the reference system Σ' has a velocity v with respect to the preferred frame Σ_o, what is the length that we measure for the rod in Σ' if the rod moves relative to Σ'? According to our definition of a moving length, we now need synchronised clocks in Σ'. Finding a solution to this problem will cost us a lot of hard work. However, the answer will be very simple and will give us a very deep insight into the physics of the considered phenomena of crystals. We will concern ourselves with this in the Chaps. 12–15. We now calculate that $q^I(x,t)$ is in fact just as much a solution of the sine-GORDON equation (88) as is $q_o^I(x)$. In order to do this, we put

$$u := \frac{x - vt}{\gamma} \ . \tag{114}$$

According to (111) then it is

$$q^I(x,t) = q_o^I(u) \ , \tag{115}$$

and we get

$$\frac{\partial^2}{\partial x^2} q^I(x,t) = \frac{1}{\gamma^2}\frac{\partial^2}{\partial u^2} q_o^I(u) \ , \qquad \frac{1}{c_o^2}\frac{\partial^2}{\partial t^2} q^I(x,t) = \frac{1}{\gamma^2}\frac{v^2}{c_o^2}\frac{\partial^2}{\partial u^2} q_o^I(u) \ ,$$

thus, because $q_o^I(u)$ fulfils the time-independent sine-GORDON equation with u in the place of x,

$$\frac{\partial^2}{\partial x^2} q^I(x,t) - \frac{1}{c_o^2}\frac{\partial^2}{\partial t^2} q^I(x,t) = \frac{1}{\gamma^2}\left(1 - \frac{v^2}{c_o^2}\right)\frac{\partial^2}{\partial u^2} q_o^I(u) = \frac{\partial^2}{\partial u^2} q_o^I(u) = \frac{D}{\sigma}\sin\left(\frac{2\pi}{a} q_o^I(u)\right)$$

and thus

$$\frac{\partial^2}{\partial x^2} q^I(x,t) - \frac{1}{c_o^2}\frac{\partial^2}{\partial t^2} q^I(x,t) = \frac{D}{\sigma}\sin\left(\frac{2\pi}{a} q^I(x,t)\right) \ ,$$

which is exactly what we wanted to prove.

We will now turn to the clock. Again, the clock moved at the uniform velocity v has to develop out of the displayed function $q_o^{III}(x,t)$ in Figs. 9.3 and 9.4 by a displacement $v\,t$ to the right, with the important condition that the sine-GORDON equation maintains its validity. If one simply displaces the function $q_o^{III}(x,t)$ by $v\,t$ to the right, we get according to (105) the function

$$\tilde{q}_o^{III}(x,t) = \frac{2a}{\pi} \arctan \frac{\sin(2\pi t/T_o)}{\cosh \frac{\pi(x-vt)}{L_o\sqrt{2}}} \,. \tag{116}$$

Using this oscillating line as the moving observer's clock we would get exactly the same oscillation period T_o and thus according to (107), register an exact equivalence with the static clock of the observer in Σ_o at $b = v\,t$. However, a moving observer's clock possessing an oscillation period of T_o giving us the function (116) does not exist. Function (116) is not a solution of the sine-GORDON equation. We can thus make a further important observation in our crystal:

It is impossible to rigidly displace the clock defining oscillating line .

This means that there is no clock in the reality of our real continuum that does not change its pace when it moves. The line form in Figs. 9.3 and 9.4, thus the function $q_o^{III}(x,t)$ undergoes a characteristic deformation in order to be able to move uniformly with the velocity v. The corresponding solution $q^{III}(x,t)$ of the sine-GORDON equation (cf. A. SEEGER [86], H. GÜNTHER [34]) is,

$$\left.\begin{aligned} q^{III} &= q^{III}(x,t) = \frac{2a}{\pi} \arctan \frac{\sin \frac{2\pi(t-vx/c_o^2)}{T_o\gamma}}{\cosh \frac{\pi(x-vt)}{L_o\gamma\sqrt{2}}} \,, \\ \gamma &= \sqrt{1-\frac{v^2}{c_o^2}} \,. \end{aligned}\right\} \tag{117}$$

Here $q^{III}(x,t)$ is in fact a $x = v\,t$ localised line form that moves with the velocity v to the right along the x-axis. This function, depicted for three different points of time in Fig. 10.4 shows characteristic differences in comparison to the function $q_o^{III}(x,t)$ in Fig. 9.3.

Do the oscillating lines depicted in Fig. 10.4 have anything to do with a clock? We must proceed with caution in order to be able to determine the oscillation period T' using the moving clock from (117). Firstly we notice that $q^{III}(0,0) = q_o^{III}(0,0)$. Here we can add that at $x = 0$ both clocks, in other words the clock U_1 of the stationary observer and the clock U_v of the moving observer have the reading 0. The moving clock moves away with the velocity v and arrives at time t at a static clock positioned at $x = vt$. This clock is the clock from (108) at $b = vt$ as depicted in Fig. 10.6 as U_2.

We will now compare. We receive the expected measure of oscillation T_o for the static clock U_2 positioned at $x = b$, if we insert in (108) the position of the maximum

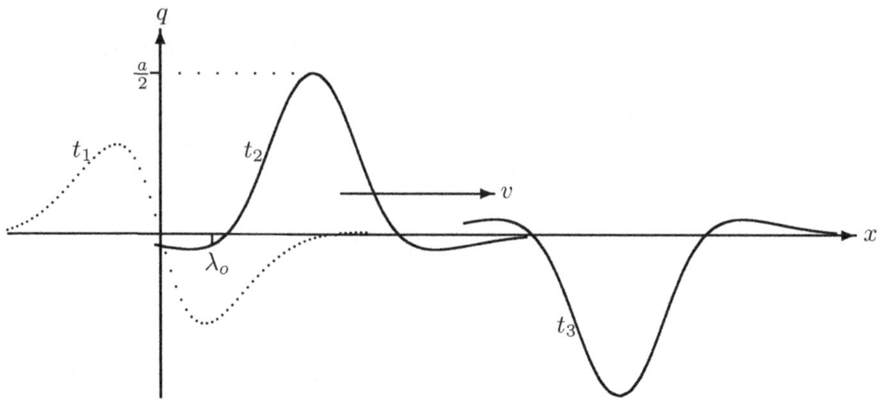

Fig. 10.4 The moving breather solution (117) $q^{III}(x,t) = \frac{2a}{\pi} \arctan \frac{\sin \frac{2\pi(t-vx/c_o^2)}{T_o\gamma}}{\cosh \frac{\pi(x-vt)}{L_o\gamma\sqrt{2}}}$ moving with $v = 0,8\,c_o$, $\gamma = 0,6$ for the same points of times as in Fig. 9.3, thus $t_1 = 0$, $t_2 = \frac{\pi\lambda_o\sqrt{2}}{2c_o}$, $t_3 = \frac{3\sqrt{2}\pi\lambda_o}{2c_o}$. Here, according to (102) $T_o = 2\pi\sqrt{2}\,\lambda_o/c_o$ and $L_o = \pi\,\lambda_o$ according to (99). The oscillating lines in Fig. 9.3 undergo a change in their period of oscillation by the motion with the uniform velocity v which we calculate in (120). Here a remarkable wavelike displacement of the oscillations can be observed

amplitude of oscillation $x = b$,

$$q_{ob}^{III}(b,t) = \frac{2a}{\pi} \arctan \frac{\sin \frac{2\pi t}{T_o}}{1} \quad \longrightarrow \quad T_o \ . \tag{118}$$

If we insert into (117) (into the equation for the moving clock U_v) the position of the maximum oscillation after time t has passed, in other words $x = vt$ we receive

$$q^{III}(vt,t) = \frac{2a}{\pi} \arctan \frac{\sin \dfrac{2\pi(1-v^2/c_o^2)\,t}{\gamma\, T_o}}{1} = \frac{2a}{\pi} \arctan \frac{\sin \dfrac{2\pi\,\gamma^2\, t}{\gamma\, T_o}}{1} \ ,$$

hence

$$q^{III}(vt,t) = \frac{2a}{\pi} \arctan \frac{\sin \dfrac{2\pi t}{T_o/\gamma}}{1} \ . \tag{119}$$

From this the period of oscillation T' of the moving clock U_v can be determined as

$$T' = \frac{T_o}{\sqrt{1-v^2/c_o^2}} \ . \tag{120}$$

The LORENTZ factor $\gamma = \sqrt{1 - v^2/c_o^2}$ is always smaller than 1. The period of oscillation T' of the moving clock is thus always larger than the period of oscillation T_o of the static clock. The unit of measure T' for time is increased by motion. Here we have found the famous time dilatation, written down using the units of measure of time, in other words the periods of oscillation of the moving and the static clocks. This is not however that what we can immediately see and compare when we observe a clock, the positions of the arms on the face of the clock.

The position of the arms on the face of the clock that are used to read the time count the number of the oscillations. This number represents the coefficient of measure of time. The arms of the moving clock always go behind the arms of a static clock. When the moving clock U_v with the hand setting t' arrives at the stationary clock U_2 with the hand setting t, then the time $\mathbf{T} = t\, T_o = t'\, T'$ has passed since it started at the clock U_1. Thus, because of (120), following is valid

$$t' = t\sqrt{1 - \frac{v^2}{c_o^2}} \ . \qquad \text{Time dilatation of the moving clock} \qquad (121)$$

The moving clock goes behind.

Once again we wish to emphasise: In order to check (121) one has to compare the readings of the single moving clock U_v to the readings of at least two static clocks. These two clocks have to have the distance $v\,t$ from each other and have to be stationary in the preferred frame Σ_o, see Fig. 10.6.

If the first static clock is not positioned at the origin of the coordinates, in other words we start at an arbitrary clock in Σ_o, we have to use Δt and $\Delta t'$ instead of the readings t and t' as depicted by the arms of the clock. We thus write

$$\Delta t' = \Delta t\sqrt{1 - \frac{v^2}{c_o^2}} \ . \qquad (121a)$$

In the Eq. (121) for the time dilatation, the coefficient of measure of time is used. The reduced coefficient of measure of time t' is that what one can immediately read from the face of the clock. Here we wish to remind the reader of the difference concerning the formula (112) for LORENTZ contraction. In (112) it is contraction of the moving unit of measure which is primarily observed. For the unit of measurement of time, the period of oscillation, time dilatation has to be formulated according to (120).

We will now prove that $q^{III}(x, t)$ is just as much a solution of the sine-GORDON equation (88) as $q_o^{III}(x, t)$. According to (117) it is

$$q^{III}(x, t) = q_o^{III}\left(\frac{x - vt}{\gamma}, \frac{t - vx/c_o^2}{\gamma}\right) .$$

We put

$$\left.\begin{aligned} u &:= \frac{x - vt}{\gamma} \,, \\ w &:= \frac{t - vx/c_o^2}{\gamma} \,. \end{aligned}\right\} \tag{122}$$

The partial differentiation with respect to x and t can be expressed using the partial differentiation with respect to u and w. After simple rearrangements we receive

$$\left.\begin{aligned} \frac{\partial u}{\partial x} &= \frac{1}{\gamma}\,, & \frac{\partial u}{\partial t} &= \frac{-v}{\gamma}\,, \\ \frac{\partial w}{\partial x} &= \frac{-v}{c_o^2\gamma}\,, & \frac{\partial w}{\partial t} &= \frac{1}{\gamma}\,, \\ \frac{\partial}{\partial x} &= \frac{\partial u}{\partial x}\frac{\partial}{\partial u} + \frac{\partial w}{\partial x}\frac{\partial}{\partial w}\,, & \frac{\partial}{\partial t} &= \frac{\partial u}{\partial t}\frac{\partial}{\partial u} + \frac{\partial w}{\partial t}\frac{\partial}{\partial w}\,, \\ \frac{\partial^2}{\partial x^2} &= \left(\frac{\partial u}{\partial x}\right)^2 \frac{\partial^2}{\partial u^2} + 2\frac{\partial u}{\partial x}\frac{\partial w}{\partial x}\frac{\partial^2}{\partial u \partial w} + \left(\frac{\partial w}{\partial x}\right)^2 \frac{\partial^2}{\partial w^2}\,, \\ \frac{\partial^2}{\partial t^2} &= \left(\frac{\partial u}{\partial t}\right)^2 \frac{\partial^2}{\partial u^2} + 2\frac{\partial u}{\partial t}\frac{\partial w}{\partial t}\frac{\partial^2}{\partial u \partial w} + \left(\frac{\partial w}{\partial t}\right)^2 \frac{\partial^2}{\partial w^2}\,. \end{aligned}\right\} \tag{123}$$

With

$$q^{III}(x, t) = q_o^{III}(u, w) \tag{124}$$

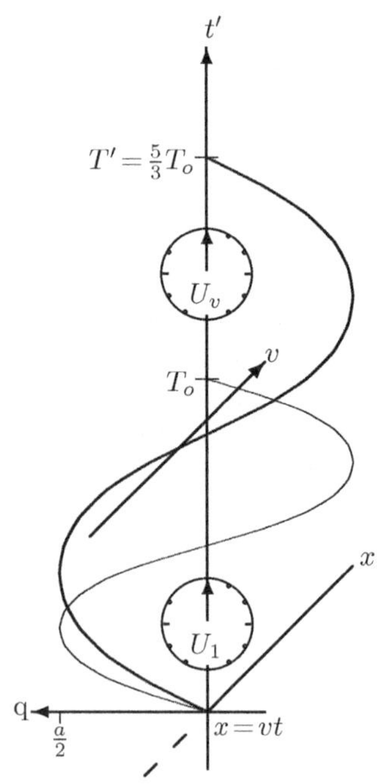

Fig. 10.5 The choice of the initial condition. When the clock U_v moving with the velocity v moves past the static clock U_1 at $x = 0$, both clocks should have their hands at zero. The clock U_1 counts the oscillations of the thin drawn line described by the function $q_o^{III}(0, t)$, see Fig. 9.4. The clock U_v counts the oscillations of the thick curve described according to (119) by $q^{III}(vt, t)$, see Fig. 10.4

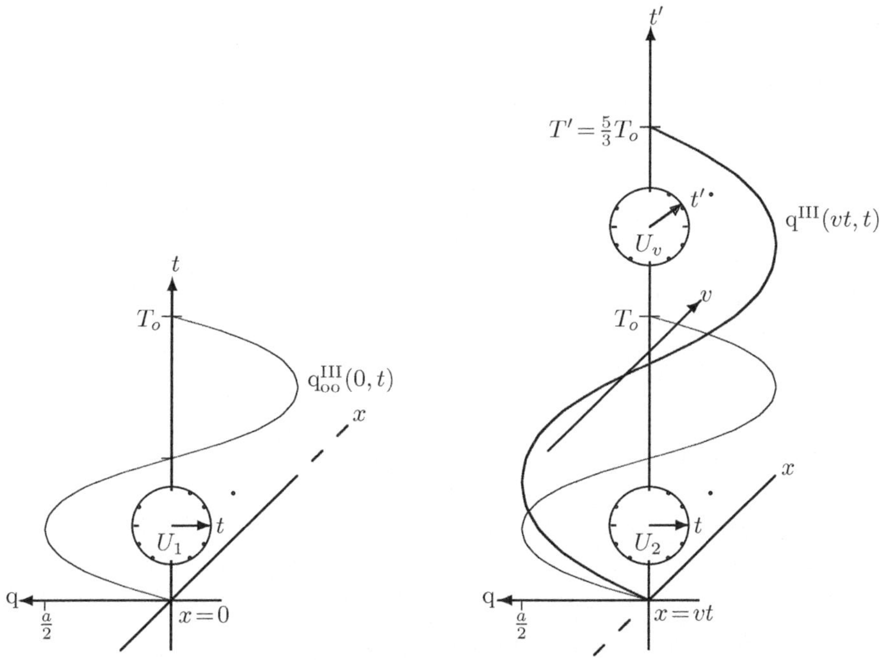

Fig. 10.6 The measuring of time dilatation. Due to the initial condition according to Fig. 10.5, the clock U_v, as defined by the moving breather solution (119), $q^{III}(vt, t) = \frac{2a}{\pi} \arctan \sin \frac{2\pi t}{T_o/\gamma}$, had the same hand setting as the static clock U_1 as it flew past: zero. The hand setting on U_v has to be compared to a second static clock U_2. We always have to observe one moving clock and at least two static clocks when determining time dilatation. The oscillations of the latter are depicted by the thin curves. The localised breather solution according to (100a) for $x = 0$, $q_{oo}^{III}(0, t) = \frac{2a}{\pi} \arctan(\sin \frac{2\pi t}{T_o})$, belongs to U_1. The localised breather solution according to (118), $q_{ob}^{III}(b, t) = \frac{2a}{\pi} \arctan(\sin \frac{2\pi t}{T_o})$, belongs to U_2. Both functions coincide. We have chosen the velocity $v = 0,8\, c_o$ for U_v. We have gauged the clocks in such a manner that the hand on U_2 shows $t = 30$ scale partitions when it meets with U_v (in other words in the position 'half past'). The hand of U_v then shows, according to (121) $t' = 30\sqrt{1 - 0,64 c_o^2/c_o^2} = 30 \cdot 0,6 = 18$ scale partitions

thus follows

$$\frac{\partial^2}{\partial x^2} q^{III}(x,t) = \frac{1}{\gamma^2} \frac{\partial^2}{\partial u^2} q_o^{III}(u,w) + 2 \frac{1}{\gamma} \frac{-v}{c_o^2 \gamma} \frac{\partial^2}{\partial u \partial w} q_o^{III}(u,w) + \left(\frac{-v}{c_o^2 \gamma} \right)^2 \frac{\partial^2}{\partial w^2} q_o^{III}(u,w),$$

$$\frac{\partial^2}{\partial t^2} q^{III}(x,t) = \frac{v^2}{\gamma^2} \frac{\partial^2}{\partial u^2} q_o^{III}(u,w) + 2 \frac{1}{\gamma} \frac{-v}{\gamma} \frac{\partial^2}{\partial u \partial w} q_o^{III}(u,w) + \frac{1}{\gamma^2} \frac{\partial^2}{\partial w^2} q_o^{III}(u,w)$$

and after a simple calculation

$$\frac{\partial^2}{\partial x^2} q^{III}(x,t) - \frac{1}{c_o^2} \frac{\partial^2}{\partial t^2} q^{III}(x,t) = \frac{\partial^2}{\partial u^2} q_o^{III}(u,w) - \frac{1}{c_o^2} \frac{\partial^2}{\partial w^2} q_o^{III}(u,w) \ .$$

We will now notice that $q_o^{III}(u, w)$ fulfils the sine-GORDON equation (88) using the variables u, w in place of x, t,

$$\frac{\partial^2}{\partial u^2} q_o^{III}(u,w) - \frac{1}{c_o^2} \frac{\partial^2}{\partial w^2} q_o^{III}(u,w) = \frac{D}{\sigma} \sin\left(\frac{2\pi}{a} q_o^{III}(u,w)\right) \ ,$$

and also find out with (124) that

$$\frac{\partial^2}{\partial x^2} q^{III}(x,t) - \frac{1}{c_o^2} \frac{\partial^2}{\partial t^2} q^{III}(x,t) = \frac{D}{\sigma} \sin\left(\frac{2\pi}{a} q^{III}(x,t)\right) \ ,$$

which is exactly what we wanted to prove.

Using the Eqs. (112) and (121) we have shown the two most spectacular kinematic effects of the Special Theory of Relativity: LORENTZ contraction and time dilatation.

However this is not yet the Special Theory of Relativity!

First of all we distinguished one inertial system Σ_o, the system in which our crystal rests. We described the sine-GORDON equation in this preferred frame and derived the rest from this equation. Thus we have only shown: When measuring-rods and clocks move against this preferred inertial frame Σ_o, then these measuring-rods are shortened and these clocks go behind.

Here a comparison with electrodynamics before the discovery of EINSTEIN's theory in 1905 would be in order. In 1889 G. F. FITZGERALD [20] postulated a hypothesis for the explanation of the MICHELSON- MORLEY experiments, "that the length of material bodies changes, according as they are moving through the ether or across it, by an amount depending on the square of the ratio of the velocities of that of light." H. A. LORENTZ [61] independently postulated a similar hypothesis in 1892. LORENTZ's endeavours concentrated on a quantitative expression for this contraction. In his paper from 1904 on the electromagnetic theory in moving systems, LORENTZ [62] came to the conclusion: "We are therefore led to suppose that the influence of a translation on the dimension (of the separate electrons and of a ponderable body as a whole) is confined to those that have the direction of the motion, these becoming k times smaller than they are in the state of rest", where k is the factor $k = \sqrt{1 - v^2/c_L^2}$. This is exactly what our Eq. (112) states (only that in our equation c_o replaces c_L from the other equation). In Chap. 16 we will make a detailed evaluation of the FITZGERALD–LORENTZ hypothesis for the Special Theory of Relativity, cf. also Chap. 12.

Just as 'our LORENTZ contraction' (112) takes place against the preferred frame Σ_o, our static crystal, in which the sine-GORDON equation is valid, H. A. LORENTZ postulated that the contraction of measuring-rods against the corresponding preferred frame Σ_o, in which the MAXWELLian equations, on which his calculations were based, were valid. The question exactly what the MAXWELLian equations for a moving reference system (against Σ_o) looked like and how we can undertake measurements from such a reference system were in 1904 just as unanswered as our question: Which equation does the internal observer moving with respect to the crystal, in other words the observer in a reference system Σ', find for the phenomena inside of a crystal, for which we found the sine-GORDON equation in the preferred frame Σ_o? We will answer this question in Chap. 15.

Before we continue with our line of thought, we will discuss a paradox that becomes visible by the present results of our discussions.

Chapter 11
A Clock Paradox

The EINSTEINian theory of relativity is based on the fact that the speed of light is a natural constant and thus has the same value in any arbitrary inertial system, $c_L = 299\,792\,458$ m/s. According to EINSTEIN's Special Relativity one and the same MAXWELLian equations are applicable for all inertial systems. We measure this velocity for every signal sent out from a light source. It makes no difference if we remain relatively at rest to the light source or if we move with the velocity v towards or away from the wavefront of the light signal. For the unprepared reader, these somewhat unbelievable facts are just as unbelievingly explained by EINSTEIN's Special Relativity: The moving clock goes behind and the moving measuring-rod is contracted. In detail, this means: We construct a number of identical clocks. A few of these are distributed and synchronised in order to obtain a perfect timing along the x-axis. We now take one of the remaining clocks and move it with a constant velocity v along the line of the distributed clocks. If we compare the hand setting t' of the moving clock to the hand setting t of the stationary clock at the moment it passes by, we see that $t' = t\sqrt{1 - v^2/c_L^2}$ where we assumed $t = t' = 0$ at $x = x' = 0$. This is what EINSTEIN's theory of relativity states. The hand of the moving clock goes behind. The only physical parameter in this equation is the speed of light c_L, which is an absolute natural constant. It, thus, makes no difference what type of clock we use in the experiment to prove this equation, as long as our methods of measurement are exact enough. In such an experiment, we should, if we move an old golden pocket watch with the velocity v or an atomic caesium clock, always exactly determine the factor $\sqrt{1 - v^2/c_L^2}$. The rules for the construction of a clock do not appear in the equation for time dilatation.[1]

[1] Clocks that are driven on by gravity, i.e. pendulum clocks, egg timers, water clocks, etc., have to be excluded in such experiments. We expressed that we intended to ignore gravity because

H. Günther, *Elementary Approach to Special Relativity*,
https://doi.org/10.1007/978-981-15-3168-2_11

Time dilatation is the 'experimentum crucis' of Special Relativity. The physical constitution or the technical construction of clocks is not found in the equation for time dilatation. We can thus use our breather clocks in our test for the dilatation of time in the same way as we examined the pace of these clocks in the last chapter. Here, we discover that the hand of a moving breather clock went behind the hands of the static breather clocks according to Eq. (121) $t' = t\sqrt{1 - v^2/c_o^2}$. Now factually, this is a dilatation of time as seen in the EINSTEINian theory of relativity, but the numerical value is completely different. In our equation, it is not the speed of light c_L, but the characteristic velocity c_o of the sine-GORDON equation that is included. This velocity c_o lies in the region of the velocities of sound and is thus smaller than the speed of light by many powers of ten, $c_o \ll c_L$. EINSTEIN's Special Relativity states, however, that only the speed of light alone can stand in the equation for time dilatation for all clocks. This is the focal point of the whole theory! Have we just shown that EINSTEIN's Special Theory of Relativity is not universally applicable or maybe even incorrect?—By no means.

Where can we find the solution to this paradox: The crucial statement we used to introduce the description of the dilatation of time was 'we construct a number of *identical* clocks'. The static breather clocks positioned in the inertial system Σ_o are oscillating lines q_o^{III} where the geometrical centres of inertia rest relative to the crystal. However, our moving breather clock is an oscillating line q^{III}, where the geometrical centre of inertia has the velocity v relative to the crystal. The observer of the EINSTEINian theory of relativity, the 'outside observer', wanting to use crystals in the construction of clocks examines the crystal with its oscillating breather lines as a whole. An outside observer stationary in Σ_o may introduce the oscillating lines q_o^{III} inside a crystal, whose centres of inertia rest relative to the crystal, in order to measure time. Moving clocks of this construction are for this observer always of such a kind that the whole crystal moves with its oscillating lines. The construction rules of such a clock for the outside observer include that the centres of inertia of the oscillating lines remain motionless with respect to the crystal. This condition must remain fulfilled when clocks move. These oscillating lines q^{III} with their centres of inertia have the velocity v with respect to the crystal, are for this observer completely different clocks. Such clocks could also have been introduced in order to measure time. The observer of the EINSTEINian Special Theory of Relativity finds it normal that both types of clocks have different periods of oscillation. In order to compare these, they would have to be gauged accordingly just as an oscillating quartz would have to oscillate more times in a second than the spring of a pocket watch. Thus, the observer would not arrive at a contradiction in his statements on Special Relativity. If the observer uses the static breather clock inside the crystal, so that the clock moves together with the crystal, he finds after comparing the hand setting $\tilde{t}$ of a breather clock moving together with the crystal with the hand settings t of such breather

(Footnote 1 continued)
this would need a complete new line of thought. This new line of thought would also play no role in our needs for solids. (For a pendulum clock moving with the velocity v away from earth, the influence of earth's gravity would decrease; however, the influence from the moon could increase).

clocks resting in Σ_o, as well as inside of the crystal his known equation for time dilatation containing the expression c_L, the speed of light $\tilde{t} = t\sqrt{1 - v^2/c_L^2}$. There is thus no contradiction. Does this mean that our Eq. (121) for the time dilatation of breather clocks is worthless? Naturally not.

We examine an infinitely extended crystal as well as the observers *inside* of this crystal. This representation of 'internal observers' as we say goes back to EINSTEIN and was used by him in a demonstration of his ideas on the General Theory of Relativity. EINSTEIN argued using his 'two-dimensional creatures', his 'flat beings' on a sphere, see for example L. INFELD [42], A. EINSTEIN [17]. Just as was used in the case above, we will use such an ideal construction in order to show to what degree the development of our theory depends on the possible methods of measurement.[2]

For our internal observers, the crystal as a whole does not move. We could also formulate it as follows. The small internal observers inside the crystal see this as an infinitely large world. They do now ask themselves about the motion of their world. Only the craziest and audacious philosophers amongst them would think of or even mention the possibility—leading to the accusation of blasphemy and insanity and probably ending up being burnt at the stake as a warlock—that their world is finite.

The outside observers of EINSTEIN's Special Relativity and the internal observers inside the crystal work with two completely different classes of reference systems. There is only one common element if they find themselves in the same preferred frame Σ_o. This is the solution to the paradox.

Let us, however, now continue with the theoretical conceptions the observers inside of the infinitely large widths of the crystal make. Here, we will consequently use only those 'mechanical instruments' defined by the sine-GORDON equation and will especially not use the speed of light. We will make sure that all the 'instruments' we use are of mechanical origin in our crystal. These are the many structures above the ground state, the energetically distinguished state of an ideal lattice.

Using A. SEEGER [86], we will not only understand this ground state (the vacuum) as an ideal lattice, but we will also count the infinitely long dislocations passing as straight lines through the lattice as the vacuum, cf. (97).

---

[2] This EINSTEINian concept of an internal observer was used in a different context in continuum mechanics by E. KRÖNER [51, 52]. E. KRÖNER refers to the problems connected to the spatial measurements inside of the crystal. The problem is thus completely different to our own. The KRÖNERian internal observers should not be mistaken with our internal observers.

Chapter 12
The Measurement of the Critical Velocity

We now arrive at the most sensitive point of the Special Theory of Relativity: The measurement of the critical signal velocity. The testing of the speed of light, the critical velocity of electromagnetic phenomena was the main reason why the violent dispute concerning the ether had arose. The unexpected result of this dispute was EINSTEIN's postulate in his paper from 1905, cf. EINSTEIN [14, 15], the principle of the universal constancy of this signal velocity, which led as a consequence to a revised formulation of the whole of physics.

Here, we will understand the quantity c_o in the sine-GORDON equation by the critical signal velocity. Primarily, we have no real picture of this velocity inside of a mechanical continuum. We are far more used to dealing with sound velocities. The sound velocities do give us an orientation of the magnitude of c_o. In order to see this, we consider the approximation of the underlying crystal replaced by an isotropic continuum. In many cases, it is exactly this isotropic continuum that delivers a useful concept in describing mechanical processes inside of a solid. There are two different sound velocities inside of an isotropic continuum (see Chap. 25). The first is the transversal polarised sound wave, where the oscillating vector **s** is orthogonal with the direction of propagation **k** of the wave. The second is the longitudinal wave, where the oscillating vector **s** is parallel with the direction of propagation **k**. Let us consider transversal sound waves, those waves that move through the medium with the transversal sound velocity c_T. This velocity c_T is closely related to the critical velocity c_o in the sine-GORDON equation. Here, we remind ourselves that an excitation propagating along the dislocation line according to the sine-GORDON equation is just as much a transversal displacement out of this dislocation direction (see Chap. 7) as is the transversal displacement of the linear chain propagating with the transversal sound velocitiy (see Chap. 6). Furthermore, the linear terms of the FRENKEL-KONTOROVA equation (77), which approximate the sine-GORDON equation, describe the linear chain of the lattice atoms in the vicinity of the dislocation line.

H. Günther, *Elementary Approach to Special Relativity*,
https://doi.org/10.1007/978-981-15-3168-2_12

This makes J. D. ESHELBY's [19] result plausible, the result that the parameters of the crystal lattice really lead to a remarkable correspondence between the critical velocity c_o of the sine-GORDON equation and the transversal sound velocity c_T. In this sense, we will use this conformity as a model presumption,

$$c_o \approx c_T \ . \tag{125}$$

Factually, the validity of the hypothesis $c_T \approx c_o$ is of no real importance for our further considerations. We can however use this fact to illustrate our point,[1] and we will arbitrarily refer, due to reasons of linguistic simplicity, to the signal moving through the crystal with the velocity c_o as the 'sound signal'. By definition, c_T is a velocity that we measure in our preferred frame Σ_o, that is, in the inertial system in which an outside observer would see that the crystal as a whole would be static. Here, we measure in fact one and the same signal velocity c_T in every direction of our, a crystal approximating three-dimensional continuum. We consider from this isotropic continuum only one single space dimension described by the choice of the x-axis. Our linear chain lies in this x-direction, which defines our dislocation line. We then need the transition to a continuum for this one direction. We only take the periodical order of the surrounding lattice atoms of the crystal into account. This led us in Chap. 8 to the sine-GORDON equation with $c_o \approx c_T$. However, isotropy is also for one-dimensional problems a property of extreme importance and will force us to make a few explanations further along in this chapter. It has the meaning of constancy for the critical velocity of both opposite directions. Firstly, we can state:

Our one-dimensional continuum is isotropic in the preferred frame Σ_o.

This one-dimensional isotropy of space becomes transparent for the sine-GORDON equation (88), which is the basis of our one-dimensional considerations, as follows: Whilst keeping the time t, we change the orientation of the x-axis; in other words, we consider a space inversion, so that all the positions can be described using new coordinates $\bar{x}$ according to

$$\bar{x} = -x \ , \quad \bar{t} = t \ . \qquad \text{Space inversion} \tag{126}$$

The transversal displacement $q = q(x, t)$ remains unchanged, in other words,

$$\bar{q}(\bar{x}, \bar{t}) = q(x, t) = q(-\bar{x}, \bar{t}) \ . \tag{127}$$

Due to the fact that only the second derivatives appear in the sine-GORDON equation, using (126) and (127) we get from (88) immediately

---

[1] Though, the transversal sound velocity c_T does in fact take the part of the relativistic critical velocity inside of an isotropic medium for straight screw dislocations (and only for these), and it even takes this part for the interactions between dislocations and elastic deformations of the surrounding lattice which is ignored here. In Chap. 26, we will illustrate this with an example.

$$\frac{\partial^2}{\partial \bar{x}^2}\bar{q}(\bar{x},\bar{t}) - \frac{1}{c_o^2}\frac{\partial^2}{\partial \bar{t}^2}\bar{q}(\bar{x},\bar{t}) = \frac{D}{\sigma}\sin\left[\frac{2\pi}{a}\bar{q}(\bar{x},\bar{t})\right]. \tag{128}$$

In other words, referred to our new coordinates, the sine-GORDON equation remains invariant. The critical velocity has one and the same value c_o for both space directions. Our one-dimensional continuum is isotropic.

If we wish to measure the velocity c_o in Σ_o, we can use our measuring-rods and clocks. Due to the isotropy in Σ_o, we have to take the following steps: We consider a length $\Delta x\ L_o$ in Σ_o. We determine the coordinates x_1 and x_2 of the end points out of reasons of simplicity as $x_1 = 0$ and $x_2 = \Delta x$. In other words, Δx is the number that informs us how many times our measuring-rod L_o fits on to the length. The statement of time t (the coordinate t in Σ_o) also informs us how many times the pendulum of our normal clock with the oscillation period of T_o swung. At the time $t_o = 0$ at $x_1 = 0$, we send a sound signal to x_2. At time Δt_1, it is reflected and arrives, due to the isotropy of signal propagation, at time $\Delta t_o = 2\Delta t_1$ at $x_1 = 0$, so that we have measured the critical velocity c_o in Σ_o (see Fig. 12.1) according to

$$c_o = \frac{2\Delta x}{\Delta t_o} = \frac{2\Delta x}{2\Delta t_1}\ . \tag{129}$$

We now consider a measuring section, a 'rod' $\Delta \mathbf{X}$ (being a corresponding structure inside of the crystal) moving with the uniform velocity v relative to the preferred frame Σ_o. This rod is at rest in a reference system Σ'. The internal observers in Σ' can measure the length of this rod simply by using their measuring-rods L' and determine the coefficient of measure $\Delta x'$. We can measure any sort of lengths in the preferred frame Σ_o as was discussed in Chap. 10 after Eqs. (112) and (113), p. 96.

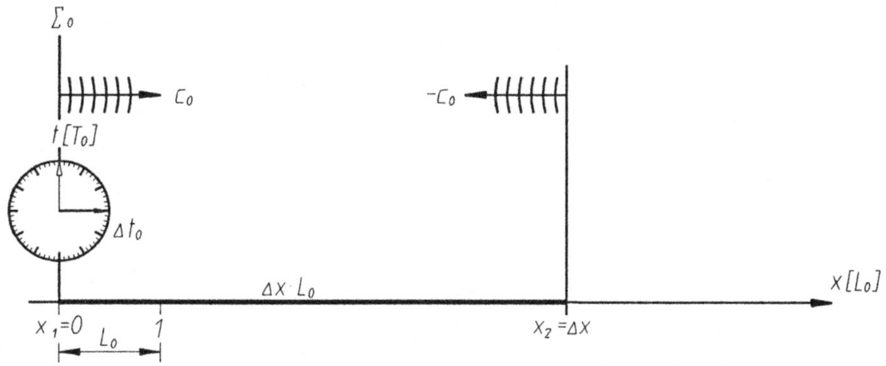

Fig. 12.1 The measuring of the critical velocity c_o in the preferred frame Σ_o. At time $t_o = 0$, a 'sound signal' is sent from $x_1 = 0$ to x_2. This signal is reflected at time Δt_1 and arrives at time $\Delta t_o = 2\Delta t_1$ at x_1. In the figure, the clock is gauged in such a way that the time Δt_o is shown at 15, in other words 'quarter past'. Taking the fact as granted that the critical velocity in our preferred frame Σ_o has the same value on its way in both directions (see the explanation in the text above), we have measured using $c_o = 2\Delta x/\Delta t_o$ the critical velocity c_o in the reference system Σ_o, see (129)

We assume that the length of the moving rod in Σ_o has the coefficient of measure Δx. Its end points $x_1(t)$ and $x_2(t)$ (for one and the same moment t) may be described in Σ_o by

$$x_1(t) = +v\,t\ , \quad x_2(t) = \Delta x + v\,t\ . \tag{130}$$

Thus, it is $\Delta\mathbf{X} = \Delta x\, L_o = \Delta x'\, L'_o$ and for the coefficients of measure Δx and $\Delta x'$ Eq. (113a) for the contraction of length of a moving rod applies,

$$\Delta x' = \frac{\Delta x}{\sqrt{1 - v^2/c_o^2}}\ . \tag{113a}$$

Once again at time $t_o = 0$, we emit a sound signal in the positive direction towards x_2. It arrives here at time $\Delta t_{\rightarrow}$, gets reflected and arrives at time $\Delta t = \Delta t_{\rightarrow} + \Delta t_{\leftarrow}$ at the left end point of the measuring section $x_1 = +v\,t$. On the first part of the journey, the point x_2 moves with the velocity v away from the sound signal, which therefore only moves towards the point with the velocity $c_o - v$. We therefore get for the time $\Delta t_{\rightarrow}$ that the sound signal needs for the first part of the journey

$$\Delta t_{\rightarrow} = \frac{\Delta x}{c_o - v}\ . \tag{131}$$

On the way back, the mark x_1 moves towards the sound signal with the velocity v; therefore, the sound signal actually moves towards the mark x_1 with the velocity $c_o + v$. We receive for the time $\Delta t_{\leftarrow}$ that the signal needs for the return journey

$$\Delta t_{\leftarrow} = \frac{\Delta x}{c_o + v}\ . \tag{131a}$$

Here, we wish to state that Eqs. (131) and (131a) strictly apply. They do not contain any approximation. This addition of the velocities c_o and v should not be confused with EINSTEIN's [14, 15] 'composition of velocities'. This will be discussed in detail in Chap. 17, see Eqs. (179) and (180) as well as Figs. 17.2 and 17.3. Summing up, we observe for the complete procedure in the reference system Σ_o a velocity $\bar{c}_o$ according to

$$\bar{c}_o = \frac{2\Delta x}{\Delta t} = \frac{2\Delta x}{\Delta t_{\rightarrow} + \Delta t_{\leftarrow}} = \frac{2\Delta x}{\dfrac{\Delta x}{c_o - v} + \dfrac{\Delta x}{c_o + v}} = \frac{2}{\dfrac{c_o + v + c_o - v}{c_o^2 - v^2}} = c_o\Big(1 - \frac{v^2}{c_o^2}\Big)\ . \tag{132}$$

What kind of velocity is this?

The quantity $\bar{c}_o$ is an effective velocity. In (132), we divided the moving length Δx, which was traversed by the signal, by the arithmetic mean of both time intervals, so that $\bar{c}_o = \frac{\Delta x}{(\Delta t_{\rightarrow} + \Delta t_{\leftarrow})/2}$. Together, the velocities $c_{\rightarrow} = \Delta x/\Delta t_{\rightarrow}$ that the signal needs to traverse the distance on the first part of the journey, where the measuring section effectively moves away from the signal with the velocity v and $c_{\leftarrow} = \Delta x/\Delta t_{\leftarrow}$ on

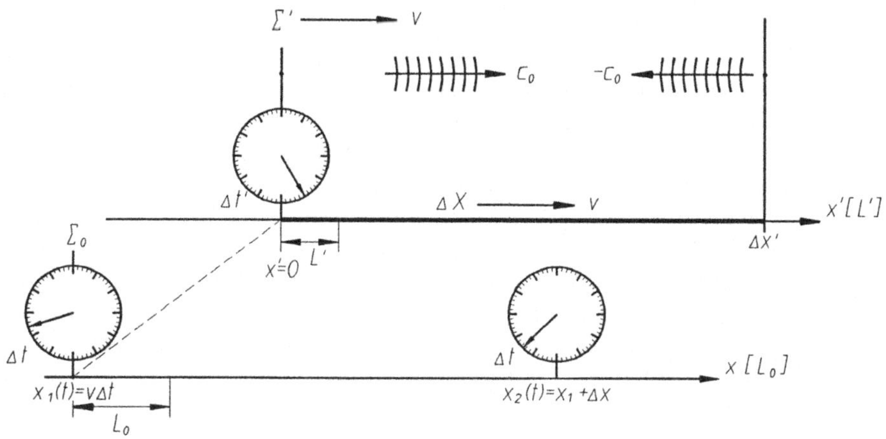

Fig. 12.2 The determination of the average velocity $\bar{c}_o$. Here, it is assumed that the section to be measured is moving with the velocity $v = 0,8\,c_o$. This results in a factor $\sqrt{1 - v^2/c_o^2} = 0,6$. Thus, $L' = 0,6 \cdot L_o$. If we take in comparison the coefficient of measure of time represented in Fig. 12.1, $\Delta t_o = 2\Delta x/c_o = 15$, we calculate using (131) and (131a) due to the fact that we use the same Δx, $\Delta t_{\rightarrow} = \Delta t_o \frac{c_o}{2(c_o - v)} = 37,5$. This is shown on the right-hand static clock of Σ_o. On the other hand, we get $\Delta t_{\leftarrow} = \Delta t_o \frac{c_o}{2(c_o + v)} = 4,2$. The total signal running time, the sum of both times, is shown on the left-hand clock of Σ_o as the coefficient of measure $\Delta t = \Delta t_{\rightarrow} + \Delta t_{\leftarrow} = 42$. In comparison, the hand setting on the clock resting in Σ' is according to time dilatation $\Delta t' = \Delta t\,\gamma = 25$. The dotted line combines two points that describe one and the same event. We remind ourselves: An event is defined by the position where it takes place and by the time when it takes place. The characterisation of an event is dependent on the reference system in which it is described, by the statement of its spacetime coordinates

the return journey, where the measuring section moves towards the signal with the velocity v result in the effective velocity $\bar{c}_o$ in (132), see Fig. 12.2, taking the mean,

$$\frac{1}{\bar{c}_o} = \frac{1}{2}\left(\frac{1}{c_{\rightarrow}} + \frac{1}{c_{\leftarrow}}\right) = \frac{1}{c_o(1 - v^2/c_o^2)} \,. \tag{132a}$$

We now consider the observer moving with the velocity v, who is thus in the reference system Σ' in which our measuring section ΔX is at rest, which we can illustrate to ourselves as a resting 'rod' in Σ'. The decisive question is: What value ${c_o}'$ does the observer in his reference system Σ' find out for the velocity with which our signal propagates? How large is the, from the preferred frame Σ_o, c_o attributed velocity for the observer moving with the velocity v relative to the crystal lattice?

We will answer these questions in two steps. We will first ask, what value ${\bar{c}_o}'$ does the observer discover in Σ' for the effective velocity that the signal needs to traverse back and forth along the measuring section $\mathbf{\Delta X}$?

We can answer this question without any further assumptions. If the observer wants to measure this velocity ${\bar{c}_o}'$, all he has to do is to copy our method where we determined the effective velocity $\bar{c}_o$, cf. (132). The only difference is that he uses the measuring-rods and clocks which are at rest relative to him. Let us observe how he proceeds.

His measuring-rod L' is according to Eq. (112) contracted in comparison with our measuring rod L_o by the factor $\sqrt{1 - v^2/c_o^2}$. Thus, the coefficient of measure $\Delta x'$ that he determines for the rod $\Delta\mathbf{X}$ resting with respect to him is larger than our coefficient of measure Δx by this factor, in other words $\Delta x' = \Delta x/\sqrt{1 - v^2/c_o^2}$, just as we have written it above in Eq. (113a), see also Fig. 12.2. Besides this, the observers clock positioned at the starting point of the measuring section and moved (in reference to Σ_o) with the velocity v together with this measuring section shows for the time interval between the event of sound signals emission and the event of its arrival the coefficient of measure $\Delta t'$ that has to be calculated using Eq. (121a) for time dilatation of moving clocks. The coefficient of measure $\Delta t'$ is smaller than the measured (with the help of two clocks) coefficient of measure in Σ_o, $\Delta t = \Delta t_{\rightarrow} + \Delta t_{\leftarrow}$ by the factor $\sqrt{1 - v^2/c_o^2}$ for the difference in time of both events, see Fig. 12.2

$$\Delta t' = \Delta t\sqrt{1 - \frac{v^2}{c_o^2}} \ . \tag{121a}$$

The moving observer in Σ' thus measures according to (113a) and (121a) a velocity $\bar{c}_o{}'$ according to

$$\bar{c}_o{}' = \frac{2\Delta x'}{\Delta t'} = \frac{2\Delta x}{\Delta t}\frac{1}{1 - v^2/c_o^2} \ .$$

We have calculated the factor $2\Delta x/\Delta t = \bar{c}_o$ in (132). Hence, the observer moving with respect to the crystal lattice with the velocity v measures for the effective velocity $\bar{c}_o{}'$ of the signal in his reference system Σ' the value

$$\bar{c}_o{}' = c_o\,\frac{1 - v^2/c_o^2}{1 - v^2/c_o^2} \ ,$$

and thus, as we have shown in 1994, cf. GÜNTHER [34],

$$\bar{c}_o{}' = c_o \ . \tag{133}$$

As remarkable as this conformity between both velocities may seem, the quantity $\bar{c}_o{}'$ measured by the observer in Σ' is an *effective* velocity. If we denote the above velocity of the signal on the first part of the way as $c'_{\rightarrow}$ and on the way back as $c'_{\leftarrow}$ we can calculate $\bar{c}_o{}'$ as shown above by taking the mean,

$$\frac{1}{\bar{c}_o{}'} = \frac{1}{2}\left(\frac{1}{c'_{\rightarrow}} + \frac{1}{c'_{\leftarrow}}\right) = \frac{1}{c_o} \ . \tag{133a}$$

Our opinion however, which should be considered cautiously, tends towards the general conditions depicted in our preferred frame Σ_o, which leads to the view that the velocity which is effective to traverse the measuring section on the way back is greater than on the way there. However, if we could be certain that the velocity for our observer in Σ' is isotropic, i.e. that the velocity has the same value for the way there and back, then this would be a sensation. We have however not shown this. The decisive question is the question about the signal velocity $c'_{\rightarrow}$ that the observer in Σ' measures for the first part of the measuring section. If we really could prove that $c'_{\rightarrow} = c_o{}'$, then because of (133a) the following would immediately result, $c'_{\rightarrow} = c'_{\leftarrow} = c_o{}'$ and we would have reached our goal. However, without further reflections, we cannot solve this problem.

In order to be able to measure a 'true' velocity (and not just an effective value for way there and back), we need two clocks, one at the start and the second at the end of the length to be measured. We have to synchronise these clocks in order to be able to measure the time needed for the signal to travel. We need, in order to be able to synchronise these clocks, a 'true' velocity and not just an effective value. We intended to determine such a velocity, but such a velocity is not at our disposal. We find ourselves in a loop. Let us recapitulate and examine our situation.

The moving observer can make pure geometrical measurements in his reference system Σ' without the help of any clocks. All he needs is a sufficient number of measuring-rods L'_o in an unbroken line, so that he can determine by counting these measuring-rods, distances and lengths in his reference system Σ'. Using this method, he can determine $\Delta x'$ according to (113a). All occurring events in the preferred frame Σ_o can be attributed with the coordinates (x, t) and the observer in Σ' can determine the space coordinates x' for these events. He can thus study geometry. The observer in Σ' cannot however determine an order of time for the events that take place somewhere and compare this order with the order of time that has been determined for these events in the reference system Σ_o. The observer can 'see' that the objects in Σ' move, but he is not able to determine a well-defined velocity for these motions. He can distribute his clocks throughout the system, but as long as he does not have a well-defined rule of how to set up these clocks, any 'measured' velocity using these clocks would be completely arbitrary.

We thus need a rule that informs us how we have to start a clock with a certain hand setting at an arbitrary position in Σ'. We will do this for a clock positioned at the beginning of our length, the origin of our coordinates Σ'. We will call this clock U_v^o. Let us label the clocks resting in Σ' generally with the subscript v. We want to determine that U_v^o together with the clock U_o at the coordinate origin of the reference system, Σ_o both have the value 0 when these clocks meet. This coincidence of the coordinate origins will be called the event O. Our (arbitrarily determined) initial condition is thus, see Fig. 12.3,

$$\Sigma_o : O \ (x = 0 \, ; \, t = 0) \quad \longleftrightarrow \quad \Sigma' : O \ (x' = 0 \, ; \, t' = 0) \ . \qquad \text{Initial condition} \tag{134}$$

Then when all clocks in Σ' have started they should all run in the same manner, because they are all of the same construction. Thus, the only question is when to start the other clocks in Σ' (except U_v^o, which is started by the initial condition). As harmless as this question may seem, this process determines the time procedure of every event, of every physical law. We must also accomplish a way of setting clocks for all possible systems of reference Σ', in other words for all possible velocities v,

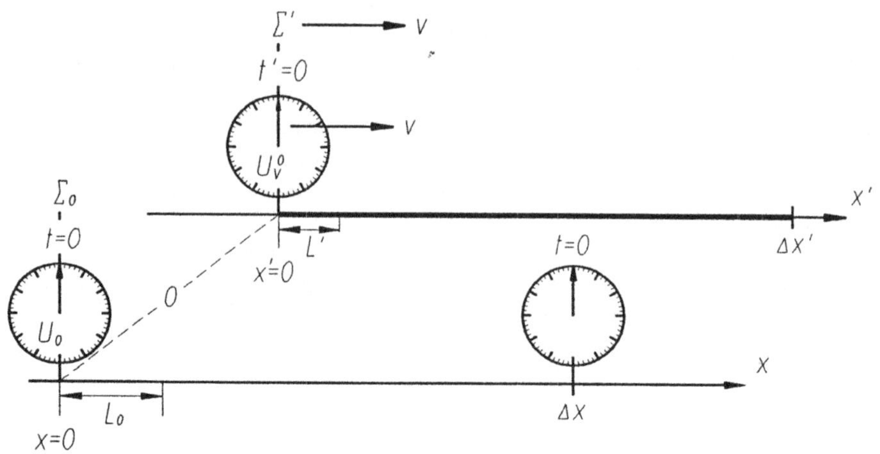

Fig. 12.3 The initial condition for the synchronisation of clocks in Σ'. The dotted line combines two points that describe one and the same event, the event O

which are measured from the preferred frame Σ_o. We would thus be well advised to look for the most simplest rules for the synchronisation of clocks in the moving systems of reference Σ', if our descriptions of physical events from out of Σ' are not to end in chaos. However, the question concerning the synchronisation of clocks in Σ' may seem to be blown out of proportion. Who forces us to describe an arbitrary process of motion from an arbitrary reference system Σ', when we know exactly how to do this from our preferred frame Σ_o ?

At least since the discovery of the connection between symmetries and conservation laws, it has become one of the most privileged tasks in physics to explain the symmetry conditions of natural events, see for example H. Genz [26] and R. Decker. The question concerning the synchronisation of clocks in moving reference systems Σ' should be formulated contrarily, namely: Is there a rule for the setting of clocks in Σ' through which these systems of reference gain equivalence with our preferred frame Σ_o ? Is there a symmetry for the description of our observed phenomena in solids that can be expressed by the equivalence of all of our observed reference systems?

In fact, this really is the case. A simple principle has been deduced from Einstein's principle of relativity, the so-called reciprocity principle, cf. V. Berzi [2] and V. Gorini. Here, our point will be the following: It is exactly this reciprocity principle that enables us to separate the synchronisation of clocks from the other part of SRT axiomatics. Let us found the rules which we will follow when setting up clocks in a moving reference system on this reciprocity principle. Because of its part in our axiomatics, we will call this principle the *elementary principle of relativity* and formulate it as follows, cf. Günther [30]:

> After an observer resting in Σ_o has measured that Σ' possesses the velocity v with respect to Σ_o, all clocks in Σ' should be set such that an observer resting in Σ', measures that the reference system Σ_o possesses the velocity $-v$ with respect to Σ'.

Here, v is an arbitrary velocity with $|v| < c_o$. We therefore use the velocity $-v$ of the preferred frame Σ_o with respect to Σ' in order to synchronise the clocks in the reference system Σ'. All other velocities observed from Σ' have thus been fixed.

Summarising our reflections, we followed H. Poincaré, who in 1898 in his famous essay 'La Mesure du Temps' called attention to the fundamental importance of a *definition* of the simultaneity for two events needed in order to describe natural processes. Poincaré [74] stated that 'Il est difficile de séparer le problème qualitatif de la simultanéité du problème quantitatif de la mesure du temps; soit qu'on se serve d'un chronomètre, soit qu'on ait à tenir compte d'une vitesse de transmission, comme celle de la lumière, car on ne saurait mesurer une pareille vitesse sans *mesurer* un temps. $\cdots$ La simultanéité de deux événements, ou l'ordre de leur succession, l'égalité de deux durées, doivent être définies de telle sorte que l'énoncé des lois naturelles soit aussi simple que possible. En d'autres termes, toutes ces règles, toutes ces définitions ne sont que le fruit d'un opportunisme inconscient."[2]

Indeed, our elementary principle is in fact nothing more than an arrangement between the observers of both reference systems in order to be able to compare the measured velocities of both systems of reference.

We now assume that the observer in Σ' has distributed and set up his clocks so that he can synchronise them according to our elementary principle of relativity. We will do exactly this for our two clocks positioned at the end points of our length $\Delta\mathbf{X}$, the stationary rod in Σ', which possess the velocity v when observed from Σ_o. One of the clocks U_v^o is positioned at the left end point of the length at the coordinate origin of Σ'. We have started this clock according to our initial condition (134). All other clocks have to follow this clock. At the right-hand side of the length is the clock U_v^*. This clock has to be synchronised with U_v^o. In order to achieve this, we designate an event as A, when the right end point of the length with the clock U_v^* finds itself at the coordinate origin of the reference system Σ_o.

Both clocks U_v^o and U_v^* then run synchronically, in the sense of our elementary principle of relativity, if and only if the hand of U_v^* finds itself at the position $t'_A = -\Delta x'/v$ when the event A takes place. When the coordinate origin of Σ_o moving, according to our principle, from Σ' with the velocity $-v$ arrives at the left-hand side of our measuring section, the coordinate origin of Σ', then the pointer of U_v^* has moved by $\Delta x'/v$ and comes to hand setting 0 just as the pointer of U_v^o does in accordance with our initial condition. Therefore, a synchronisation, in other words an simultaneity in Σ' has been defined for both clocks, according to which the velocity $-v$ for the reference system Σ_o was ascertained. This is exactly what we wanted to achieve, see Fig. 12.4. If we take (113a) for $\Delta x'$ into consideration, then the event A is described by the observer in Σ' in the following manner,

[2] 'It is difficult to separate the qualitative problem of simultaneity from the quantitative problem of the measurement of time; no matter whether a chronometer is used, or whether account must be taken of a velocity of transmission, as that of light, because such a velocity could not be measured without *measuring* a time. $\cdots$ The simultaneity of two events or the order of their succession, the equality of two durations, are to be so defined that the enunciation of the natural laws may be as simple as possible. In other words, all rules, all these definitions are only the fruit of an unconscious opportunism'. Quoted from the English translation in Poincaré [74].

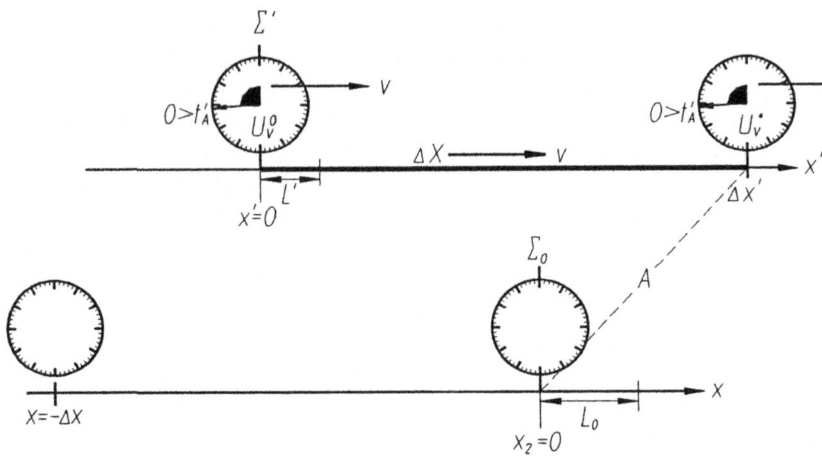

Fig. 12.4 The synchronisation of both clocks U_v^o and U_v^* in Σ'. Here, we have defined event A in such a fashion that the right end point of the length resting in Σ' has the coordinate $x_2 = 0$ in Σ_o. We assume, as in Fig. 12.2, a velocity of $v = 0,8\,c_o$ with a factor $\gamma = 0,6$. We also gauge the clocks, as in Fig. 12.1, according to $\Delta t_o = 2\Delta x/c_o = 15$. For the hand setting t'_A, we can calculate with the help of (135) a coefficient of measure $t'_A = -\Delta t_o \frac{c_o}{2v\gamma}$. The dotted line combines both points that once again describe one and the same event, event A. The time coordinate of this event in Σ_o still has to be determined

$$\Sigma' : A \left(x'_2 = \Delta x' = \frac{\Delta x}{\sqrt{1 - v^2/c_o^2}} \;, \quad t'_A = \frac{-\Delta x}{v\sqrt{1 - v^2/c_o^2}} \right) . \tag{135}$$

The observer in Σ_o discovers that the event A takes place as awaited at $x_2 = 0$. This is where his clock U_o is positioned. Because the length $\Delta\mathbf{X}$ in Σ_o accordingly possesses the coefficient of measure Δx, the observer registers the coordinate $-\Delta x$ for the left end point moving towards him with the velocity $+v$ as shown in Fig. 12.4. The left end point of the length arrives at his position after the time $\Delta x/v$, where the pointer of his clock U_o positioned at the coordinate origin has the hand setting 0. At the time of event A, the pointer of the clock U_o is at $x_2 = 0$, thus at the time coordinate of $t_A = -\Delta x/v$ so that,

$$\Sigma_o : A \; (x_2 = 0 \,, \quad t_A = \frac{-\Delta x}{v}) \,. \tag{136}$$

This results in Fig. 12.5.

The coincidence of both coordinate origins makes up the event O with the space and time coordinates according to our initial condition (134). According to (136), the time Δt_A passes between event A and event O, the coincidence of both coordinate origins, in Σ_o,

$$\Delta t_A = \frac{\Delta x}{v} \,. \tag{137}$$

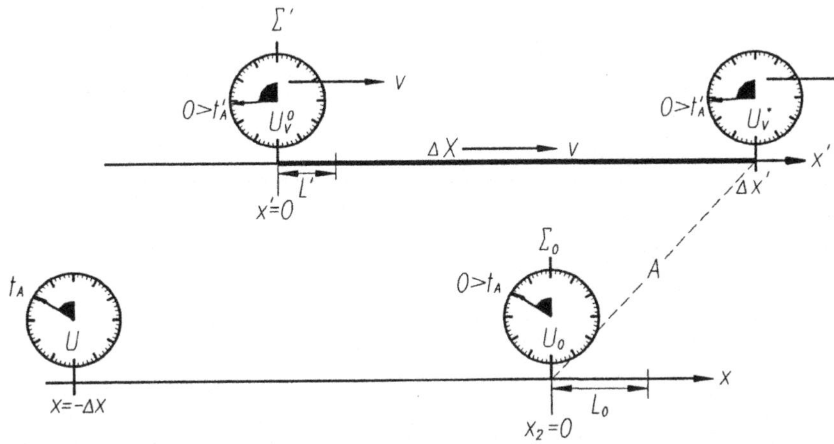

Fig. 12.5 At event A the clocks U_v^* and U_o meet. The positions of the hand settings can be directly compared. The coefficient of measure t_A for the time shown by U_o is calculated using (135) and (136) with $t'_A = 15,6$ as $t_A = \gamma t'_A = -9,4$. Once again, the dotted line combines both points describing event A. Both of the left end points cannot be combined by the dotted line because we have illustrated two different events which will be described below

The hand of the moving clock U_v^* with respect to Σ_o, moves more slowly due to the time dilatation (121a) namely by the value $\Delta t'_A$,

$$\Delta t'_A = \Delta t_A \sqrt{1 - \frac{v^2}{c_o^2}} = \frac{\Delta x}{v}\sqrt{1 - \frac{v^2}{c_o^2}} \ . \tag{138}$$

The hand of U_v^* finds itself for the coincidence of both coordinate origins at the position $t'_B = t'_A + \Delta t'_A$, so that with (135) and (138)

$$t'_B = \frac{-\Delta x}{v\sqrt{1 - v^2/c_o^2}} + \frac{\Delta x}{v}\sqrt{1 - \frac{v^2}{c_o^2}} = \frac{-\Delta x}{v}\left(\frac{1}{\sqrt{1 - v^2/c_o^2}} - \sqrt{1 - \frac{v^2}{c_o^2}}\right)$$

and thus

$$t'_B = \frac{-\Delta x \, v/c_o^2}{\sqrt{1 - v^2/c_o^2}} \ . \tag{139}$$

The event B is described by both reference systems as follows,

$$\Sigma' : B \ \left(x'_2 = \Delta x' \ , \quad t'_B = \frac{-\Delta x \, v}{c_o^2\sqrt{1 - v^2/c_o^2}}\right) \ , \tag{140}$$

$$\Sigma_o : B \ (x_2 = \Delta x \ , \quad t_B = 0) \ . \tag{141}$$

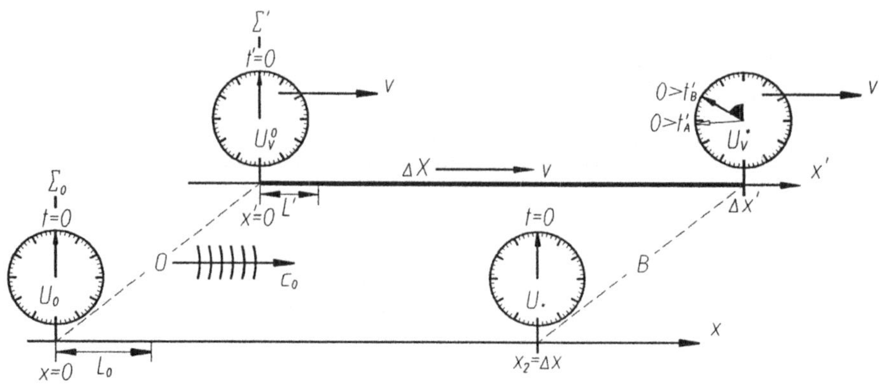

Fig. 12.6 The signal emission. When $t = 0$ and $x = 0$, then $t' = 0$ and $x' = 0$. This is our event O. The hand setting of U_v^o shows 0. At $t = 0$, the Σ' clock U_v^* in the reference system Σ_o positioned at the right side of the measuring section has the coordinates of this end point, $x_2 = \Delta x$. Event B has according to (141) the coordinates $x = \Delta x$ and $t = 0$ in Σ_o and according to (142) the coordinates $x' = \Delta x$ and t'_B in Σ'. We have calculated the coefficient of measure t'_B for the hand setting of U_v^* in (139). Using the values $v = 0,8\,c_o$, thus $\gamma = 0,6$ and with the coefficient of measure $\Delta t_o = 2\Delta x/c_o = 15$, we receive $t'_B = -\Delta t_o \frac{v}{2c_o\gamma} = 10$. We once again have points describing one and the same event. These have been combined by the dotted lines. Here, these are the events O and B. In accordance with Eqs. (140) and (141) we can directly see the following from the illustration: Both events O and B that take place simultaneously in Σ_o do not take place simultaneously in Σ'. At the event B, the observer in Σ_o states that the signal at the coordinate origin starts 'now', as depicted in the figure, but the observer in Σ' contradicts this and states that the signal is not started simultaneously with event B, but at the later point of time $t' = 0$. For the observer in Σ', event B takes place at $t'_B < 0$. There is no 'now' that makes sense for both observers

The pointers of the Σ' clocks, U_v^o and U_v^* at the uniform time $t_o = 0$ of the reference system Σ_o are shown in Fig. 12.6. Here, two events, O and B are depicted.

The length Δx of the measuring section in Σ_o is of course arbitrary. This means that (139) generally describes the hand settings of the pointers $\Delta t' = t' = t'(\Delta x, 0)$ of those clocks of Σ' that belong to the uniform time $t = 0$ of Σ_o. For $t = 0$, $\Delta x = x$ because the starting point of the measuring section is positioned at the coordinate origin. We can therefore also write for (139)

$$t'(x, 0) = \frac{-x\, v/c_o^2}{\sqrt{1 - v^2/c_o^2}} \ . \tag{142}$$

This is the equation for the synchronisation of the Σ' clocks as observed for time $t = 0$ from Σ_o, see Fig. 12.7.

If read generally, Eq. (139) describes the difference $\Delta t'$ of the hand settings of two clocks resting in Σ', that take the positions x_o and $x_o + \Delta x$ respectively at the uniform time t_o in Σ_o according to

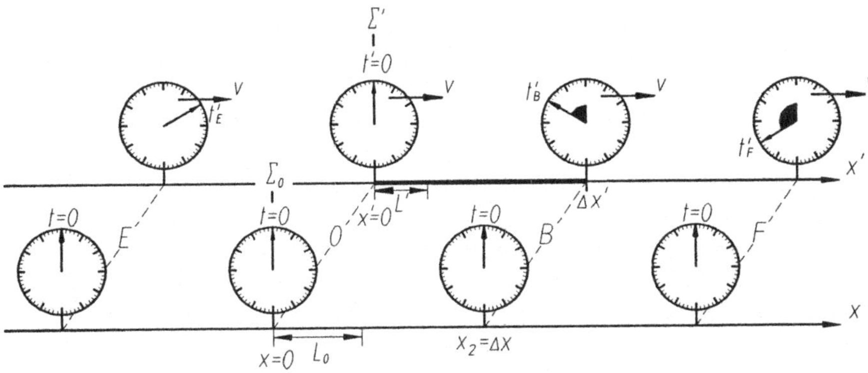

Fig. 12.7 Assuming the elementary principle of relativity, we calculate the hand settings of the synchronised clocks in Σ' at the uniform time in the reference system Σ_o of $t = 0$. Points belonging to one and the same event are once again combined by the dotted lines. Events E and F have been added to the events O and B shown in Fig. 12.6. The hand settings of the Σ' clocks are calculated according to Eq. (139). Here, we once again consider the case where $v = 0,8\,c_o$, thus $\gamma = 0,6$ and gauge the clocks in the same way as in the previous figures, so that the coefficient of measure $\Delta t_o = 2\Delta x/c_o$ has the value $\Delta t_o = 15$, which takes up the quarter past position. Using Eq. (139) in the reference system Σ', we calculate for the events E, O, B, F the corresponding points of time the coefficients of measure $t'_E = 10$, $t_o{}' = 0$, $t'_B = -10$, $t'_F = -20$

$$\Delta t' = \frac{-\Delta x\, v/c_o^2}{\sqrt{1 - v^2/c_o^2}} \,. \tag{142a}$$

Because of the minus sign in the numerator, for $\Delta x > 0$ and $v > 0$, the clock at $x_o + \Delta x$ has to be, as opposed to the clock at x_o, set back, see Fig. 12.7. This also leads us to a key sentence of the Special Theory of Relativity: Two events, here the events O and B, that take place simultaneously in one reference system, here according to (134) and (141) at $t = 0$ in Σ_o do not take place simultaneously in a moving reference system Σ' (with respect to Σ_o), because in Σ' at time $t' = 0$ for the event O and at $t'_B < 0$ according to (141) for the event B. This can be seen directly in Fig. 12.6. The magnitude of the difference $t'_B - t_B$ is determined by the velocity c_o. We can therefore state:

> We assume the elementary principle of relativity. As a consequence, the simultaneity of two events is a property that is only valid for that reference system in which it was measured. The magnitude of the deviation from simultaneity is dependent on the critical signal velocity.

We can read in formula (139) the following: Only for the limiting case of a reference system independent, infinitely large signal velocity could a reference system independent absolute simultaneity develop. Notice that here the elementary principle of relativity is taken for granted. NEWTON's description of gravity, the universal attraction of masses, contains such a presumption of simultaneous interaction. NEWTON's theory of gravity includes the hypothesis that the mass forces of attraction propagate through space with an infinitely large velocity, so that the force is simultaneously omnipresent. Such a description is mathematically consistent, however when examined more closely, collides with our conception of the world and can be falsified experimentally.

As quoted above, the criticism of the opinion of an absolute, a proiri given, reference system independent simultaneity as promoted by NEWTON's theory of gravitation was undertaken by H. POINCARÉ [73, 74] in 1898. A quantitative definition for simultaneity is found for the first time in EINSTEIN's [14, 15] paper from 1905. Here, the relativity of simultaneity is a consequence of the principle of the universal constancy of the critical velocity (there the speed of light c_L) including the definition of simultaneity with the help of this principle,[3] cf. also our discussion in Chap. 16. We however have taken an alternative way and receive the same result as a consequence of the LORENTZ contraction and time dilatation of measuring-rods and clocks, respectively, that move with respect to the preferred frame Σ_o and under the condition of our elementary principle of relativity which enables a synchronisation of clocks in the moving reference systems Σ'. This synchronisation makes easier a comparison of velocities in the different reference systems. Nevertheless, nobody can force us to apply even thus method for synchronisation of clocks in the 'moving' frames Σ'.

Due to the fact that the consequences arising from the synchronisation of clocks according to our elementary principle of relativity in connection with the time dilatation of moving clocks and the LORENTZ contraction of moving lengths are so fundamental for our lines of thought, we once again want to stray a little.

If we just ignore our elementary principle of relativity we could following POINCARÉ choose a different rule by which we could start our clocks in Σ'. This could be for example the following:

We once again consider the point of time $t = 0$, defined in our preferred frame Σ_o by the positions of the 'static' clocks. In Σ', we distribute sufficiently many 'moving' clocks. We start all of these clocks at the position $t' = 0$ when they meet the static clocks of Σ_o, which also have the hand setting $t = 0$, as illustrated in Fig. 12.8. Then, by definition, two events that take place simultaneously in Σ_o also take place simultaneously in the moving reference system Σ'. Our need for simultaneity would thus be fulfilled. The Σ' clocks still underlie the principle of time dilatation, so that admittedly at $t = 0$ in Σ_o all clocks including the ones in Σ' would show $t' = 0$. However, for an arbitrary point of time t, $t' = \gamma t$ is valid. If we insert $x - vt$ in place of Δx according to (130) into Eq. (113a), then we get the coordinates $x' = (x - v\,t)/\gamma$. We thus receive as a total, including the inversion,

[3]The correction of, NEWTON's theory of gravitation could not be achieved just by taking this fact into consideration. The consistent incorporation of universal gravitation in physics throws up new fundamental questions which were answered by, A. EINSTEIN's General Theory of Relativity from 1915. The measurable effects of this theory are under 'earthly conditions' relatively minute. The depiction of special relativistic phenomena allows us to drop the premise of gravitation. This we have done throughout this book.

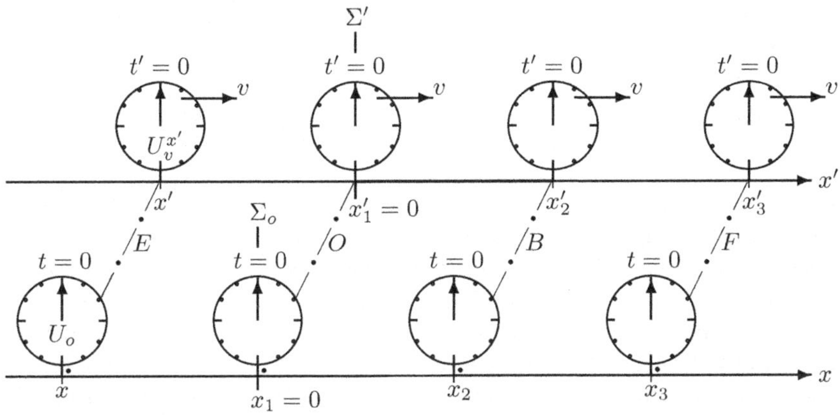

Fig. 12.8 The definition of an absolute simultaneity. We consider the point of time $t = 0$, defined in our preferred frame Σ_o. In Σ' we distribute sufficiently many 'moving' clocks. We start all of these clocks at the position $t' = 0$ when they meet the static clocks of Σ_o, which also have the hand setting $t = 0$. (A dotted line combines two points that describe one and the same event, here the events E, O, B and F.)

$$\left.\begin{array}{ll} x' = \dfrac{x - v\,t}{\gamma} \quad , & x = \gamma\, x' + \dfrac{v}{\gamma}\, t' \, , \\ t' = \gamma\, t \quad , & t = \dfrac{1}{\gamma}\, t' \, . \end{array}\right\} \text{Absolute simultaneity: REICHENBACH transformation} \qquad (143)$$

It was H. Reichenbach [81–83], who predicted the existence of such a transformation with upholding of an absolute simultaneity and without calling the physical laws of SRT into question, cf. our discussion in Chap. 16. Therefore, we have named Eq. (143) after H. Reichenbach, cf. also Günther [38].

Notice that we started with a well-defined simultaneity in the preferred frame Σ_o assuming there an isotropic signal propagation. After that, we could introduce the absolute simultaneity (143) in a 'moving' reference system Σ'. Due to the factor γ, time intervals are different between two events when measured from Σ_o and Σ' respectively. Therefore, the absolute simultaneity (143) must not confused with absolute time in Newtonian mechanics, where we have $t' = t$, cf. also the next chapter. Using Eq. (143), we can immediately determine the measured velocity of the coordinate origin of Σ_o from Σ', namely $v' = x'/t' = -v/\gamma^2$ for $x = 0$. If the observer in the preferred frame Σ_o determines a velocity v for the system Σ', the observer in Σ' measures for the reference system Σ_o a velocity $v' = \frac{-v}{1-v^2/c_o^2}$. Generally, an object, moving through the positions $x' = x'(t')$, has the velocity $u' = dx'/dt'$ in Σ'. We ask for the velocity $u = dx/dt$, that is measured for the same object from out of the reference system Σ_o. Using Eq. (143) we find,

$$\frac{\Delta x}{\Delta t} = \frac{\Delta x'\,\gamma^2 + \Delta t'\,v}{\gamma} \cdot \left(\frac{\Delta t'}{\gamma}\right)^{-1} = \frac{\Delta x'}{\Delta t'}\gamma^2 + v \ .$$

Hence, a composition of velocities, which is based on the REICHENBACH transformation (143) is valid,[4]

$$u = u'\gamma^2 + v \quad \longrightarrow \quad u' = \frac{u - v}{\gamma^2} \,. \qquad \text{REICHENBACH's composition of velocities} \tag{144}$$

Whilst the observer in Σ_o can only determine velocities smaller than c_o (one should think about the moving measuring rods and clocks), the observer in Σ' registers velocities u' that, because of the factor $1/\gamma^2$, can become arbitrarily large if his velocity v in comparison to Σ_o is sufficiently near to c_o. If one however wishes to accept these peculiarities as an expression for the different nature of these reference systems and also the distinction of our reference system Σ_o, then one cannot sustain this statement. We will at once see this: Using (143), we would have introduced a description into the reference system Σ' that blocks our view of an important symmetry of the observed phenomena in solids. Nevertheless, this description is correct and reflects the reality.

Notice that as well as our clock paradox of the last chapter is nothing but an illustration for POINCARÉ's ideas. We here as well refer to H. REICHENBACH [83], Chap. 2, 'We could arrange the definition of simultaneity of a system K in such a manner that it leads to the same results as that of another system K' which is in motion relative to K;...', which is exactly what we have done in (143). We will come back to this question and its consequences, e.g. concerning the contraction of moving length's in Chap. 16.

We will now leave absolute simultaneity and deal with the determination of the velocity $c_o{}'$ in Σ' using that synchronisation procedure, based on our elementary principle of relativity for the Σ' clocks. In connection to this, we will be able to show in the next two chapters, that this reference system (and with it all other systems moving with a uniform velocity with respect to Σ_o) is completely equivalent to Σ_o. Here, the important point is that we can prove the *existence* of such an instruction for the synchronisation, with which we receive such a symmetry of the reference systems. In the appendix, Chap. 26, we will give a further example for the above-quoted ideas of POINCARÉ and REICHENBACH. There we will explain that the so-called structural eigen strains, which will be generated by arbitrary dislocations in a solid with arbitrary symmetry, fulfil one and the same D'ALEMBERTian wave equation in all reference systems Σ'. This is, in its generality, a surprisingly simple result.

Let us now move on to the velocity $c_o{}'$ that the observer measures in Σ'.

For this purpose, we send the signal attributed with the velocity c_o by Σ_o on to the measuring section with the intention to determine the value $c_o{}'$ in Σ'. We start the signal at the mutual coordinate origin, our event O. The observers in Σ_o and Σ' register the same starting time $t_o = t_o{}' = 0$ at the starting point of the measuring section $\Delta\mathbf{X} = \Delta x\, L_o = \Delta x'\, L$ being at rest in Σ'. The travel time $\Delta t'_*$ of the signal in the reference system Σ' up to the end point of the measuring section can be read

[4]Notice that the single difference of (144) to the famous EINSTEINian composition of velocities (194) is the definition of simultaneity so that no contradiction can arise with experiments.

immediately form the hand setting of the synchronised clock U_v^* in Σ'. The arrival of the signal at the clock U_v^* will be called event C. We will therefore determine the hand setting $\Delta t'_*$ of the clock U_v^* when event C takes place. From Σ_o, we can determine that the hand of U_v^* for $x = \Delta x$ and $t = 0$ is, according to (139), at the position t'_B, as was calculated in our example, see Fig. 12.6. Calculated from Σ', the signal does not move yet, because t'_B has a negative value and the signal only starts to move at $t_o{}' = 0$. Once again, observed from Σ_o, the travel time of the signal from the beginning to the end point of the measuring section has, according to (131), the coefficient of measure $\Delta t_{\rightarrow} = \frac{\Delta x}{c_o - v}$. Because of time dilatation, the hand of the moving clock U_v^* only moves forward during this Σ_o time by $\Delta t'_{\rightarrow} = \Delta t_{\rightarrow}\sqrt{1 - v^2/c_o^2}$, so that at the end it is positioned at $\Delta t'_*$ according to $\Delta t'_* = t'_B + \Delta t'_{\rightarrow}$, thus with (139),

$$\Delta t'_* = \frac{-\Delta x\, v/c_o^2}{\sqrt{1 - v^2/c_o^2}} + \Delta t_{\rightarrow}\sqrt{1 - \frac{v^2}{c_o^2}} \; . \tag{145}$$

Hence, the event C, the arrival of the signal at U_v^* is described by both reference systems as follows,

$$\Sigma' : C \left(x'_2 = \Delta x' \, , \quad t'_C = \Delta t'_* = \frac{-\Delta x\, v/c_o^2}{\sqrt{1 - v^2/c_o^2}} + \Delta t_{\rightarrow}\sqrt{1 - \frac{v^2}{c_o^2}} \right) \, , \tag{146}$$

$$\Sigma_o : C \; (x_2 = \Delta x \, , \quad t_C = \Delta t_{\rightarrow}) \; . \tag{147}$$

The signals arrival at the clock U_v^* has been shown in Fig. 12.9. With (113a) and (145) we determine the velocity $c_o{}'$ in Σ',

$$c_o{}' = \frac{\Delta x'}{\Delta t'_*} = \frac{\Delta x/\sqrt{1 - v^2/c_o^2}}{\Delta t_{\rightarrow}\sqrt{1 - v^2/c_o^2} + \dfrac{-\Delta x \cdot v/c_o^2}{\sqrt{1 - v^2/c_o^2}}} = \frac{\Delta x}{\Delta t_{\rightarrow}(1 - \dfrac{v^2}{c_o^2}) - \dfrac{\Delta x \cdot v}{c_o^2}} \, ,$$

and, if we take into account $\Delta x/\Delta t_{\rightarrow} = c_o - v$ according to (131) we get

$$c_o{}' = \frac{c_o - v}{1 - v^2/c_o^2 - (c_o - v)\, v/c_o^2} = c_o^2 \frac{c_o - v}{c_o^2 - v^2 - c_o v + v^2} = c_o^2 \frac{c_o - v}{c_o(c_o - v)} \, ,$$

and thus, see GÜNTHER [30],

$$c_o{}' = c_o \; . \tag{148}$$

This is in fact a remarkable result which contradicts everyday knowledge concerning velocities. If the critical velocity c_o is isotropic in just one reference system—we assumed isotropy in the preferred frame Σ_o—then it is also isotropic in all other reference systems Σ' moving with an arbitrary constant velocity v with respect to Σ_o, because of the behaviour of the measuring-rods and clocks. We will observe one and the same numerical value for this velocity c_o in all reference systems Σ' if we synchronise the clocks in the moving systems according to our elementary principle of relativity:

Two observers moving towards each other with a uniform velocity v measure one and the same critical velocity c_o.

This is the core of the Special Theory of Relativity.

Expressed differently: The result of a measurement of the critical velocity c_o does not depend on whether we move towards or away from the signal emitter, with the velocity v. The equivalence to Special Relativity for our physical spacetime with the speed of light c_L is thus perfect. For our internal observers inside of our crystal, the critical velocity c_o (or the transversal sound velocity c_T as we state with certain justification) is an 'absolute natural constant'. We can thus repeat this by saying:

The critical velocity c_o of the sine-GORDON equation is, for the internal observers inside of a crystal, an 'absolute natural constant'.

This is nothing else than EINSTEIN's [14, 15] postulate, which we quoted on p. 12: 'Any ray of light moves in the 'stationary' system of coordinates with a determined velocity c whether the ray be emitted by a stationary or a moving body. Hence

$$\text{velocity} = \frac{\text{light path}}{\text{time interval}}$$

where time interval is to be taken in the sense of the definition in Sect. 1'.

Here, we wish to note: We discovered the relativity of simultaneity before the universal constancy of the critical velocity c_o. This was based on our primary knowledge of the behaviour of our measuring-rods and clocks and with the help of the synchronisation regulations according to our elementary principle of relativity. Once the universal constancy of c_o has been accepted, we can show the relativity of simultaneity for the internal observers just as we can show it according to EINSTEIN [14, 15] using light.

Our observer in Σ' observes that two c_o-signals emitted from both end points of our measuring section towards the central point were emitted at exactly the same point of time, if they meet each other exactly in the centre of the measuring section. For an observer in Σ_o, the signal coming from the left, in other words moving in the direction of motion, needs according to Eq. (131) the time $\Delta t_{\rightarrow} = \frac{\Delta x/2}{c_o - v}$ to reach the center of our measuring section $\Delta x/2$, whereas the signal coming from the right needs only the time $\Delta t_{\leftarrow} = \frac{\Delta x/2}{c_o + v}$. The signal coming from the right must thus be emitted at a later point of time than the signal coming from the left, if they are to meet in the central point. According to the observer in Σ_o, the signals are thus not emitted simultaneously. The reference systems Σ_o and Σ' show themselves as complete equivalent. Observed from Σ_o, we can state:

The coordinates x' and t' are the coefficients of measure of space and time measurement in the reference system $\Sigma'(x', t')$, with the units of measure L' and T'.

The traditional method of deriving Special Relativity with its a priori postulated universal constancy of critical velocity for all observers moving towards each other with a uniform velocity has been turned upside down. In fact, we have done nothing more than to consequently continue developing the FitzGerald–Lorentz contraction hypothesis (cf. FitzGerald [20], Lorentz [61, 62], Einstein [15]) as well as applying Poincaré's ideas for synchronising clocks, until we received the complete Special Theory of Relativity. We have thus proved that this contraction hypothesis does not stand in contradiction to Special Relativity as is sometimes believed, e.g. in the textbook of R. Pathria [72] on relativity we read, 'Although the Lorentz–FitzGerald hypothesis explains the null result of Michelson- Morley experiments reasonably well, its ad hoc nature and hypothetical character prevent it from being convincing enough. In fact, we shall find that in the theory of relativity there is no place at all for such a hypothesis; the explanation of the null results of Michelson- Morley experiments is actually contained in the postulate on the 'constancy of the speed of light for *all* inertial observers'. S. P. Puri [79] writes 'This result[5] differs from the contraction hypothesis postulated by Lorentz and FitzGerald to explain the negative results of Michelson- Morley experiment, since Eqs. (1–36) gives a symmetrical relation between two measuring sticks in relative motion, whilst hypothesis required a change in length for a single rod depending on its actual velocity through a real fixed aether'. Even in R. Becker's [1] traditional standard textbook on electrodynamics incorporating Special Relativity, the FitzGerald–Lorentz hypothesis is commented on in the following way: 'Although the contradiction hypothesis is to be regarded as an immediate forerunner of the Theory of Relativity, it should be emphasized that the concept, when standing by itself, contradicts the fundamental Principle of Relativity. Thus, if the moving observer compares the scale moving with him with a scale at rest, he can obviously confirm, that his own scale is actually shortened. In principle, then, we should have available a means of experimentally determining the state of absolute rest simply by observing which of various more or less rapidly moving unit-length scales possessed the greatest length'.

Where is the mistake in this argumentation? The author fails to see that the observer sitting on the moving measuring rod, thus the observer in Σ' is not in the position to measure a length being at rest in Σ_o, in reference to him a moving length, if he did not synchronise his clocks according to a well-defined set of regulations. This length is in fact, according to definition, nothing more than the difference of the *simultaneous* coordinates of its end points in Σ'. However, due to the fact that H. A. Lorentz's moving observer does not have a set of regulations for the synchronisation of clocks, neither the timed order of events, nor the length of a moving ruler is defined in Σ'. It is however correct, that FitzGerald and Lorentz believed at that time in an ether that defined an absolute reference system and that all moving lengths were contracted with respect to this reference system. Such behaviour has been shown by our kinks moving relative to the crystal lattice as discussed in Chap. 10. However, we now know that this ether falls out of the theory if we consequently continue using this hypothesis. In order to achieve this, the contraction of the measuring-rod has to be supplemented with the dilatation of time of a moving clock. Here, it is also taken for

[5] "This result" is Puri's formula (1–36), which is identical with our contraction formula (112).

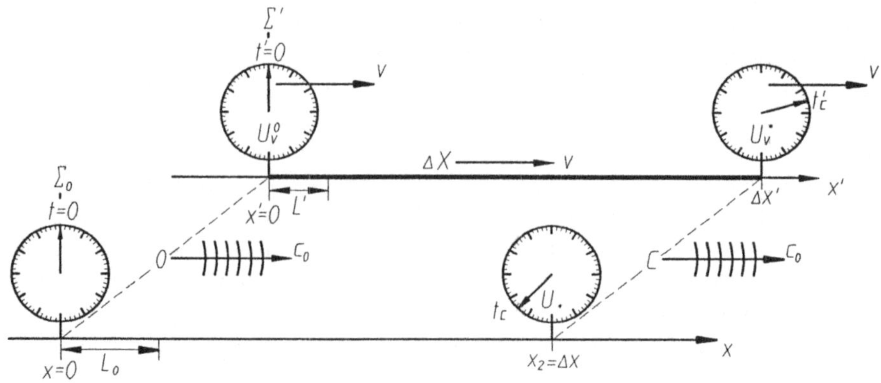

Fig. 12.9 All clocks are synchronised. The arrival of the signal at the right-hand end point of the measuring section $\Delta \mathbf{X}$ is the event C. The observer in Σ_o measures for this event the time $t_C = \Delta t_{\rightarrow}$, whereas the observer in Σ' sees the hand setting $t'_C = \Delta t'_*$ on his clock U^*_v. We once again consider the case $v = 0,8\,c_o$, thus $\gamma = 0,6$ and gauge the clocks so that we receive the value $\Delta t = 15$ for the coefficient of measure $\Delta t_o = 2\Delta x/c_o$. Using this result we calculated in Fig. 12.2 the value $\Delta t_{\rightarrow} = 37,5$, so that the hand of U_* shows $t_C = 37,5$ in Σ_o. We calculate using $t'_B = -10$ from Fig. 12.6, compare also with (145) the value $t'_C = \Delta t'_* = t'_B + \gamma\,\Delta t_{\rightarrow} = 0,6 \cdot 37,5 - 10 = 12,5$ for the hand setting of U^*_v in Σ'. If we take $\Delta x = \frac{\Delta t}{2}\,c_o$ into consideration, thus $\Delta x' = \frac{1}{\gamma}\,c_o = \frac{15}{0,6\cdot 2}\,c_o$, then the observer in Σ' registers for the velocity $c_o{}'$ that the signal possesses between the events O and C, the value $c_o{}' = \Delta x'/\Delta t'_* = \frac{15}{0,6\cdot 2\cdot 12,5}\,c_o = c_o$

granted that the clock moves relative to a preferred frame Σ_o. Up to this point, we do not in fact find anything concerning relativity. However, this relativity is developed if we synchronise the clocks in the reference system Σ' based on our elementary principle of relativity. We can also now determine a timed order of events in Σ', as well as determine lengths of moving rulers in Σ' and compare them with each other. In Chap. 15, see Figs. 12.9 and 17.2, we explicitly showed this.

Based on our axiomatic system, we attained the Einsteinian principle of relativity in all at once and thus complete Special Relativity. On the other hand, according to H. Reichenbach [83], Chap. 2, it is really possible to 'arrange the definition of simultaneity of a system K in such a manner that it leads to the same results as that of another system K' which is in motion relative to K;...', which is exactly what we have done in (143). As we will discuss in Chap. 16, in this case the above-quoted conclusion of R. Becker is correct. The moving length as measured from the reference system Σ' is extended, i.e. the length of a rod resting with respect to the lattice, is extended as measured from a reference system which is moving with respect to the lattice. However, what does this mean?

At first, we violate H. Poincaré's principle of simplicity. As we showed above, the elementary principle of relativity is out of order in this case. If the system Σ' has the velocity v with respect to Σ_o, the velocity v' for Σ_o as measured from Σ' is $v' = \frac{-v}{1-v^2/c_o^2}$.

The question arises, what is the matter with Einstein's famous principle of relativity, quoted on p. 12, the physical equivalence of all inertial systems? Indeed, this symmetry, which becomes obviously by synchronising clocks according to the elementary principle of relativity is a little bit hidden here. Nevertheless, it survives due to the fact that any inertial frame is equally qualified to take the part of the preferred frame, i.e. in the procedure of absolute simultaneity the reference systems Σ_o and Σ' could be replaced by each other. Then the length of a rod resting in Σ' would be extended as measured from the reference system Σ_o. Both situations are indistinguishable, which most easily could be seen with the help of EINSTEIN's formulation of SRT. It would be not very useful to formulate the fundamental laws of physics with the help of an absolute simultaneity.

We were in the physically enviable position of being able to construct measuring-rods and clocks as solutions of a physical equation, the sine-GORDON equation, which made their behaviour calculable. This is obviously a far more simple situation than that of the physicists at the turn of the century, who faced the question of electrodynamics in moving reference systems. The author counts himself to the mass of those who draw any conclusions out of the vicious circle of the never-ending debate on the speed of light, however, not the conclusion that clocks change their pace and that lengths shorten. Notice that there is no difference in derivation of Special Relativity for our physical spacetime with c_L or for its miniature version in the crystalline solid with c_o. We will return to the axiomatic systems of the Special Theory of Relativity in Chap. 16.

We will now make it clear for ourselves that all of the consequences of Special Relativity apply to our crystal without restriction and what this means for every single case. Of special interest are the circumstances of our mechanically based Special Relativity, with which we can reach the boundaries of its validity more easily and that we can get a physical overview of these conditions. In the framework of SRT, the crazy phenomenon of tachyons arises. We will be able to make simple statements in our lattice model of relativity in the Chaps. 28–30.

Chapter 13
The LORENTZ Transformation

We consider an observer in the reference system Σ' moving with respect to our reference system Σ_o with the velocity v. The measuring-rods and clocks used in Σ' are described by us from out of Σ_o using the solutions $q^I(x,t)$ and $q^{III}(x,t)$ of the sine-GORDON equation. In order to make measurements, the observer in Σ' distributes these measuring-rods and clocks and synchronises these clocks according to our elementary principle of relativity. Because we now know that the observer measures the 'universal' velocity c_o (or the equivalent sound velocity c_T) for the propagation of a 'sound signal'; he could proceed with this synchronisation using the velocity c_o, as in the traditional description of Special Relativity. The initial points of his space and time measurement are determined by the initial conditions (134), the coincidence of the coordinate origins of Σ_o and Σ' that we have described as event O with $(x,t) = (x',t') = (0,0)$.

Our measuring section, $\Delta \mathbf{X} = \Delta x\, L_o = \Delta x'\, L'$, rests in Σ'. Its left end point lies in the coordinate origin of Σ' with $x_1' = 0$ and thus has in Σ_o the coordinates $x_1 = v\,t$. The length of the measuring section is arbitrary. We can thus write for the right end point in Σ' the coordinate $x_2' = \Delta x' = x'$ and in Σ_o the coordinates $x_2 = x = \Delta x + v\,t$, compare to (130), so that with $\Delta x' = x'$ and $\Delta x = x - v\,t$, we get from Eq. (113) for the LORENTZ contraction

$$x'(x,t) = \frac{x - v\,t}{\sqrt{1 - v^2/c_o^2}} \tag{149}$$

with arbitrary x and t. The hand setting $t'(x,0)$ of the clocks of Σ' for an arbitrary position x and at the uniform time $t = 0$ of Σ_o was calculated using our elementary principle of relativity in Eq. (142). We know how these clocks run. They go behind our clocks in Σ_o by the factor $\sqrt{1 - v^2/c_o^2}$. The clock that was at x at the Σ_o time $t = 0$ and had the hand setting (142) finds itself at time t at the position $x + v\,t$. It has moved on by $t\sqrt{1 - v^2/c_o^2}$ and thus has the hand setting

H. Günther, *Elementary Approach to Special Relativity*,
https://doi.org/10.1007/978-981-15-3168-2_13

$$t'(x+v\,t,t) = \frac{-x\,v/c_o^2}{\sqrt{1-v^2/c_o^2}} + t\sqrt{1-\frac{v^2}{c_o^2}} = \frac{-x\,v/c_o^2+t\,(1-v^2/c_o^2)}{\sqrt{1-v^2/c_o^2}}$$
$$= \frac{t-(x+v\,t)\,v/c_o^2}{\sqrt{1-v^2/c_o^2}}\,.$$

Here, x and t are completely arbitrary numbers. We therefore know the function of two variables $t'(x,t)$,

$$t' = \frac{t-x\,v/c_o^2}{\sqrt{1-v^2/c_o^2}}\,. \tag{150}$$

We can calculate for every event that took place at time t and at the position x in Σ_o, with the help of the Eqs. (149) and (150), which time t' and which position x' the observer in Σ' registers for this same event. These are the famous LORENTZ transformations as discovered in 1887 by W. VOIGT [96] (up to a transformation of the primed units of measure with the common factor $\sqrt{1-v^2/c_o^2}$), in 1904 by H. A. LORENTZ [62] and independently in 1905 by A. EINSTEIN [14, 15]. Their complete physical interpretation was left up to EINSTEIN. We will come to this in the next two chapters. In literature, the term LORENTZ transformation is generally found in connection with the speed of light, see J. A. SCHOUTEN [85], H. GOENNER [27]. We will keep this term in our mechanical model with the critical velocity c_o of the sine-GORDON equation and thus use SEEGER's [86] terminology. These transformations, the mathematical core, of Special Relativity are completely shown as

$$\left.\begin{aligned} x' &= \frac{x-vt}{\sqrt{1-v^2/c_o^2}}\,, \\ t' &= \frac{t-v\,x/c_o^2}{\sqrt{1-v^2/c_o^2}}\,. \end{aligned}\right\} \quad \text{LORENTZ transformation} \tag{151}$$

We also take notice of the inversion of these transformations that gives us the same equation except that v has been replaced with $-v$,

$$\left.\begin{aligned} x &= \frac{x'+vt'}{\sqrt{1-v^2/c_o^2}}\,, \\ t &= \frac{t'+v\,x'/c_o^2}{\sqrt{1-v^2/c_o^2}}\,. \end{aligned}\right\} \tag{151a}$$

One can see by inserting (151a) in (151) that the inversion is correct.

In the passing to the limit $c_o \longrightarrow \infty$, all relativistic effects disappear and we get from (151) for the connection between the coordinates (x,t) of an event observed in one inertial system with the coordinates (x',t') of the same event observed from another inertial system the simple GALILEI transformation of NEWTONian mechanics,

$$x' = x - v\,t \;, \quad t' = t \qquad \text{GALILEI transformation} \tag{152}$$

with the inversion

$$x = x' + v\,t' \;, \quad t = t' \,. \tag{152a}$$

If one replaces the variables x and t in the NEWTONian inertia term $m\,d^2x(t)/dt^2$, considered here in one space dimension, with the variables x' and t' according to (152a), then $m\,d^2x'(t')/dt'^2$ follows. The differential equation of NEWTONian mechanics has the same form in all inertial systems. Measuring-rods keep their length; clocks run synchronically, whether moved or not, if only the critical velocity c_o for signal propagation is infinitely large.

Questions linked to 'infinitely large velocities' will be considered during our discussion on tachyons and causality in the Chaps. 20 and 28–30.

Chapter 14
The Linear Approximation of Special Relativity

As we have seen in the preceding chapters, the physical contents of Special Relativity shortly can be summarised in this way: We assume LORENTZ contraction[1]

$$L' = L_o \sqrt{1 - \frac{v^2}{c_L^2}} \tag{7}$$

time dilatation

$$t' = t \sqrt{1 - \frac{v^2}{c_L^2}} \tag{8}$$

as well as a well defied synchronisation of clocks in a frame Σ_o. At the end, this must be completed by the synchronisation of clocks in the 'moving frames' Σ'. In principle, this synchronisation is arbitrary and can be described by a synchron function $\phi(x) = t'(x, 0)$. If we introduce absolute simultaneity, we assume a synchron function ϕ according to, cf. Fig.12.8,

$$\phi(x) = t'(x, 0) = 0 \tag{153}$$

with the result of REICHENBACH transformation,

$$x' = \frac{x - vt}{\gamma} , \qquad t' = \gamma t . \tag{143}$$

[1] Here, the velocity c stands for the speed of light c_L in the case of Special Relativity of our spacetime and for c_o of sine-GORDON equation in the case of our mechanical model.

H. Günther, *Elementary Approach to Special Relativity*,
https://doi.org/10.1007/978-981-15-3168-2_14

On the other hand, we can realise the elementary principle of relativity with the help of a synchron function ϕ according to, cf. (142) and (150),

$$\phi(x) = t'(x,0) = \frac{-x\,v/c^2}{\sqrt{1-v^2/c^2}}\ , \tag{154}$$

with the result of LORENTZ transformation

$$x'(x,t) = \frac{x - v\,t}{\sqrt{1-v^2/c_L^2}}\ , \qquad t'(x,t) = \frac{t - x\,v/c_L^2}{\sqrt{1-v^2/c_L^2}}\ . \tag{151}$$

The synchron function (154) is nothing but EINSTEIN's simultaneity derived with the help of the universal constancy of the speed of light.

In the following, we want to neglect all non-linear terms in v/c, so that $\gamma = \sqrt{1-v^2/c^2} = 1$.

Let us notice GALILEI transformation

$$x' = x - v\,t\ , \qquad t' = t \tag{152}$$

with the synchron function (153), $\phi(x) = t'(x,0) = 0$.

Since no linear terms in v/c are liberated in REICHENBACH transformation (143), its linear approximation is GALILEI transformation (152). The REICHENBACH transformation is well adapted for this transition, since both transformations make use of one and the same definition of simultaneity by using the same synchron function (153).

On the other hand, the linear approximation resulting from LORENTZ transformation (151) is

$$x'(x,t) = x - v\,t\ , \qquad t'(x,t) = t - \frac{v}{c}\frac{x}{c} \qquad \text{Linearised in } v/c \text{ LORENTZ transformation} \tag{155}$$

with the synchron function

$$\phi(x) = t'(x,0) = \frac{-x\,v}{c^2}\ , \tag{156}$$

which is the linear approximation of (154). We see:

> The linear approximation, in v/c of LORENTZ transformation includes a definition of simultaneity, which is the linear approximation of EINSTEINs simultaneity. All physical effects of Special Relativity have disappeared.

In other words, (155) is different from GALILEI transformation (152), since the linarisation (156) of the relativistic synchron function (154) is different from absolut simultaneity $\phi = 0$ of GALILEI transformation.

The *definition* of simultaneity is a matter of convenience. All physical effects of Special Relativity as LORENTZ contraction (7) and time dilatation (8) start with non-linear terms in v/c. It is a matter of convenience to change the definition of simultaneity in the linarised LORENTZ transformation (154) from (156) to absolute simultaneity (153) with the result of GALILEI transformation (152).[2]

[2]The linear, in v/c approximation (155) of LORENTZ transformation corresponds to the linear, in v/c approximation of the transformation formula for the so-called four-vector k_i normal to an electromagnetic wave. Notice that also the transformation formula for the MINKOWSKI tensor of the electromagnetic field has a non-trivial linear, in v/c approximation. Therefore, the linearised transformation (155) works very well, if we consider wave phenomena as, e.g. aberration, cf. the footnote on Chap. 19.

Chapter 15
The Principle of Relativity: The Lost Crystal

We now make a discovery. We have already written down the LORENTZ transformation (151) without having given the matter any great thought, in fact we only used it in mathematical calculations. In Chap. 10 we introduced a new variable u according to (114) for a moving kink $q^I(x,t)$, so that the moving kink went over, according to (115) into the function $q_o^I(u)$, so that $q_o^I(u)$ fulfils in the variable u the static sine-GORDON equation,

$$\frac{\partial^2}{\partial u^2} q_o^I(u) = \frac{D}{\sigma} \sin\left(\frac{2\pi}{a} q_o^I(u)\right). \tag{157}$$

We had, for the moving breather $q^{III}(x,t)$ according to (122), introduced new variables u and w instead of x and t so that the function $q^{III}(x,t)$ changed into the breather function $q^{III}(u,w)$, so that $q_o^{III}(u,w)$ fulfils the sine-GORDON equation in the variables u, w,

$$\frac{\partial^2}{\partial u^2} q_o^{III}(u,w) - \frac{1}{c_o^2}\frac{\partial^2}{\partial w^2} q_o^{III}(u,w) = \frac{D}{\sigma} \sin\left(\frac{2\pi}{a} q_o^{III}(u,w)\right). \tag{158}$$

However, (114) is already half of the transformation (151) if we write x' in place of u, and Eq. (122) is completely identical to the LORENTZ transformation (151) if one replaces u with x' and w with t'. These facts are generally applicable. We can, using exactly the same mathematical steps as with variables u and w in Chap. 10, calculate, using the variables x' and t', that if the function $q(x,t)$ fulfils the sine-GORDON equation

$$\frac{\partial^2}{\partial x^2} q(x,t) - \frac{1}{c_o^2}\frac{\partial^2}{\partial t^2} q(x,t) = \frac{D}{\sigma} \sin\left(\frac{2\pi}{a} q(x,t)\right), \tag{159}$$

H. Günther, *Elementary Approach to Special Relativity*,
https://doi.org/10.1007/978-981-15-3168-2_15

then the function $\tilde{q}(x', t') = q(x(x', t'), t(x', t'))$ fulfils the equation

$$\frac{\partial^2}{\partial x'^2}\,\tilde{q}(x', t') - \frac{1}{c_o^2}\frac{\partial^2}{\partial t'^2}\,\tilde{q}(x', t') = \frac{D}{\sigma}\,\sin\left(\frac{2\pi}{a}\,\tilde{q}(x', t')\right)\,, \tag{160}$$

where $x = x(x', t')$ and $t = t(x', t')$ are inserted according to the transformation (151a).

Equation (160) is nothing else than the sine-GORDON equation for a moving observer in Σ', whose measurement results are given in the coordinates x' and t'. For this, one says that the sine-GORDON equation is LORENTZ invariant. The moving observer in Σ' thus discovers especially the solutions (157) and (158) of his Eq. (160),

$$\left.\begin{aligned} &\frac{\partial^2}{\partial x'^2}\,q_o^I(x') = \frac{D}{\sigma}\,\sin\left(\frac{2\pi}{a}\,q_o^I(x')\right)\,, \\ &\frac{\partial^2}{\partial x'^2}\,q_o^{III}(x', t') - \frac{1}{c_o^2}\frac{\partial^2}{\partial t'^2}\,q_o^{III}(x', t') = \frac{D}{\sigma}\,\sin\left(\frac{2\pi}{a}\,q_o^{III}(x', t')\right)\,, \end{aligned}\right\} \tag{161}$$

which define in the well-known way a natural measuring-rod L_o and an oscillation period T_o for his space and time measurements. The physical state that the observer in Σ_o considers as the moving kink $q^I(x, t)$ is considered by the observer in Σ' as the static kink $q_o(x')$. And vice versa, the physical state considered by the observer in Σ' as a moving kink $q^I(x', t')$ is considered by the observer in Σ_o as the static kink $q_o^I(x)$.

One can state precisely the same for the breather solutions. There is no physical difference between the observers in Σ_o and Σ' anymore. The two internal observers of our crystal move relatively towards each other with the velocity v or $-v$, respectively, and discover one and the same physical law, the sine-GORDON equation, with the same physical parameters and thus the same solutions; in other words, the observers discover the same physical states. Although both observers classify each and every state differently (e.g. the moving kink and the static kink), every physical state discovered by one observer is also discovered by the other observers and vice versa. This is nothing else than the principle of relativity formulated in 1905 by A. EINSTEIN [14, 15] which we quoted in Chap. 2:

> The laws by which the states of physical systems undergo changes are not affected, whether these changes of state be referred to the one or the other of two systems of coordinates in uniform translatory motion.

Although the mathematical equations of the LORENTZ transformation (151) were formulated before EINSTEIN by W. VOIGT [96] and H. A. LORENTZ [62], which were derived from certain symmetry properties of equations as we have seen in Chap. 10, it was left to A. EINSTEIN to clarify the relation to space and time measurement in reference systems moving relative to each other on the basis of his principle of relativity.

We have thus found, via the physical properties of our measuring-rods and clocks, what was in those days a theoretical principle used as the fundament of a whole theory. We also discovered the second EINSTEINian postulate, the constancy of critical velocity, in the same manner in Chap. 12. Let us once more summarise this:

From law (79) for the motion of dislocations relative to the lattice, we can derive the sine-GORDON equation (88) with its critical velocity c_o defined by the parameters of the lattice, in the same way as we derived the wave equation (61) from Eq. (31) for a linear chain. The periodical potential of the lattice, in which the linear chain of dislocation masses is embedded, delivers the sine term to us. Due to this non-linearity, the sine-GORDON equation possesses soliton solutions, with which we can define natural measuring-rods and clocks. We thus arrive at a remarkable conclusion:

> The contraction of length and the time dilatation of moving measuring-rods or clocks, respectively, are reduced to elastic interactions in the linear chains of dislocations in the lattice.

Thus, the circle is closed. The mechanical oscillations, with which we started in Chap. 4, give us a physical mechanism for explaining the elementary effects of Special Relativity: Length contraction and time dilatation as registered by the internal observer in a crystal can be explained using HOOKE's law for the constituents of this crystal,, i.e. the atoms that remain invisible for the internal observers—a curious connection.

We will repeat this so that no misunderstanding can arise: The effects of EINSTEIN's Special Relativity of our physical spacetime caused by the speed of light have nothing at all to do with elastic interactions of the atoms in a crystalline solid. We will come back to this point later on.

Using our elementary principle of relativity, we are now able to define simultaneity in the 'moving reference system', that makes the LORENTZ invariance of the sine-GORDON equation visible. Here, we note together with POINCARÉ [73, 74] that this definition of simultaneity is not altogether necessary; however, using it easily allows for a special symmetrical formulation of the physical laws. Once this has been done, we derive the constancy of the critical velocity for all observers moving uniformly towards each other, which then leads to the whole EINSTEINian principle of relativity, the complete equivalence of all reference systems moving uniformly towards each other.

Here, one could protest, saying that we are always only talking about the sine-GORDON equation, whereas the physics inside of a crystal deals with several other states that are not described by this equation. Correct is that we do in fact strictly limit the range of states to those structures described by the sine-GORDON equation, that as we know, describe the deviation from the 'vacuum state' of the lattice (compare to Chap. 8). The number of these states is nevertheless immensely large. The non-linear character of the sine-GORDON equation belongs to the comprehensive area of soliton physics, where new solutions are being made using new methods. For the first view of these, see the paper of A. SEEGER [86]. In the light of this, limiting our states does not seem so narrow.

Dislocation structures belong to the area of plasticity or micro-plasticity, respectively, that always also lead to elastic deformations of the lattice, that we ignored here. This coupling of dislocations with the elastic deformations of the lattice is determined by the totality of the parameters of HOOKE's tensor. This means for the general case, 21 material parameters and generally lead to a very complicated coupled system of differential equations for elastic deformations under the influence of general dislocation structures. One can however show that these deformations, originally derived from *arbitrary* dislocation distributions, finally containing only one single parameter can be determined using simple wave equations[1] and also fulfil the LORENTZ symmetry (cf. GÜNTHER [35]). We therefore have another immensely large class of states in a crystal that also underlie the special principal of relativity with the velocity c_o or sound velocity. A few further explanations will be made in the appendix, in Chaps. 26 and 27. We can now also reasonably say that the class of states at the disposal of our internal observers is a cosmos of its own, for which Special Relativity with the critical velocity c_o of the sine-GORDON equation applies (or respectively, cum grano salis also with the transversal sound velocity c_T).

The un-hoped derivation of the famous EINSTEINian principle of relativity from our humble considerations of NEWTONian mechanics has made us oversee something, we lost our crystal during this process! The crystal has disappeared! There is now nothing in the world of our internal observers capable of detecting or finding our crystal. Our internal observer is not capable of detecting the slightest trace of the crystal, even with the help of the most clever and complicated experiments. Every single of these internal observers in any single reference system measures one and the same velocity c_o. Using his measuring-rods and clocks does not help him any further. The observer in Σ_o still states that the clock resting in Σ' goes behind and that measuring-rod resting in Σ' is shorter. Thus he is compelled to state that only Σ_o is distinguished as the absolute resting reference system and that Σ' possesses the velocity v.

This conclusion however is null and void, since we know that the same sine-GORDON equation with the same physical parameters applies for the observer in Σ' as it does for the observer in Σ_o. Thus, the observer in Σ' discovers the kink and the breather solutions with the length L_o and the oscillation period T_o. The length that was observed from Σ_o as being shorter is registered in Σ' as the normal length L_o, and the oscillation period that we observed as dilated is registered in Σ' as the normal period T_o. The observer in Σ' registers that the measuring-rods and clocks at rest in Σ_o possess the velocity $-v$. He thus discovers that the clock resting in Σ_o goes behind by the LORENTZ factor and that the measuring-rod is shortened correspondingly. The sign of the velocity plays no role, because all of these effects only depend on the ratio v^2/c_o^2.

Here, we stop, reconsider and ask, because of the consequences of these conclusions: Is this in fact really so? Have we not shown explicitly in Chap. 10 that the kink moving with respect to the crystal is shortened, shortened in comparison with the kink resting with respect to the crystal. It is obviously the resting kink that possesses the largest geometrical extension. The sketch in Fig. 10.2 is, in principle, experimentally provable. The breather solution moving with respect to the crystal

[1]Here, we wish to note that these equations go over to the LORENTZ invariant equations with the transversal sound velocity c_T, in the special case of straight screw dislocations in an isotropic medium, cf. GÜNTHER [34, 35].

oscillates slower than the breather solution resting with respect to the crystal, which we have shown in Fig. 10.5. This is also, in principle, experimentally provable. From this follows that the reference system Σ_o of the static crystal is distinguished and that both the LORENTZ contraction and time dilatation only take place, in reality, relative to the crystal?

Such an evaluation of the situation always silently requires one thing that all measurements are naturally carried out using the measuring instruments of the outside observer, as are all other measurements in physics. We can indeed do it this way and also arrive at the same evaluation as seen above but maybe different measuring methods would be better suited for the observed phenomena.

What does *'in reality'* mean? We do not intent to start a philosophical discussion here, however we intend to concentrate alone on what happens inside of a crystal, what functions inside of a crystal, what has its reality inside of a crystal. If we however solely concentrate on the crystal and the considered class of phenomena, then the question concerning the symmetry of these phenomena is a problem of primary importance. The mathematical methods of physical examination of such phenomena depend heavily on symmetry. Connected to this is the question concerning the description of phenomena using reference systems moving relative to each other, which is of elementary interest. If however we intend to describe these observed phenomena using reference systems moving relative to the crystal lattice, then we must first decide which measuring-rods and clocks are best suited for this purpose and we must also ask for a meaningful definition of a velocity in these reference systems? We have answered this question using our elementary principle of relativity, see Chap. 12, as well as Fig. 12.4: *'If the observer in Σ_o registers that Σ' moves with the velocity v, then the observer in Σ' should observe that Σ_o moves with the velocity $-v$'*. Everything else is deduction.

Let us once again make it clear that our measuring procedure, based purely on mechanical phenomena in a crystal, tells us that the observer in Σ' does in fact register that a breather clock resting relative to the crystal oscillates slower than the breather clock resting relative to Σ'. We can derive these facts from Fig. 12.6.

In order to do this, we first have to consider the event B with $t_B = 0$ for the Σ_o clock U_* and with $t'_B = \frac{-v\Delta x}{c_o^2\gamma}$ for the Σ' clock U_v^*. After time $\Delta t = \Delta x/v$, U_* with the hand setting Δt lies opposite the Σ' clock U_v^o, whose hand then shows the time $t'_S = \frac{\Delta x}{v}\gamma$, because of the time dilatation of the clock which moves with respect to Σ_o. Thus whilst the time $\Delta t = \Delta x/v$ on one of the Σ_o clocks has run out, the observer in Σ' determines a time period $\Delta t'$ (which he has to read from both clocks U_v^* and U_v^o) according to $\Delta t' = t'_S - t'_B = \frac{\Delta x}{v}\gamma - \frac{-v\Delta x}{c_o^2\gamma} = \frac{\Delta x}{v}(\gamma + \frac{v^2}{c_o^2\gamma}) = \frac{\Delta x}{v}\frac{c_o^2\gamma^2+v^2}{c_o^2\gamma} = \frac{\Delta x}{v}\frac{c_o^2-v^2+v^2}{c_o^2\gamma} = \frac{\Delta x}{v}\frac{1}{\gamma}$, and thus

$$\Delta t = \Delta t'\gamma\,. \tag{162}$$

This is Eq. (121a) for the time dilatation of a moving clock, but with swapped positions as was determined by the observer in the reference system Σ' moving relative to the reference system Σ_o. In other words, this means that the observer in Σ' registers that the breather clock resting relative to the crystal goes behind:

> The moving clock goes behind, even if it rests by chance with respect to the crystal and only moves relative to the observer, who himself moves relative to the crystal.

This only seemingly contradicts the information in Fig. 10.6. The relativity of time dilatation for the internal observers in connection with the discussion concerning the twin paradox in Chap. 16 is once again depicted in Fig. 17.2.

We also want to see how it can be that the static kink resting relative to the crystal is registered by the observer in Σ' as shorter than the kink resting relative to Σ' by the LORENTZ factor γ, which is once again apparently contradictory to what we have shown in Fig. 10.2. Here, one has to keep in mind that we have only shown momentary snapshots, as registered by the internal observer in the reference system Σ_o. These observations are also made by us 'outside' observers. We can peer into a crystal from the outside and measure, using the measuring-rods and clocks that we constructed using our 'outside' physics. Important symmetrical properties of the observed phenomena as measured by the internal observers remain invisible to us when we use these outside measuring-rods and clocks. Momentary snapshots as registered by the internal observer in the reference system Σ' have been depicted in Fig. 12.4 and the lower picture of Fig. 17.1.

How can it be possible for the observer in the reference system Σ' (that moves relative to Σ_o with the velocity v) to measure a LORENTZ contracted length of a rod being at rest in Σ_o (thus being at rest relative to the crystal)? Let us once again make this clear: If we measure the length of a rod in the reference system in which it rests, then it plays no role when we note the coordinates of its end points, because these do not change. Due to the fact that the rod from Fig. 12.8 rests in the reference system Σ', $\Delta x'$ is the coefficient of measure of its length, even though both Σ' clocks show different time readings. However, if the rod moves, we need the coordinates of both end points at one and the same point of time, so that we can calculate the coefficient of measure of the moving length from the coordinate differences, also see our explanations about LORENTZ contractions at the end of Chap. 10.

Let us consider a rod $\mathbf{X}$ resting in Σ_o, as we have shown in Fig. 12.1, with the end points $x_1 = 0$ and $x_2 = \Delta x$. The coordinate difference Δx is thus the coefficient of measure of its length with the measuring-rod L_o and it is $\mathbf{X} = \Delta x\, L_o = \Delta\tilde{x}'\, L'$, so that it has to be $\Delta\tilde{x}' = \Delta x/\gamma$ because of $L' = \gamma\, L_o$. What is $\Delta\tilde{x}'$? We consider the end points of the rod at time $t = 0$ in Σ_o. These are the events O and B with their coordinates $(x = 0,\ t = 0)$, or $(x_2 = \Delta x,\ t_B = 0)$ in Σ_o. We have already described the synchronisation of clocks of the reference system Σ' for this case in Fig. 12.6. In Σ', O has the coordinates $(t' = 0,\ x' = 0)$ and event B has the coordinates, as seen in Fig. 12.6 and using Eq. (140), $x' = \Delta x/\gamma,\ t'_B = \frac{-v\Delta x}{c_o^2\gamma}$. What is $\Delta\tilde{x}'$?

Although in this case the number $\Delta\tilde{x}'$ also states how often the measuring-rod L' fits into the positions, which are defined by the events O and B in Σ'. Both of these events however are the end points of a moving rod in Σ' at two different points of time. The quantity $\Delta\tilde{x}'$ is thus not the length of the moving rod $\mathbf{X}$ with respect to Σ' anymore, but only the distance between the events O and B, that are only simultaneous in Σ_o and whose coordinate difference makes up the coefficient of measure for the length of the rod **X**. In order to find out which length the observer in Σ' determines for the rod **X**, we need the coordinates of its end points at one and the same point of time in Σ', let us say at $t' = 0$. For the left-hand end point, we once again have $x' = 0$ (this is event O). The right-hand end point finds itself at event B, thus at time $t'_B = \frac{-v\Delta x}{c_o^2 \gamma}$ at $x' = \Delta\tilde{x}' = \Delta x/\gamma$. Thus, the right-hand end point finds itself at time $t' = 0$ at the position $\Delta x' = \Delta\tilde{x}' + v \cdot t'_B = \frac{\Delta x}{\gamma}(1 - v^2/c_o^2)$. This $\Delta x'$ is the coefficient of measure for the length of the resting rod **X** in Σ_o measured in Σ'. Compared to the coefficient of measure Δx of its length determined in the reference system Σ_o, the following is valid

$$\Delta x' = \gamma \cdot \Delta x \ . \tag{163}$$

For the rod $\mathbf{X} = \Delta x\, L_o$ resting in the reference system Σ_o, the moving length $\Delta x'\, L'$ is measured in Σ'. In other words, the measuring-rod L' has to be used exactly $\gamma\, \Delta x$ times in Σ' in order to fill out the distance in Σ' corresponding to the moving length in question. This situation has been shown in Fig. 15.1, using an arbitrary Δx.

Perhaps the statement in Eq. (163) becomes clearer if we consider the case $\Delta x = 1$. In other words, $\mathbf{X} = L_o$. Here, the length of the measuring-rod resting in the reference system Σ_o is determined from the reference system Σ' using the measuring-rod L' resting in Σ'. This situation has been shown and described in Fig. 15.2. The observer in Σ' registers that the measuring-rod L_o that from his point of view is in motion with the reference system Σ_o can be covered using the fraction $\gamma\, L'$ of his measuring-rod. As curious as this fact may seem from the outside—because from the outside we always register that L' is smaller than L_o and that it can become arbitrarily small if the velocity v of Σ' relative to Σ_o sufficiently approaches c_o—for the internal observer, who gauges his velocities using our elementary principle of relativity, the world looks completely different (Fig. 15.1).

Equation (163) is Eq. (113a) for the LORENTZ contraction of a moving rod, with exchanged roles, so that the observer in the reference system Σ' moving relative to the reference system Σ_o measures the rod **X** resting in the reference system Σ_o:

> The moving rod is contracted, even if it rests, by chance, with respect to the crystal and only moves relative to the observer, who himself moves relative to the crystal.

The hasty conclusion of an internal observer in Σ_o, that the reference system Σ' moves with respect to the crystal, would be disputed and reversed using exactly the same argumentation. It thus remains a fact that the crystal cannot be proven by anything in the world of the internal observers.

However, no one will want to conclude from this that we have proven the crystal to be non-existent. Anyone who, against reality, still wishes to believe this only has to take a chunk of crystal and bang it against his head until he has assured himself of the opposite. It is important for the success of this reflective experiment that one uses ones own head and not the head of another. We as outside observers do not have any difficulties in accepting the existence of crystals. The internal observers however need a more subtle line of thought. They have to answer the question concerning

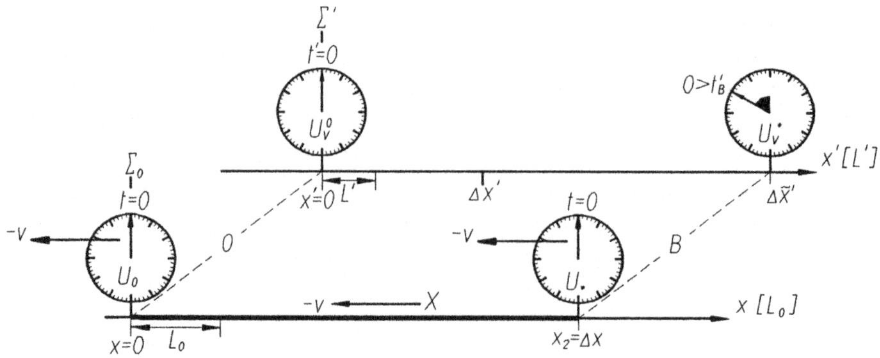

Fig. 15.1 The relativity of the LORENTZ contraction. The rod $\mathbf{X} = \Delta x\, L_o$ rests in the reference system Σ_o. Here, the coefficient of measure of its length is thus Δx. The observer in Σ' determines for this, from his point of view, moving rod the coefficient of measure as $\Delta x'$, he thus measures the distance $\Delta x'\, L'$. He discovers according to (163) $\Delta x' = \gamma \cdot \Delta x$. Using the assumed velocities from the earlier illustrations, $v = 0,8\, c_o$, thus $\gamma = 0,6$, he discovers that his measuring-rod L' has to be used exactly $\gamma\, \Delta x$ times in order to determine the length in question. We have used the hand setting $t'_B = -10$ from (43) for event B

the ether, a question that has caused many of the most important physicists of the nineteenth and twentieth centuries immense headaches. Let us remind ourselves (see Chap. 1) what A. EINSTEIN [12, 13] in 1920 had to say about the question concerning ether, 'To deny the ether is ultimately to assume that empty space has no physical qualities whatever. The fundamental facts of mechanics do not harmonise with this view'. The empty space is, in the eyes of the internal observers, the vacuum state of this crystal, in other words the infinitely extended, ideal crystal with its infinitely long straight dislocations, a crystal containing, as we can observe from outside of this crystal, undeniable, manifest physical properties. It is however difficult for the internal observers to make out this ether as the space in which they experiment and for which the EINSTEINian principle of relativity is valid.

In fact the situation is when considered from the outside somewhat curious. The tiny small energies that are found in a kink, or a breather, or any solutions of the sine-GORDON equation can be precisely calculated by the internal observers in the crystal. In comparison, the 'tonnes' of weight that the atoms and molecules of the lattice represent remain unnoticed. These internal observers fly past these giant molecules and atoms with the grace and precision of a circus artist. However, using this to derive that the ether does not exist would go too far. It would be like stating, after having constructed an artful method of navigation past the lattice obstacles, that this lattice does not exist. We therefore see:

The ether is space, and in our mechanical continuum it is the lattice.

Let us clarify this again: Where in the atom lattice of a crystal do the properties of a physical vacuum lie? This is already found in the oscillating masses of NEWTONian mechanics. In Chap. 4, we were able to determine for a single oscillating mass, *'Both* *characteristics of a harmonic oscillation, its angular velocity ω and its total energy U, do not allow us to draw a conclusion about the mechanical state of the oscillating system'.* We discovered, as we considered only two of these elastic coupled masses that the transportation of energy U occurs in the same manner as it does with waves. Such a transportation of energy through a complete system of elastic coupled masses disguises the physical nature, as well as the number of these masses. We have seen how this was established in axioms of NEWTONian mechanics in Chaps. 4 and 5.

An elastic wave is nothing else than a disturbance of the ideal lattice structure in the form of elastic deformations. This disturbance contains an energy that moves along the lattice with a characteristic velocity defined by the constants of the lattice. We have seen in Chap. 7 that a lattice can have elastic and plastic deformations.

Now come to the important step. We examine *structural disturbances made up of plastic deformations above an ideal arrangement of lattice atoms*. In Chap. 8, we showed how we could register these structural disturbances using the postulates of NEWTONian mechanics in order to derive the sine-GORDON equation. It is the solutions of this equation, the physically real small local deviations from the ideal lattice, for which this lattice itself possesses the properties of a vacuum. In Chaps. 22 and 23, we will further show that these areas are attributed the properties of a physical particle with respect to its vacuum, the atomic lattice.

Thus, the axiomatic system of NEWTONian mechanics already supplies us with a model for the phenomena of a physical ether or of a physical space, respectively. It would of course be a grievous error to conclude that every ether is of NEWTONian origin. The only thing we have shown is that NEWTONian mechanics supplies us with an ether possessing a characteristic velocity c_o of plastic disturbances moving through the crystal. This has nothing to do with electrodynamics. However, we now know just how one could imagine an ether model, so that this ether does not possess any state of motion at all. The explanation is obviously simple. The instruments of measurement with which we quantitatively determine any motion are built up out of structures on the ether, just as the particles themselves that we want to measure are and cannot thus per construction respond to this ether—*the eye cannot see itself* (Fig. 15.2).

And the ether for the speed of light c_L?—That is our physical space. In this physical space, we are the internal observers. What however does this mean? The quantum phenomena are an obvious reminder that we should not imagine the motion of matter in this, our space in an all too naive manner. Particles and space cannot be thought of separately. This conclusion was made by the philosophers of ancient Greece—without the help of science—maybe this was the reason?[2] Every new result in elementary particle physics is always a new view into the structure of our space, the structure of the vacuum as today's physicists would state. If we use the modern term for space or ether, we see that the 'physical vacuum' has been the object of strenuous physical discussion for many years. This is rather difficult for us. We ourselves constitute the internal observers who cannot escape from this space—just

[2]For a detailed discussion of these questions, we refer to C. F. v. WEIZSÄCKER's [98] essays in his book 'The unity of Nature'.

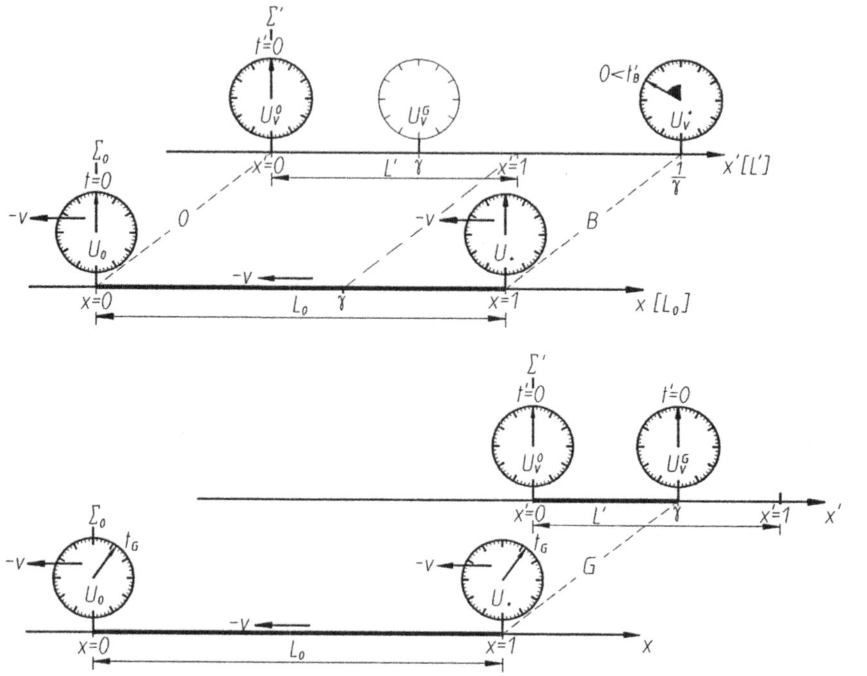

Fig. 15.2 The relativity of the LORENTZ contraction for the measuring-rods. The observer in Σ_o 'sees' that the length L' of the measuring-rod moving with the velocity v, which thus is resting in the reference system Σ' is shorter than the measuring-rod L_o resting with him, according to our Eq. (112) by the factor γ, $L' = \gamma\, L_o$. With $v = 0,8\, c_o, \gamma = 0,6$. In the lower illustration, the middle dotted line indicates that the measuring-rod L' in Σ_o possesses the coefficient of measure γ. The point with the coefficient of measure γ is shown on the x-axis. In the upper picture, we consider both events O and B that take place simultaneously in Σ_o. In order to use the same hand settings in Fig. 12.6 of the synchronised clocks in the reference system Σ', U_v^o and U_v^*, that correspond to the events O and B, we need to assume that our measuring-rod L_o in Σ_o is just as long as the distance between the clocks U_o and U_*, which is what we want to assume. The clock U_* has the coordinate $x = 1$ in Σ_o. The points belonging to one and the same event, have once again been connected by dotted lines. The space coordinate of the clock U_v^* of Σ' at the event B is then $\Delta\tilde{x} = 1/\gamma$. The hand of the clock U_v^* is then position at $t'_B < 0$ according to (141). This results, in our numerical example, as calculated in Fig. 12.9 as $t'_B = -10$. In order to determine the length of the moving measuring rod L_o in Σ' (by using the measuring-rod L' resting in Σ'), we need those space coordinates $\Delta x'$ in Σ', at which the right end point of the measuring-rod L_o passes by at time $t' = 0$. According to (163), we receive for this (due to the fact that for the measuring-rod $\Delta x = 1$) $\Delta x' = \gamma$. This is our event G with ($x'_G = \gamma, t'_G = 0$). We have added this into our illustrations with $x' = \gamma = 0,6$. Event G is shown once more in the lower illustration. Because the clock U_* belonging to the reference system Σ_o shows a time dilatation when observed from the reference system Σ', according to Eq. (162), we register for event G in Σ_o ($x_G = 1, t_G = -t'_B\, \gamma$). In our numerical example, the hand of U_* is positioned at $t_G = 6$ at the event G. The clock U_o at the left end point of the measuring-rod L_o has moved on during this and cannot be connected to U_v^o using a dotted line. The distance between both clocks U_v^o and U_v^G in Σ' is the length L_v that the internal observer in Σ' determines for the measuring-rod L_o resting in Σ_o, thus $L_v = \gamma\, L'$. In the reality of the internal observer of our crystal, the measuring-rod L_o resting in Σ_o is, when observed from Σ', contracted by the same value as the rod L' resting in Σ' when observed from Σ_o

as the internal observers inside of our crystal are bound to their space. Thus we find ourselves together with the most clever and refined mathematicians, physicists and engineers in experimental atomic research centres and try with the utmost effort and utmost energy to poke our vacuum a little in order to, with a bit of luck knock a bit out of this unbelievingly large source.

Physically, on the basis of our crystal lattice, the theory of relativity reaches one of the limits of its validity when it comes to length contractions and their corresponding LORENTZ factors $\gamma = \sqrt{1 - v^2/c_o^2}$ leading to distances lower than the lattice parameter of the crystal. Principally, such distances cannot be measured using kinks on dislocations, because the field theoretical description of the kink with the help of the sine-GORDON equation due to the used continuum approximation (85) is simply based on the geometrical structures of these lines built up of a large number of lattice atoms. We are thus in a position to be able to predict the moment when Special Relativity loses its validity on the crystal lattice. We will return to this in Chap. 21.

Here in this connection, it is remarkable that we know of limiting lengths in our 'outside physics', so that geometric distances smaller than these lengths become problematic, or cannot physically be meaningfully determined at all. An absolute lowest value for space distances according to today's science was already introduced in 1906 by M. PLANCK. This elementary length l_o named after him is made up of three fundamental parameters, PLANCK's constant $\hbar$, the gravitational constant f and the speed of light c_L. Together, these three parameters define PLANCK's elementary length $l_o = \sqrt{f\hbar/c_L^3} = 1,6 \cdot 10^{-33}$ cm. W. HEISENBERG even proposed the COMPTON wavelength λ_c of a nucleus with the mass m, $\lambda_c = \frac{2\pi\hbar}{mc_L} = 1,32 \cdot 10^{-13}$ cm, a value twenty times larger than PLANCK's length, as the lowest possible limit past which no meaningful geometric measurements could be made, cf. also H. TREDER [94, 95].

A space lattice made up of physical constituents whatever type they may be, whose next neighbours are positioned at a distance equivalent to the HEISENBERG, or even the PLANCK length, could indeed according to our considerations supply us with the physical background needed not only for Special Relativity but also for further elementary properties of matter. Here, we explicitly repeat that the hypothetical lattice 'constituents' of our physical space is under no circumstances inertial masses of our physical space such as atoms or elementary particles. The masses of our physical space would be structures on such a lattice, so that the basic kinematic properties of our masses, being that masses move relative to each other with a certain inertia, could in principle not be attributed to the constituents of such a lattice. Nevertheless, such a lattice would be physical real and would assert fundamental influence on the physics of our matter—this would correspond to the statement made by A. EINSTEIN [12, 13] in 1920 concerning the ether problem, 'To deny the ether is ultimately to assume that empty space has no physical qualities whatever. The fundamental facts of mechanics do not harmonise with this view'. The hypothetical discussion using a lattice for our physical space has nothing to do with the introduction of a classical ether. It is just a vague idea of a possible physical background of our relativistic spacetime structure. In order to become more involved in this subject, we would need to include the quantum theory and maybe even the theory of gravity.

Chapter 16
Two Axiomatic Systems for Special Relativity

We will now turn to the axiomatic construction of the Special Theory of Relativity. Here, we will consider our physical spacetime with its distinguished critical velocity, the speed of light c_L. The starting point of our considerations in any case is the reference systems in which we experiment and describe the results of these experiments. Starting out from the oldest physical discipline, mechanics, we arrive at the distinction of inertial systems as seen in Chap. 4.

In mathematical terms, the problem can be described as follows. The equations of NEWTONian mechanics are identical for all inertial systems, if the GALILEI transformation (152) is valid for the coordinates (x, t) and (x', t') of an event in two arbitrary inertial systems Σ_o and Σ', respectively,

$$\left.\begin{array}{l} x' = x - vt \ , \\ t' = t \ , \end{array}\right\} \quad \longleftrightarrow \quad \left.\begin{array}{l} x = x' + vt' \ , \\ t = t' \ . \end{array}\right\} \tag{152}$$

If a scalar solution $f = f(x - c_L t)$ of the wave equation

$$\frac{\partial^2 f(x,t)}{\partial x^2} - \frac{1}{c_L^2}\frac{\partial^2 f(x,t)}{\partial t^2} = 0 \tag{1}$$

is observed in Σ_o, then this describes the propagation of a plane wave signal in the x-direction with the speed of light c_L (see for example Fig. 16.1). Accepting the GALILEI transformation (152) results in $f = f(x - c_L t) = f(x' + vt' - c_L t') = f[x' - (c_L - v)t']$ for the same signal when observed from Σ' and thus

$$\frac{\partial^2 f}{\partial x'^2} - \frac{1}{(c_L - v)^2}\frac{\partial^2 f}{\partial t'^2} = 0 \ . \tag{164}$$

H. Günther, *Elementary Approach to Special Relativity*,
https://doi.org/10.1007/978-981-15-3168-2_16

The observer in Σ' would thus determine, according to (164) that the (light) signal in Σ' moves in the x'-direction with the velocity $c_L - v$. However, as we already know, MICHELSON's experiments proved exactly this to be false.

Having in mind the analysis of POINCARÉ, at first we have to observe that the axiomatics of Special Relativity in any case decomposes into two different statements:

(A) A *definition* of simultaneity.
(B) An *axiomatic statement* about certain physical facts.

Then, we need to distinguish between two fundamentally different approaches to the axiomatic formulation of Special Relativity:

I. We ask for a general principle of symmetry that governs the objects of our physical search in all inertial systems.

II. We primarily ask: How do our measuring-rods and clocks behave in a single and therefore preferred inertial frame?

I. This is the EINSTEINian, the royal path to the Special Theory of Relativity. Its explicit axiomatic fundamental is the EINSTEINian principle of relativity that we quoted in Chap. 2, (and at the beginning of the last chapter). In order to arrive, following this path, together with EINSTEIN at Special Relativity, we demand that MAXWELL's equations for the description of electromagnetic processes have to be identical for all inertial systems. This would have the consequence that all physical parameters of this theory, here the speed of light c_L, would have one and the same numerical value in all inertial systems. This important conclusion from the application of his principle of relativity on the MAXWELL theory was especially formulated by EINSTEIN as the universal constancy of the speed of light (quotation in Chap. 2, p.7).

With the help of the universal constancy of the speed of light, EINSTEIN introduced a synchronisation of clocks in an arbitrary inertial system and deduced the relativity of simultaneity, LORENTZ contraction, time dilatation, the LORENTZ transformation, the dependence on velocity of inertial masses, as well as the inertia of energy. These effects would also function identically in every inertial system.

Constructing Special Relativity with the help of MAXWELL's electrodynamics does not mean that electromagnetic phenomena have a dominant meaning for the whole of physics. Any other theory could be taken up this position, e.g. the theory of weak interactions or simply mechanics. Electrodynamics was just the historical cardinal point for the discovery of Special Relativity—not more, not less. The basis of theoretical physics is the Special Theory of Relativity.

We do not intend to illustrate EINSTEIN's path and therefore refer the reader to his original papers EINSTEIN [14, 16] (cf. the English translation in EINSTEIN [15]). The simple synchronisation procedure according to EINSTEIN is described in Chap. 3. After that a synchronisation of clocks in any physical textbook was explained along these lines (except LIEBSCHER's [57] approach, see below), so that the unprepared reader could get the impression this procedure would be physically compelling. Though H.

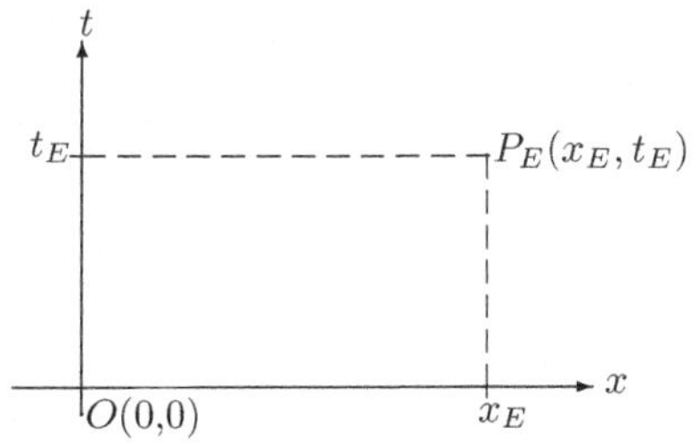

Fig. 16.1 Every event E in the inertial system Σ_o is attributed a point P_E with the coordinates x_E and t_E in the corresponding coordinate system of MINKOWSKI's spacetime continuum

REICHENBACH[1] criticised this opinion in 1920, hardly a physical lecture really took notice of this. It is characteristic for the approach to relativity according to EINSTEIN that A) the definition of simultaneity and B) the axiomatic demand of a certain physical fact, here the constancy of the speed of light, cannot be separated from each other. Nevertheless, EINSTEIN's axiomatics includes a *definition*, though very meaningful, but a definition which also could be replaced by another one. Hence, the conclusions are dependent on this definition, the relativity of simultaneity and its consequences are. This means, using another definition for synchronisation we could arrive at another simultaneity. There is a price for this, symmetry. We return to this question below, when explaining our second approach, where we will give an example.

H. MINKOWSKI [66] has earned special honours for his mathematical formulation and formalisation of Special Relativity; see MINKOWSKI's paper from 1908 in LORENTZ [60]. Without MINKOWSKI's formulation of Special Relativity, modern relativistic theories would practically be non-existent. All presentations of Special Relativity

with the help of MINKOWSKI's formalism are based alone on the EINSTEINian principle of relativity. Applied to the above considered signal propagation, we could formulate this principle mathematically as follows:

We are looking for those transformations of

$$\left.\begin{aligned} x' &= x'(x,t)\ , \\ t' &= t'(x,t)\ , \end{aligned}\right\} \quad \longleftrightarrow \quad \left.\begin{aligned} x &= x(x',t')\ , \\ t &= t(x',t')\ , \end{aligned}\right\} \tag{165}$$

which ensure that the same scalar equation for the propagation of waves in Σ' follows from the wave equation (1) of the observer in Σ_o, thus

$$\frac{\partial^2 f(x',t')}{\partial x'^2} - \frac{1}{c_L^2}\frac{\partial^2 f(x',t')}{\partial t'^2} = 0\ , \tag{166}$$

and not Eq. (164) that we received when we applied the GALILEI transformation (152).

[1] cf. REICHENBACH [83], Chap. 2, 'One mistake results from the derivation of the relativity of simultaneity from the different states of motion of various observers. It is true that one *can* define simultaneity differently for different moving systems, $\cdots$ but such a definition is not *necessary*'.

H. MINKOWSKI recognised that this could be given a geometric formulation in the following way: The three-dimensional space and the time can be combined to a four-dimensional spacetime continuum. Every physical event E at the location x at time t corresponds to exactly one point P in the spacetime continuum. In order not to overcomplicate our considerations, we will suppress the two space dimensions y and z. A certain inertial system Σ_o corresponds to a certain coordinate system (x, t) in MINKOWSKI's spacetime; see Fig. 16.1. In this coordinate system, we *define* a distance s of a point $P(x, t)$ from the coordinate origin, the point $O(0, 0)$, according to

$$s^2 := x^2 - c_L^2 t^2 \ . \tag{167}$$

The transition to a different inertial system Σ' means in the spacetime continuum the transition to a new coordinate system (x', t'), in which the point P attributed to the event E has the coordinates x' and t'. Here, we assume that both coordinate systems share the same coordinate origin,

$$O(x = 0\,,\ t = 0) = O(x' = 0\,,\ t' = 0) \ . \tag{168}$$

This corresponds to the initial condition (134) of inertial systems.

H. MINKOWSKI recognised that the definition of the 'true' spacetime transformation made by the demand for the invariance of the wave equation (1) is equivalent to the demand for the invariance of the above defined distance s according to (167) with respect to these transformations. According to MINKOWSKI, EINSTEIN's principle of relativity (including EINSTEIN's definition of synchronisation) could also be mathematically formulated as follows:

In the coordinate system (x', t'), we calculate a distance s' from the coordinate origin according to

$$s'^2 = x'^2 - c_L^2 t'^2 \ . \tag{169}$$

Searched for are those transformations (165) for which the form (167) for the distance s is retained, in other words

$$\begin{aligned} &x^2 - c_L^2 t^2 = x'^2 - c_L^2 t'^2 \ , \qquad \text{hence} \\ &s^2 = s'^2 \ . \end{aligned} \tag{170}$$

This mathematical formulation of the principle of relativity is also called *the invariance of the line element* and can also be formulated for the distance Δs between two arbitrary points $P_1(x_1, t_1)$ and $P_2(x_2, t_2)$ according to

$$\begin{aligned} &\Delta x^2 - c_L^2 \Delta t^2 = \Delta x'^2 - c_L^2 \Delta t'^2 \ , \qquad \text{hence} \\ &\Delta s^2 = \Delta s'^2 \ . \end{aligned} \tag{170a}$$

Hcrc, $\Delta x = x_2 - x_1$, $\Delta t = t_2 - t_1$ and for the same points in the primed coordinates $\Delta x' = x_2' - x_1'$, $\Delta t' = t_2' - t_1'$. If one once again takes all three space coordinates, then the demand for the invariance of the line element runs as follows:

$$\Delta x^2 + \Delta y^2 + \Delta z^2 - c_L^2 \Delta t^2 = \Delta x'^2 + \Delta y'^2 \Delta z'^2 - c_L^2 \Delta t'^2 \ . \qquad (170b)$$

The spacetime continuum for which a distance s is defined, according to (167), is called MINKOWSKI *space*. For a detailed discussion of MINKOWSKI geometry, we refer the mathematical interested reader to G. L. NABER [67]; see also A. P. FRENCH [23] and for an elementary geometrical representation, LIEBSCHER [57].

Let us make it clear for ourselves that the LORENTZ transformations (151) are filtered out from the transformations (165) by the demand made by (170).

We remind ourselves of elementary mathematics:

Let us write

$$y := c_L t \ .$$

If we have a plus instead of a minus sign in Eq. (170), hence

$$x^2 + y^2 = x'^2 + y'^2 \ , \qquad (171)$$

then the primed coordinate axes in the x-y-plane, described by CARTESIAN coordinates, are created from the unprimed by an angular rotation of φ,,

$$\left.\begin{aligned} x' &= \ \ x \cos\varphi + y \sin\varphi \ , \\ y' &= -x \sin\varphi + y \cos\varphi \ . \end{aligned}\right\} \qquad (172)$$

We can now reduce Eq. (170) to (171) by introducing a purely imaginary coordinate T into the x-t-plane according to

$$T := i\, c_L t \ , \quad \text{where} \quad i^2 = -1 \ . \qquad (173)$$

Thus, (170) becomes $x^2 + T^2 = x'^2 + T'^2$ and from the solution of this equation we receive with the purely imaginary angle $i\varphi$

$$\begin{aligned} x' &= \ \ x \cos(i\,\varphi) + T \sin(i\,\varphi) \ , \\ y' &= -x \sin(i\,\varphi) + T \cos(i\,\varphi) \ . \end{aligned}$$

We now take into consideration the relationship between the hyperbolic and the trigonomic functions, namely $\cos(i\varphi) = \cosh\varphi$, $\sin(i\varphi) = i \sinh\varphi$, replace T with (173) and receive

$$\left.\begin{aligned} x' &= x\,\cosh\varphi - c_L t\,\sinh\varphi\,, \\ c_L t' &= -x\,\sinh\varphi + c_L t\,\cosh\varphi\,. \end{aligned}\right\} \tag{174}$$

One can also immediately verify that Eq. (170) is fulfilled if (174) is valid. One only has to observe that for every angle φ the relationship $\cosh^2\varphi - \sinh^2\varphi = 1$ applies.

In order to be able to physically interpret (174), we have to replace the parameter φ, which has not yet been physically interpreted, with another non-interpreted parameter v according to

$$\tanh\varphi := \frac{v}{c_L}\,. \tag{175}$$

Because of

$$\cosh\varphi = \frac{1}{\sqrt{1-\tanh^2\varphi}}\,, \quad \sinh\varphi = \frac{\tanh\varphi}{\sqrt{1-\tanh^2\varphi}}\,,$$

the following applies

$$\cosh\varphi = \frac{1}{\sqrt{1-v^2/c_L^2}}\,, \quad \sinh\varphi = \frac{v/c_L}{\sqrt{1-v^2/c_L^2}}\,. \tag{176}$$

We insert (176) into (174) and find after a few calculations for the transformations (165) the equations

$$x' = \frac{x - v\,t}{\sqrt{1-v^2/c_L^2}}\,, \quad t' = \frac{t - v\,x/c_L^2}{\sqrt{1-v^2/c_L^2}} \tag{177}$$

that let the wave equation (166) and the line element (170), respectively, invariant.

The critical case $c_L \longrightarrow \infty$ delivers the GALILEI transformation (152), $x' = x - vt, t' = t$ and thus also the interpretation of the parameter v: v is the uniform velocity with which the inertial system Σ' moves with respect to the inertial system Σ_o. Equation (177) is nothing else than the LORENTZ transformation (144).

The mathematically elegant formulation of the principle of relativity by MINKOWSKI is used in many theoretical presentations as a reason to choose the unit of measure for time, so that the coefficient of measure for the speed of light is exactly 1. The invariance of the line element is then simply written as $x^2 - t^2 = x'^2 - t'^2$. As comfortable as such an arrangement for mathematical purposes may seem, it can lead to the erroneous view that the speed of light is just a numerical constant like the numbers 1, 7 or π and not a natural parameter, whose inalterability alone underlies the judgement of measurements as well as the definition of synchronisation!

Another approach to Special Relativity based on the principle of relativity is the inertial masses' dependency on velocity. The inertial mass m of an object moving with the velocity v relates to the inertial mass m_o of the same object in its state of rest as

$$\frac{m}{m_o} = \sqrt{1 - \frac{v^2}{c_L^2}} \ . \tag{10}$$

This we already described in Chap. 3, and we will discuss it in detail in Chaps. 22–23. The procedure is now primarily to postulate this law, in accordance with the first part of EINSTEIN's principle of relativity, Chap. 2, p.7, for every inertial system and to define with the help of this a simultaneity. As a consequence, the complete Special Relativity is fulfilled. This method was developed in LIEBSCHER's book [57]. Here, the reader can find a detailed geometrical illustration of MINKOWSKI's method. Notice that the possibility for the definition of simultaneity with the help of the laws of mechanics also was envisaged by REICHENBACH [83].

II. We will call this the LORENTZian way to Special Relativity, because it was in fact founded by H. A. LORENTZ, even though LORENTZ did not complete this way. LORENTZ's role as a pioneer of the Special Theory of Relativity is honoured in every serious depiction. However, this LORENTZian way is so completely different from the above discussed EINSTEINian approach that it has been misinterpreted and misunderstood. Even until today, this approach has been defined in standard textbooks even as an impasse as we already mentioned at the end of Chap. 12. It is especially the statement made by LORENTZ that 'the influence of a translation on the dimensions (of the separate electrons and of a ponderable body as a whole) is confined to those that have the direction of the motion, these becoming k times smaller than they are in the state of rest', with $k = \sqrt{1 - v^2/c_L^2}$ that sometimes receives false interpretation. It is not the case that LORENTZ stated that even an observer who is moving with respect to the ether must determine that the resting rod with respect to his position, therefore a moving rod with respect to the ether would be shorter than the rod stationary in the ether, which would be a moving rod with respect to the observer by the above factor. This is not correct. Correct is that LORENTZ always upheld the notion of an ether, without actually postulating mechanical properties for this ether. LORENTZ' idea of an ether is thus very close to EINSTEIN's. LORENTZ [63] wrote in 1914 in his paper 'Consideration Élementaire sur le Principe de Relativité', 'On sait, en premier lieu, que, dans la théorie de M. Einstein, on ne parle pas plus de l'éther. C'est une question sur laquelle j'aurai à revenir, mais qui, a vrai dire, ne me semble pas très importante. Pour le moment, nous supposerons l'existence d'un tel milieu qui sera en repos pour l'observateur A. Cela implique que pour lui la lumière se propagera avec une vitesse déterminée c, qui est toujours la même, indépendamment d'un mouvement éventuel de la source qui l'émet ou d'un miroir qui la réfléchit'.[2]

Apart from LORENTZ, it was POINCARÉ who stated the principal importance of the contraction hypothesis and placed this hypothesis on the same level together with

[2] We know that one does not speak of the ether in the EINSTEINian theory anymore. This is a question to which I will return later, however, to be honest, this question is not that important. Firstly we will assume the existence of such a medium, a medium that is static with respect to observer A. This implies that light propagates for him with the distinguished velocity c. This velocity is always the same, independent of any possible motion of the source of the emitter, or the mirror that reflects the light.

the principle of relativity, in other words he made both principles equal. Although POINCARÉ [75] was wrong, when in 1909 he believed he had to postulate both, the principle of relativity (which according to EINSTEIN, cf. p.7, is formulated as two hypotheses) and: 'One needs to make still a third hypothesis, ... A body in translational motion suffers a deformation in the direction in which it is displaced'.[3] POINCARÉ, however, oversaw that the LORENTZ contraction was a deduction made from the principle of relativity. We will just see however that one can as well start out from the hypothesis of FITZGERALD- LORENTZ contraction, complete it to a closed axiomatic system and then deduce the EINSTEINian principle of relativity with its universal constancy of the speed of light from this. Therefore, POINCARÉ was correct in his evaluation that both principles were equal. The main point is that the *definition* of simultaneity in this approach to relativity is now strictly separated, what we are now going to explain.

The completed LORENTZian axiomatic system of the Special Theory of Relativity from the outset is composed of two independent postulates. As we explained at the beginning of this chapter, these are (A) a definition of simultaneity and (B) an axiomatic statement about certain physical facts. These two postulates dissect the EINSTEINian, universal, abstract principle into two basic assumptions that are independent from each other. Let us start with the physical facts that we will demand according to LORENTZ:

1. In a *preferred* inertial system Σ_o, a moving rod suffers a length contraction as this was proposed by LORENTZ [62] in 1904, cf. also LORENTZ [60], according to

$$L' = L_o\sqrt{1 - \frac{v^2}{c_L^2}} \ . \tag{112}$$

In addition, we have to postulate that a clock moving with respect to this *preferred* system Σ_o goes behind according to Eq. (121) for time dilatation,

$$t' = t\sqrt{1 - \frac{v^2}{c_L^2}} \ . \tag{121}$$

In order to be able to measuring these facts, synchronised clocks should be available in this preferred reference system Σ_o. For this, we demand the propagation of light being isotropic in Σ_o. Then, we found the synchronisation of clocks in Σ_o on this isotropy. Notice that even this synchronisation of the Σ_o-clocks is a definition: Clocks are synchronised in such a way that light propagation is observed isotropically.

This postulate (121) was also proposed by LORENTZ, although not in the sense of an independent axiom. Both postulates (112) and (121) were seen by LORENTZ [63] as the elementary experimental statements of the Special Theory of Relativity. In his lectures from the year 1906 published in 1909 in the book 'The theory of Electrons', LORENTZ [64] writes, 'Then on account of the different rates of a moving

[3] Quoted from the translation in PAIS [71].

and a stationary clock, we shall have continually .... $t' = \frac{1}{k} t$'. (Notice that here the definition is $k = 1/\sqrt{1 - v^2/c^2}$).

These postulates for the measuring-rods and clocks in a preferred frame Σ_o can be extended to an independent and complete axiomatic system for spacetime if we add a definition for synchronisation of clocks for every single inertial system. Notice that we are now completely free to do this. It makes sense here to use a symmetry principle:

2. In order to define simultaneity in every single inertial system, we assume that the elementary principle of relativity should be valid as formulated in Chap. 12, p.118, cf. GÜNTHER [30]:

> After an observer resting in Σ_o has measured that Σ' possesses the velocity v with respect to Σ_o, all clocks in Σ' should be set such that an observer resting in Σ', measures that the reference system Σ_o possesses the velocity $-v$ with respect to Σ'.

With the help of these two postulates **1.** and **2.**, we found in Chap. 13 (cf. also GÜNTHER [36]) the coordinates (x', t') of an event E for the reference system Σ', which in the reference system Σ_o has the coordinates (x, t). Hence, we get simply by replacing the critical velocity c_o with the speed of light c_L,

$$x'(x,t) = \frac{x - v\,t}{\sqrt{1 - v^2/c_L^2}} \quad , \quad t'(x,t) = \frac{t - x\,v/c_L^2}{\sqrt{1 - v^2/c_L^2}} \quad . \tag{151}$$

This again is LORENTZ transformation (177). The equivalence of both axiomatic approaches has therefore been proven. In Chap. 12, we have also explicitly shown how, with the help of our postulates **1.** and **2.**, we can immediately deduce the EINSTEINian principle of the universal constancy of critical velocity, cf. GÜNTHER [36].

The definition of simultaneity according to our elementary principle of relativity needs no further explanation. It is a definition, which is exceedingly simple. One can however ask whether the synchronisation of clocks in all inertial systems can really be achieved without contradiction using this principle. This question has not yet been discussed. The reference systems Σ' and Σ'' possess the velocities u and v with respect to the reference system Σ_o. The clocks in Σ' and Σ'' are then synchronised according to the elementary principle of relativity, so that a velocity measurement is defined in these reference systems. The observers in Σ' measure for the velocity of the system Σ'' a value w'', whilst the observers in Σ'' measure w' for the system Σ'. Our elementary principle of relativity only makes sense if its statements are valid for Σ' and Σ'', thus if $w'' = -w'$ is fulfilled. Is this so? Well, have derived in Chap. 12 that the critical velocity has one and the same value in all reference systems, therefore EINSTEIN's principle of relativity is valid and following this in any arbitrary reference systems that the weaker elementary principle of relativity is also valid, thus $w'' = -w'$, cf. also the discussion in BERZI et al. [2–4]. Here, we notice that the validity of the equation $w'' = -w'$ can also be directly proven without having to use the universal constancy of critical velocity.

However, from where do we take the properties of our measuring-rods and clocks in a preferred inertial frame postulated in the first postulate? The easiest method would be if we could simply refer to precision experiments. In fact today, it is possible to do just this, because the LORENTZ contraction of bodies, as well as the time dilatation of clocks moving with respect to our laboratory can be proven with breathtaking accuracy. Such a starting point however remains theoretically unsatisfying. It would be better to have an exact mathematical theory, whose solutions can be identified as particles and using these, we could calculate and measure those postulated properties of a LORENTZ contraction and of time dilatation. In the frame of the classical field theory, a non-linear theory gives us stabile particle configurations, so-called solitons, as solutions. Every non-linear system of equations that retains its mathematical form after its variables were changed according to the LORENTZ transformation (151) are LORENTZ invariant as one would say allows soliton solutions. These equations suffice as a model for particles with those properties demanded by the first postulate. A mathematically simple example for this would be the sine-GORDON equation for a single space dimension which we discussed in detail. In a solid, the speed of light is replaced with a critical velocity c_o, whose dimension lies in the regions of sound velocities. For Special Relativity of our physical spacetime with the speed of light, the one-dimensional sine-GORDON equation does not achieve much, because here, as in opposition to a solid, we do not know of any physical objects that could be identified using the solutions of this one-dimensional equation. Today, the sine-GORDON equation for three-dimensional space with three-dimensional solutions corresponding to the kink and breather solutions discussed by us increasingly occupies the thoughts of physicists. Here, we wish to refer to the papers of G. LEIBBRANDT [55, 56]. If such solutions of the three-dimensional sine-GORDON equation can be identified using physical objects, then one could, according to the LORENTZ method, also construct a three-dimensional Special Relativity with the help of this equation in the same fashion as we did this for one space dimension in Chaps. 9–15.

In our case, where we have a continuum approximation of the crystal lattice, we can traverse one step further. We have shown what the physical background for the sine-GORDON equation that constitutes the fundamental of the explanation of Special Relativity in this continuum according to the LORENTZ method looks like. The elastic coupled components of a periodical lattice structure constitute the background, which underlies the laws of NEWTONian motion. With the ideal lattice as a vacuum, the objects of Special Relativity are realised by localised structural imperfections of this lattice.

2.* On the other hand, after having started with **1.** we may replace our postulat **2.** of the elementary principle of relativity with another postulate for synchronisation of clocks in the systems Σ'. We replace the postulate **2.** with an 'absolute simultaneity' according to

$$t'(x, 0) = 0 \tag{153}$$

as depicted in Fig. 12.8. The result of **1.** and **2.*** is now REICHENBACH transformation (143), which replaces LORENTZ transformation (177) resulting from **1.** and **2.**,

$$\left.\begin{aligned} x' &= \frac{x - vt}{\gamma} \ , \quad x = \gamma x' + \frac{v}{\gamma} t' \ , \\ t' &= \gamma t \ , \qquad t = \frac{1}{\gamma} t' \ . \end{aligned}\right\} \begin{array}{l} \text{Absolute simultaneity} \\ \text{REICHENBACH transformation} \end{array} \qquad (143)$$

According to (143), two events are observed at equal time t' in Σ' if and only if they are observed at equal time t in Σ_o. Simultaneity now is a reference system independent property. Nevertheless, due to the factor γ *time intervals* remain dependent on the reference system where they are measured.

Let us remember that in the twenties, H. REICHENBACH [81–83] analysed the axiomatic system of the Special Theory of Relativity. We only will concentrate on two points without actually becoming involved in extended philosophical discussions.

As already mentioned above, REICHENBACH, in his studies on the problem of time and especially on simultaneity, expressively warns us to be careful when using these terms and criticises the larger part of the explanations made in various presentations of Special Relativity concerning the relativity of simultaneity, cf. REICHENBACH [83], Chap. II 'We could arrange the definition of simultaneity of a system K in such a manner that it leads to the same results as that of another system K' which is in motion relative to K;...'. Equation (143) is an explicit mathematical realisation for this thesis.

Notice that the special construction used in Chap. 12 to define a simultaneity in moving reference systems on the basis of our elementary principle of relativity was not considered by REICHENBACH.

The situation however behaves in a more complicated manner when taking REICHENBACH's argumentation concerning LORENTZ contraction versus EINSTEIN contraction into consideration, cf. also our discussion in Chap. 12, p.129. REICHENBACH [83], Chap. 2, writes, 'It would be advisible, therefore, not to use the same name for the two "contractions". There is an *Einstein contraction*, which results from the relativity of simultaneity and compares the length of the moving rod with the length of the rod at rest; and there is a *Lorentz contraction*, which compares the length of a rigid rod that satisfies the Michelson experiment with the length of the rod as defined in the classical theory. It is a coincidence that both have the same contraction factor $\sqrt{1 - v^2/c^2}$ and this is probably the reason that the two contractions have been so frequently confused. Their meanings are different'.

The length of a moving object measured in the reference system Σ_o, in other words the length L' of an object that has the velocity v with respect to Σ_o, must first be defined. If the same object rests with respect to Σ_o, then its length L_o is simply the difference of the end point coordinates. Here, time plays no role. It is of no importance when these coordinates are measured. This is however different with the moving length L'. This length is also defined by the difference of the coordinate end points, however at one and the same point of time t in Σ_o. It makes no sense to measure the coordinate of the right-hand side end point of a moving object 10 minutes later than the coordinate of its left-hand side end point and then to proclaim the difference of coordinates as its length.

We therefore have to firstly decide what we wish to understand when we use the term, 'length of a moving rod'. Here, we use the generally accepted definition: *The length of a moving rod is equal to the difference of its coordinate end points at one and the same point of time.*

This automatically takes a further definition as granted: the definition of simultaneity.

Let us start with a primary preferred frame Σ_o, where a simultaneity is defined on the basis of an isotropic propagation of light. Let us take LORENTZ contraction of a moving length with respect to Σ_o as granted. Σ' are moving frames with respect to Σ_o. A moving length, with respect to Σ', has a well-defined meaning if a simultaneity is defined in Σ'. Consider a rof $\mathbf{X} = \Delta x\, L_o$ at rest in Σ_o with its coefficient of measure Δx for its length in Σ_o. We will envisage two possibilities.

1. Firstly, we follow POINCARÉ [73, 74], (cf. also Chap. 12): 'The simultaneity of two events or the order of their succession, as well as the equality of two time intervals, must be defined in such a way that the statements of the natural laws be as simple as possible'. We can realise POINCARÉ's demand with the help of our elementary principle of relativity. In the last chapter, cf. the Figs. 12.9 and 15.1, we have seen this. Once the definition of simultaneity is based on that principle the observer in the reference system Σ' moving relative to the reference system Σ_o measures a length contraction for the rod resting in the reference system Σ_o, which is identical with the LORENTZ contraction measured for a moving rod in Σ_o. The coefficient of measure $\Delta x'$ for the length of the rod $\mathbf{X} = \Delta x\, L_o = \Delta x'\, L'$ as determined in the reference system Σ' is $\Delta x' = \gamma\, \Delta x$ with $\gamma = \sqrt{1 - v^2/c^2}$. The moving rod is observed shortened, even if it rests, by chance, with respect to the crystal and only moves relative to the observer, who himself moves relative to the crystal. According to REICHENBACH, the contraction of the rod moving relative to Σ' is called EINSTEIN contraction. However, this EINSTEIN contraction is dependent on the definition of simultaneity in Σ', as we will see now.

2. Let us suppose that the clocks in Σ' are synchronised according to the 'absolute simultaneity' as described by (143),

$$x' = \frac{x - vt}{\gamma} \quad , \quad t' = \gamma t \ .$$

Consider again the rod $\mathbf{X} = \Delta x\, L_o$ at rest in Σ_o with the coefficient of measure Δx for its length in Σ_o. The measuring-rods L' at rest in Σ' are shortened, if their length is measured in Σ_o, $L' = \gamma\, L_o$. Hence, the coefficient of measure $\Delta x'$ for the length of the rod $\mathbf{X}$ as measured with L', $\mathbf{X} = \Delta x'\, L'$, is $\Delta x' = \Delta x/\gamma$. Here, we made use of the fact that the coordinates of the end points have one and the same time coordinate t in Σ_o if they have one and the same time coordinate $t' = \gamma t$ in Σ'. REICHENBACH's EINSTEIN 'contraction' in truth now is an extension. The moving length $\Delta x'$ of the rod $\mathbf{X}$ as observed from the reference system Σ' is extended in this case of synchronisation. This means, due to the 'absolute simultaneity' the statement concerning the lengths of moving and resting rods has an absolute meaning. Nevertheless, there is no contradiction between the two statements 'the rod is contracted' or 'the rod is extendcd'. This arises from the different *definitions* of simultaneity. If we consider a definition of simultaneity which does not fulfil the elementary principle of relativity with REICHENBACH, it is important to differentiate between LORENTZ contraction and EINSTEIN contraction.

In the context of our LORENTZian way **II.** to Special Relativity, we understand REICHENBACH's argumentation in this way. We demand LORENTZ contraction as a fundamental physical law for a primary preferred frame Σ_o. EINSTEIN contraction is a property of the inertial frames Σ' moving with respect to Σ_o. It depends on the *definition* of simultaneity. Here, we have to point at that. *There is* a definition of simultaneity so that all inertial frames have equal rights. Especially, the primary preferred frame Σ_o is an arbitrary inertial frame and so LORENTZ contraction becomes identical with EINSTEIN contraction.

From the LORENTZian way **II.** to Special Relativity, the following conclusion should be emphasised:

> The FITZGERALD- LORENTZ contraction hypothesis does not contradict the basic principle of relativity.

The negative assessment of the LORENTZ hypothesis, which is sometimes still shared today, cf. our discussion in Chap. 12, p.129, corresponds with a misunderstanding of REICHENBACH's thesis. Here, we wish to state that H. MINKOWSKI did not encourage this error. Starting out from EINSTEIN's principle of relativity, MINKOWSKI shows in LORENTZ [60], 'that the Lorentzian hypothesis is completely equivalent to the new conception of space and time, which, indeed, makes the hypothesis much more intelligible'. This passage from MINKOWSKI's famous paper was obviously forgotten. This could have had something to do with the fact that REICHENBACH [83] himself, who explicitly discussed MINKOWSKI's work and ideas, to our knowledge, did not explicitly mention this MINKOWSKI's statement. And EINSTEIN himself? He writes, discussing the MICHELSON experiment, cf. EINSTEIN [17], Chap. XVII, 'Lorentz and FitzGerald rescued the theory from this difficulty by assuming that the motion of the body relative to the ether produces a contraction of the body in the direction of motion, the amount of contraction being just sufficient to compensate for the difference in time mentioned above. Comparison with the discussion in Section XII shows that also from the standpoint of the theory of relativity this solution of the difficulty was the rigth one'.

If we replace the velocity of light c_L with the critical velocity c_o of sine-GORDON equation, all considerations concerning Special Relativity can be illustrated directly using solids. This applies to the explanation of the relativistic effects as to the solution of the well-known relativistic paradoxa and thus leaves no room for interpretation.

Chapter 17
The Twin Paradox

The twin paradox is certainly one of the most provocative consequences with which Special Relativity confronts us. It is a challenge to everybody's 'common sense'. One can however also see this from various viewpoints. The unpretentiousness and indigence of our common sense which we so desperately need and the loose ground on which our indisputable facts of everyday life are built on crumbles away in the face of the twin paradox. If one can be certain about one fact, then that twins always have their birthdays on exactly the same day, even if they have completely different lifestyles and have lived completely different lives. However, this is not exactly true.[1] What is the problem here? Both twins, let us call them brother A_o and twin A go on a journey. To be more precise, brother A_o finds himself in the reference system Σ_o and his twin A in the reference system Σ', so that A moves away from A_o with the uniform velocity v and correspondingly that A_o moves away from A with the velocity $-v$. Both of them should be equipped with a sufficient number of our described clocks and measuring-rods. At their point of departure found at the common coordinate origin, event O, where $(x = 0, t = 0)$ in Σ_o and also $(x' = 0, t' = 0)$ in Σ', the clocks they both carry, let us determine them as the clock U_o^A of brother A_o and the clock U^A of his twin A that they both show the reading zero. This situation corresponds to our Fig.12.3, which we have reproduced here with different labelling, because of the different objective; see Fig. 17.1.

Using these 'personal clocks', the twins measure the time since their point of departure at event O. The personal clocks, the pulsating breathers, show how much

[1]From now on, we will talk about persons. Here, we have the situation inside of the crystal in front of us, in other words, we consider the internal observers. For us, clocks are breather solutions of sine-.GORDON equation which have the possibility of moving through the crystal with the velocity $|v| < c_o$. The number of oscillations of such a breather determines its 'personal life span'. We could replace the twin brothers anytime with 'twin breathers'.

H. Günther, *Elementary Approach to Special Relativity*,
https://doi.org/10.1007/978-981-15-3168-2_17

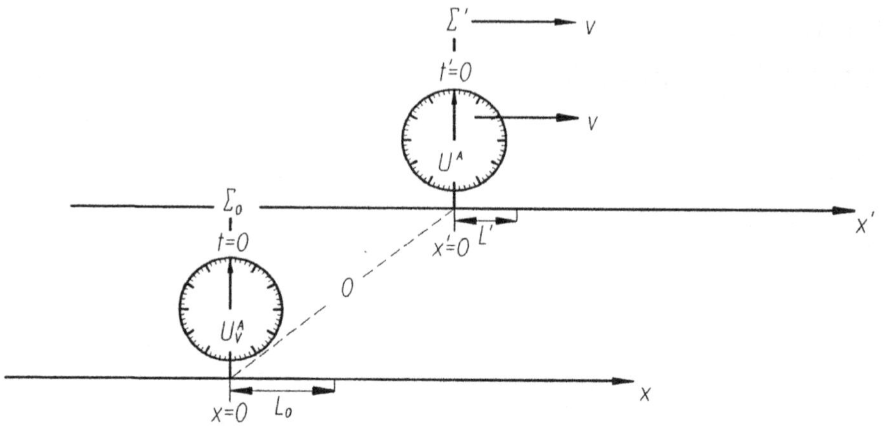

Fig. 17.1 Event O. The departure of the twins from the common coordinate origin. Brother A_o has the clock U_o^A, and the twin A has the clock U^A with him. Brother A_o observes that the measuring-rod L' of his twin A is shorter than his measuring-rod L_o by the factor γ. We once again consider the case $v = 0,8\,c_o$, thus $\gamma = 0,6$, which we have depicted as $L' = 0,6\,L_o$. We have not plotted the contraction in length that the twin A measures. He determines the same factor γ for the moving length $L_v = \gamma\,L'$ of the, from his point of view, moving measuring rod L_o; see Fig.15.2

older each twin has become since their departure. The space and time statements of an event are once again shown in Σ_o with the help of unprimed coordinates (x, t) and in Σ' by using primed coordinates (x', t'). We recapitulate as follows: The coordinates are always the *coefficients of measure* of lengths and time. The hands of the clocks show the coefficients of the measure of time. These following stated coordinates of time can thus be immediately read from the clocks.

In order to suppress as many objections as possible concerning the twin paradox, we will discuss this problem in detail from various viewpoints. A simple, special solution of this paradox can be found on the pp. 173–175.

Brother A_o reasons in his reference system Σ_o: My twin brother A with his clock U^A is moving with the velocity v. After the time $t_p = x_P/v$ according to my clock U_o^A, my twin has arrived at the location x_P in Σ_o. We will call this the event P with the coordinates in Σ_o according to

$$\Sigma_o \;:\; P\left(x_P,\; t_P = \frac{x_P}{v}\right). \tag{178}$$

The hands of the clock U^A of twin A are thus positioned, according to our Eq.(121) at the time of $t'_P = t_P\sqrt{1 - v_2/c_o^2}$. In the reference system Σ', event P thus has the coordinates

$$\Sigma' \;:\; P\left(x'_P = 0,\; t'_P = \frac{x_P}{v}\sqrt{1 - \frac{v^2}{c_o^2}}\right). \tag{179}$$

Thus, when twin A compares the time t'_P shown on his clock to the time t_P shown on the clock in Σ_o opposite from him, let us call this clock U_o^P, and he notices that his clock U^A goes behind the clock U_o^P by the amount $t_P - t'_P$. The telegram of his twin brother, that he, brother A_o is now the elder of the two (and thus has the sole claim to their dad's hereditary farm) can be confirmed. However, A is not too happy about this and thus turns the tables and also argues that: The personal clock U_o^A of my twin brother A_o moves with the velocity $-v$. If I as the slandered younger twin A read the hand setting t'_P of my clock U^A in Σ', then brother A_o with his clock U_o^A is at location $x'_R = -v\, t'_P$ in my system. We will call this event R with $t'_R = t'_P$, thus with (179)

$$\Sigma' : \quad R\left(x'_R = -x_P\sqrt{1-\frac{v^2}{c_o^2}}\;,\; t'_R = \frac{x_P}{v}\sqrt{1-\frac{v^2}{c_o^2}}\right). \tag{180}$$

Because brother A_o moves with the velocity $-v$ with respect to myself, the hand of his clock U_o^A must, with respect to the clock in my system Σ' exactly opposite from him, let us call this clock U^R, have the delayed position $t_R = t'_R\sqrt{1 - v_2/c_o^2}$ according to Eq. (162). In Σ_o, we therefore find for event R

$$\Sigma_o : \quad R\left(x_R = 0,\; t_R = \frac{x_P}{v}\left(1 - \frac{v^2}{c_o^2}\right)\right). \tag{181}$$

Thus, deducing from this, brother A_o's clock when compared to the clock U^R in the reference system Σ' showing the time of twin A goes behind by the amount $t'_R - t_R$. Twin A therefore excitedly sends a telegram back to his brother stating that he is the younger of them both, and that he can confirm this by comparing the hand setting of his clock U_o^A to the hand settings of the clocks in Σ', where he is presently located (which, as known to him, show the time of A). Thus, brother A_o has lost his claim to the farm. This argument eventually leads both brothers to accept that both of their opposite statements concerning their clocks are correct, and that they do not contradict each other, because different clocks were used in the comparison. Therefore, coming to a conclusion about an absolute younger or older made no sense (and was probably only attempted in order to obtain the sole claim to dad's farm). Exactly, this situation is used in order to derive at Eq. (162). We will recapitulate this in Fig. 17.2.

Up to now, we have no paradox. However, brother A_o is not happy with what he has so far achieved and continues forging new plans. In the equation dealing with the hand setting of twin A's personal clock, only the velocity v is squared. If he could therefore convince twin A to reverse his direction, let us say back to event P using the velocity $-v$ and therefore back towards A_o, then A's personal clock would, according to the same equation, still go behind A_o's personal clock. We will call the event according A_o's plan, where the two brothers meet, made possible by A's reverse in direction, event Y. After a further time span of t_P (shown by brother A_o's clock U_o^A), brother A_o's clock U_o^A would have the hand setting of $t_Y = 2t_P$. For event Y, we can register in the reference system Σ_o

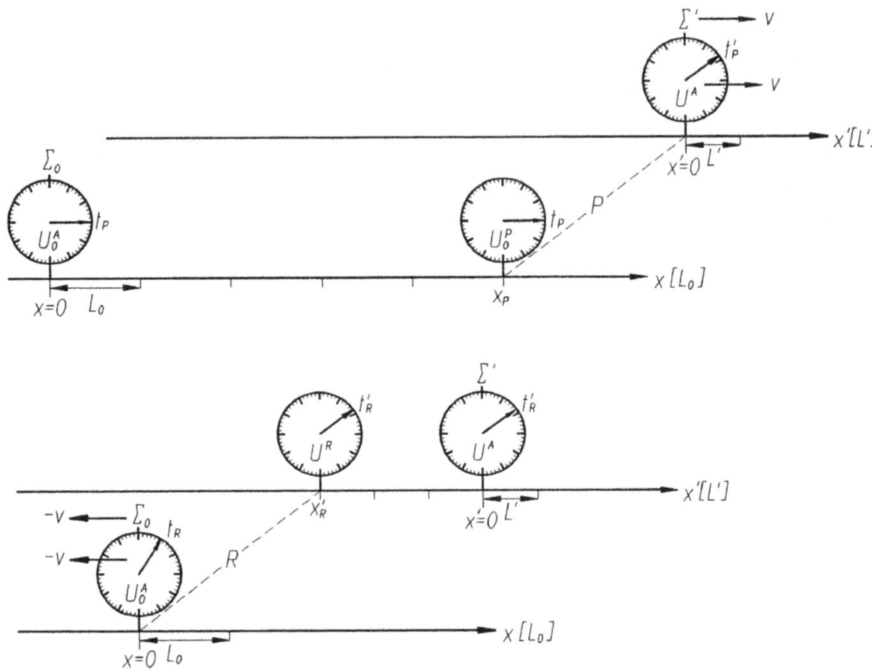

Fig. 17.2 Both of the twins see that: The personal clock of the other goes behind his own synchronised clocks in his reference system by the factor γ. We calculate using $v = 0,8\,c_o$, thus $\gamma = 0,6$. The event P is illustrated in upper picture. Twin's A clock U^A is at $x = x_P$. We have chosen the coefficient of measure $x_P = 5$ for this. The hands of all clocks in Σ_o have the uniform setting of $t_P = x_P/v$, where the display chosen shows 'quarter past' (= 15 scale marks), thus $t_P = 15$. We have inserted this into the picture for the clocks U_o^A and U_o^P. The hand setting of twin A's clock U^A is then, according to Eqs. (121) and (179), $t'_P = \gamma\, t_P$, therefore $t'_P = 9$. (If the traversed distance of $5\,L_o$ using a 'small' measuring-rod L_o is in fact somewhat small, then the used clock only has to show sufficient small time intervals using a scale mark in order to be positioned at 15 at event P). Event R is illustrated in the lower picture. Brother A_o's clock U_o^A is according to Eq. (180), when observed from the reference system Σ', positioned at $x'_R = -x_P\ \gamma = -0,6\,x_P$, thus at $x' = -3$ (in the picture the measuring-rod L' is used three times). All clocks in Σ' have the uniform position $t'_R = t'_P = 9$. We have inserted this into the picture for the clocks U^A and U^R. The hand of brother A_o's clock U_o^A then shows, according to Eqs. (162) and (181), the setting of $t_R = \gamma\, t'_R$, thus $t_R = 5,4$ scale marks

$$\Sigma_o\ :\quad Y\left(x_Y = 0,\quad t_Y = 2\,\frac{x_P}{v}\right)\,. \tag{182}$$

According to time dilatation (121), the hand of twin A's clock U^A once again goes behind by the factor γ, and thus has the setting $2\,\frac{x_P}{v}\,\sqrt{1 - v^2/c_o^2}$ when the two brothers meet.

By an immediate comparison of the clocks at one and the same location, both brothers would see that twin A is younger than brother A_o and thus, that A_o has sole claim to the farm. Brother A_o therefore sends a sanctimonious message to A stating that his, the twin's A version is the more valid. He should return as fast as possible in order to hand over the farm to the older brother. Time is pressing. How does twin A reply? The equation for the time delay of clocks only contains the square of the journey's velocity v. If he, twin A, moves with $-v$ with respect to brother A_o, then A_o moves with $+v$ with respect to A. Then, A_o's clock U_o^A that already went behind on the first leg of the journey (with respect to A) should still go behind. Following from this, when the brothers meet, the clock U_o^A should still go behind the A's clock U^A and not vice versa. However, one of both cases cannot apply. The hand setting of a clock is a fact that cannot be ignored.

> One single hand in two different positions—that would be paradox!

In fact, somewhere along the way an error has occurred, an error that we want to find. Correct without doubt, according to (121) is that it makes no difference for the time delay of twin A's moving clock whether he travels away from brother A_o with the velocity $+v$, or if he moves towards him with the velocity $-v$. Fact is that when they meet, the hand of U^A goes behind the hand of U_o^A by the amount $2\,\frac{x_P}{v}\sqrt{1 - v^2/c_o^2}$. Thus, the mistake must lie in twin A's argumentation.

If we examine Eqs. (121) and (162) for the time dilatation of moving clocks, then we see that these depend on the square of velocity v. Let us go back one step and consider the regulations for synchronising clocks in a reference system Σ' moving with the velocity $+v$ with respect to a reference system Σ_o. This is Eq. (142) as shown in Fig. 12.7. This synchronisation completely depends on the direction of the velocity v! Reversing the velocity v forces the clocks to be set in the reverse direction. Let us recapitulate as follows:

> The time dilatation of a moving clock is solely dependent on the square of its velocity. The synchronisation regulation of a clock changes its sign when the direction of velocity changes.

This asymmetry in the synchronisation of clocks plays a decisive role in the clarification of the twin paradox.

If twin A wants to return to his brother A_o, he must leave his reference system Σ' and enter a new reference system, let us call it Σ''_-, whose clocks are synchronised so that Σ''_- has the velocity $-v$ with respect to Σ_o and that this synchronisation contrarotates the one in the reference system Σ' used by twin A in his statements concerning the time flow. For twin A, it is now obliging to use the new synchronisation regulation for his time comparison with Σ_o. For brother A_o, who only needs the time dilatation of U^A, this new synchronisation plays no role, because we assumed that twin A would be returning with the velocity $-v$.

The situation is reversed if twin A remains in his reference system Σ' and his brother A_o leaves his reference system Σ_o in order to enter a new reference system, let us say Σ'' and rushes after his twin A, so that twin A registers the velocity of A_o travelling towards him as $+v$. The velocity $+v$ with respect to Σ' is now decisive for the synchronisation of clocks in Σ'', and this synchronisation contrarotates the one in the reference system Σ_o, which was decisive for brother A_o's statements concerning the time span. For brother A_o, the new synchronisation regulation is valid when comparing time with Σ'. Now, both brothers meet in Σ' at $x' = 0$. If the reverse in direction of brother A_o takes place at event R, thus at time t'_R in Σ', then he meets twin A at $2t'_R$. This is our event Z, which in Σ' together with (121) results in

$$\Sigma' : \quad Z\left(x'_Z = 0,\; t'_Z = 2\,\frac{x_P}{v}\sqrt{1-\frac{v^2}{c_o^2}}\right) . \tag{183}$$

Now, brother A_o has in fact firstly moved away from his twin with the velocity $-v$ and then, after entering the new reference system Σ'' moved back towards him with the velocity $+v$. During the total time $t'_Z = 2\,\frac{x_P}{v}\sqrt{1-v^2/c_o^2}$ in Σ', the clock U_o^A goes behind by the factor γ, so that its hand stands at $2\,\frac{x_P}{v}\,(1-v^2/c_o^2)$ when they meet at event Z and thus goes behind the hand setting of U^A by the amount $2\,\frac{x_P}{v}\sqrt{1-v^2/c_o^2}\left(1-\sqrt{1-v^2/c_o^2}\right)$. Now, brother A_o really is the younger of the two.

The confusion caused by the paradox arises when one ignores that the synchronisation of clocks in the reference Σ''_- with the velocity $-v$ with respect to Σ_o, or the synchronisation of clocks in Σ'' with the velocity $+v$ with respect to Σ', contrarotates the previous synchronisation of the brother who just reverses his direction. Due to the fact that our clocks represent oscillating breather solutions of the sine-Gordon equation, we are in the position of being able to follow these experiments, at least theoretically, inside of a crystal. Thus, the oscillating breather will have made fewer oscillations after having reversed its direction of motion back to its point of departure than the breather that remains in its original position.

These explanations have not yet completely satisfied us. It becomes far more intriguing if we try to calculate the time comparison for the meeting, from the view of the returning twin, because this twin and his clock successively find themselves in two different reference systems. Here, we wish to discuss in detail that in this story it is brother A_o who is the twin who changes his reference system during his journey. Brother A_o will thus leave his reference system Σ_o at time t_P in order to enter a new reference system so that he can rush back to his twin brother A. From twin A's point of view, who always remains in his reference system Σ', brother A_o makes a 180 degrees change in direction at exactly t'_P in Σ'. The hand setting of the clock U_o^A carried by brother A_o must therefore be successively synchronised to the clocks in different reference systems. We will represent this special hand setting of this clock by using a tilde, $\tilde{t}$.

In order to control what is happening, we will also include a neutral observer B_o, who is positioned in the reference system Σ_o and who thus always measures the coordinates (x, t).

We will firstly simplify our question in order to make the problematic of this twin story more transparent. Twin A is suspicious. 'He wants me to come to him'. He then thinks, 'if so, then this can only be in his interest and not mine. He should come to me'. He sends a message to his brother: 'As I am the one who gets the farm you should at least receive a nice journey with all expenses paid. It really is very nice here. Take the most expensive super-train existing. It virtually moves with the speed of sound c_T (a few cm/s slower, but this cannot be measured). We can thus meet as soon as possible and can bring everything concerning the farm to a conclusion'. (Here of course we assume that $c_T = c_o$).

Brother A_o takes the super-train, and his clock U_o^A reads $\tilde{t}_P = t_P = x_P/v$. We will call this event T with the coordinates in Σ_o according to

$$\Sigma_o \, : \quad T\left(x_T = 0, \;\; t_T = \frac{x_P}{v}\right) . \tag{184}$$

In Fig. 17.3, we have illustrated events T and P together in order to distinguish them from the already defined events P and R in Fig. 17.2.

Twin A is located at x_P (compare to the upper plot of Fig. 17.2) at time x_P/v in the reference system Σ_o. The observer B_o in Σ_o observes the following: Twin A rushes away with the velocity v, brother A_o rushes after him with the velocity c_T. Thus, A_o approaches to his twin brother with the velocity $(c_T - v)$ and the time t_c that B_o in Σ_o measures for the super-train from its starting point up to the meeting point is $t_c = \frac{x_P}{c_T - v}$. The settings of the personal clocks of both twins result in the following calculation:

The hand setting of brother A_o's clock U_o^A was at $\tilde{t}_P = x_P/v$ when he stepped into the train. He then moves with the velocity $u \approx c_T$, which results in the LORENTZ factor $\sqrt{1 - u^2/c_o^2} \approx 0$ for the pace of his clock, because we assumed that $u \approx c_T = c_o$. The hand of brother A_o's clock therefore does not move (or not enough to count) due to this highest possible velocity u of the super-train. We define the happy meeting reunion of the brothers after the journey in the super-train as event S. The hand setting of brother A_o's clock U_o^A therefore still has the hand setting of $\tilde{t}_S = \tilde{t}_P = x_P/v$,

$$\tilde{t}_S = \frac{x_P}{v} \; . \qquad \text{Hand setting of the clock } U_o^A \text{ at the point of reunion} \tag{185}$$

Observer B_o also discovers that: Whilst brother A_o stepped on to the train, twin A was at the location x_P in Σ_o, and that his clock U^A had the hand setting of $t'_P = \frac{x_P}{v}\sqrt{1 - v^2/c_o^2}$ according to Eq. (179). Twin A moved at the velocity v during A_o's time of journey. The hand setting of U^A changed, during the above measured journey time of brother A_o in the super-train by B_o in the reference system Σ_o, from $t_c = \frac{x_P}{c_T - v}$ by the amount $t'_c = \frac{x_P}{c_T - v}\sqrt{1 - v^2/c_o^2}$ due to time dilatation (121). Thus, the hand setting of A's clock U^A is at $t'_S = t'_P + t'_c$ when the twins meet. If we insert $c_T = c_o$, we find

$$t'_S = \frac{x_P}{v}\sqrt{1 - \frac{v^2}{c_o^2}} + \frac{x_P}{c_o - v}\sqrt{1 - \frac{v^2}{c_o^2}} = \sqrt{1 - \frac{v^2}{c_o^2}}\left(\frac{x_P}{v} + \frac{x_P}{c_o - v}\right) ,$$

$$t'_S = x_P\left(\frac{1}{v} + \frac{1}{c_o - v}\right)\sqrt{1 - \frac{v^2}{c_o^2}} = \frac{x_P}{v}\frac{c_o}{c_o - v}\sqrt{1 - \frac{v^2}{c_o^2}} = \frac{x_P}{v}\sqrt{\frac{c_o^2 - v^2}{(c_o - v)^2}} ,$$

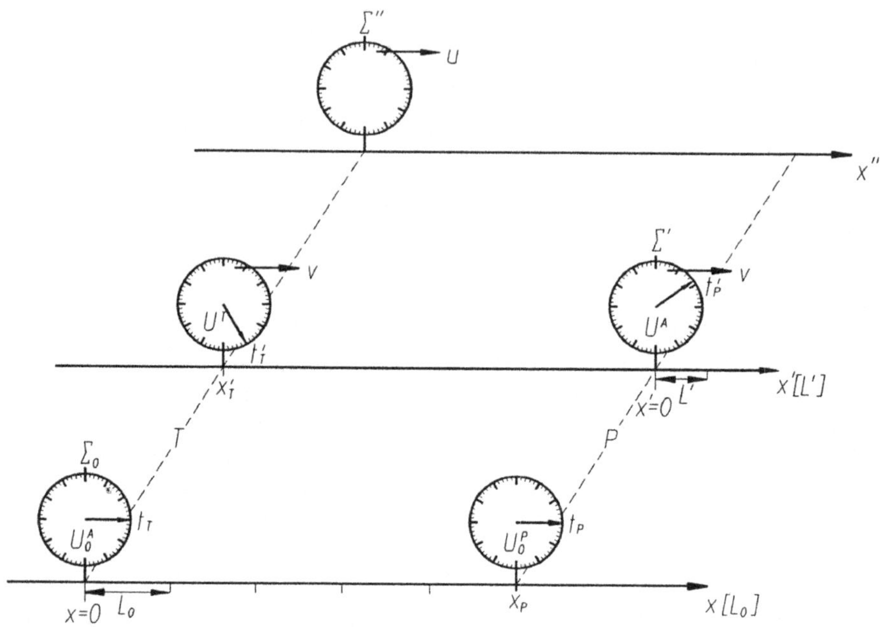

Fig. 17.3 Event T. Brother A_o who is in the reference system Σ_o at $x = 0$ moves to the reference system Σ'' exactly then when his clock U_o^A shows the coefficient of measure for time $t_T = t_P = x_P/v$. We once again choose $v = 0,8\,c_o$, therefore $\gamma = 0,6$ and gauge the clock so that $t_P = 15$ scale marks. The reference system Σ' moves with the velocity v with respect to the system Σ_o. The difference in coordinates x_P determined at one and the same point of time t_P in Σ_o is the coefficient of measure of a length moved with the velocity v in Σ_o, therefore stationary in Σ', its coefficient of measure in Σ' is x'_P. Due to LORENTZ contraction (113a), it is $x'_P = -x_P/\gamma$. The space coordinate for the event T in Σ', taking the sign into consideration is therefore $x'_T = -x_P/\gamma$ (as we also find in (188) using a different approach) and thus $x'_P\,L' = x_P\,L_o$, as plotted. As a comparison, we have included in the reference system Σ_o the event P that occurs simultaneously with event T with the coefficient of measure $x_P = 5$; see the upper picture in Fig. 17.2. The time for event T in the reference system Σ' is shown by the clock U^T located at x'_T. Since the twin's moment of departure from the common coordinate origin, brother A_o has been moving with the velocity −v and is therefore located at event T, as twin A calculated, at $x'_T = -v\,t'_T$. We therefore calculate for the hand setting t'_T of the clock U^T at event T $t'_T = x'_T/(-v) = x_P/(v\,\gamma) = t_T/\gamma$, cf. (188), which in our example leads to the hand setting $t'_T = 25$ scale marks. The numerical value of the velocity u of the reference system Σ'', measured from Σ_o, into which brother A_o transfers has not yet been included in our considerations. We will later assume a value for this velocity u just lower than the critical velocity c_o. (Dotted lines once again combine spacetime points belonging to one and the same event)

thus,

$$t'_S = \frac{x_P}{v}\sqrt{\frac{c_o + v}{c_o - v}}\,. \qquad \text{Hand setting of the clock } U^A \text{ at the point of reunion} \qquad (186)$$

and with (185)

$$\frac{t'_S}{\tilde{t}_S} = \sqrt{\frac{c_o + v}{c_o - v}} \ . \tag{187}$$

Obviously, it is $\sqrt{\frac{c+v}{c-v}} > 1$. Observer B_o therefore registers in his reference system Σ_o, $t'_S > \tilde{t}_S$. The hand setting of brother A_o's clock, the brother who rushed after the other, goes behind.

Twin A in his reference system Σ' has in the meanwhile also calculated this. Seen from Σ', A_o enters the super-train at time t'_T. Up to this point of time, A_o's clock U_o^A moved away with the velocity $-v$ and thus has the hand setting of $\tilde{t}_T = t_T = t'_T \sqrt{1 - v^2/c_o^2}$ when entering the super-train (event T) due to time dilatation. Therefore, because of (184), event T in the reference system Σ' takes place at $t'_T = \frac{x_P}{v\sqrt{1-v^2/c_o^2}}$. Because of the speed at which the train travels $u \approx c_o$, $\tilde{t}_T$ is also the hand setting $\tilde{t}_S$ of A_o's clock U_o^A at the point of arrival. Therefore, as we already know,

$$\tilde{t}_S = t'_T \sqrt{1 - \frac{v^2}{c_o^2}} = \frac{x_P}{v} \ . \qquad \text{Hand setting of the clock } U_o^A \text{ at the point of reunion} \tag{185a}$$

Since departure, brother A_o has been moving with the velocity $-v$ and thus finds himself, with respect to his twin A, during event T at $x'_T = -v\, t'_T$. We thus have the following coordinates in Σ' for the event T,

$$\Sigma' : \quad T\left(x'_T = \frac{-x_P}{\sqrt{1 - v^2/c_o^2}},\ t'_T = \frac{x_P}{v\sqrt{1 - v^2/c_o^2}}\right) . \tag{188}$$

Since brother A_o is actually on the super-train, he meets his twin brother A at $x' = 0$ after a further travelling time of $t'_d = -x'_T/c_o = \frac{x_P}{c_o\sqrt{1-v^2/c_o^2}}$, so that twin A's clock U^A should show the hand setting of $t'_S = t'_T + t'_d$ when they both meet (event S), thus

$$t'_S = \frac{x_P}{v\sqrt{1 - v^2/c_o^2}} + \frac{x_P}{c_o\sqrt{1 - v^2/c_o^2}} = \frac{x_P}{\sqrt{1 - v^2/c_o^2}}\left(\frac{1}{v} + \frac{1}{c_o}\right) = \frac{x_P}{v\, c_o}\, \frac{c_o + v}{\sqrt{1 - v^2/c_o^2}}$$

and once again,

$$t'_S = \frac{x_P}{v}\sqrt{\frac{c_o + v}{c_o - v}} \ . \qquad \text{Hand setting of the clock } U^A \text{ at the point of reunion} \tag{186}$$

Twin A can therefore fully confirm the results made by observer B_o. This consequently leads to the unbelievable fact that:

Twin A is older than his arriving twin brother A_o.

The later therefore aged less during his journey (which should compensate him for losing the farm)—stays young and drives fast.

This result is from twin A's point of view of course self-evident, because it is brother A_o who moves. And it is the moving clock that goes behind. The velocity with which A_o departs from A and the then following arbitrary velocity chosen by A_o to return to twin A should make no difference to the sign of this effect, of course from out of the view of A in the reference system Σ'. However, what does this case look like if it is observed from the reference system Σ_o, the reference system in which our observer B_o controls the personal clocks of both twin brothers?

We will accept that it is once again twin A who moves with the constant velocity v with respect to the reference system Σ_o. Brother A_o sits and watches, together with observer B_o who makes all of his measurements in the reference system Σ_o. The observer B_o registers: After time $t_T = t_P = x_P/v$, brother A_o changes into the reference system Σ'' and moves behind his twin brother A with the velocity $u > v$. This 'changing of reference systems' is the displayed event T in Fig. 17.3 with its coordinates in Σ_o according to (184) and in Σ' according to (188). Twin A is located at x_P during event T in Σ_o. From observer B_o's point of view, brother A_o needs $\Delta t_u = \frac{x_P}{u-v}$ from the point when he changed reference systems until the point where he catches up with his brother A_o. Once more, Δt_u is the time measured from out of Σ_o that brother A_o spent in Σ''. We will understand further down the reference system that moves with the special velocity u according to (196), with respect to Σ_o, by Σ''. The twin's meeting is defined, if we have the case of a general velocity u, as the event $S(u)$. From both of the times in Σ_o

$$\left.\begin{aligned} t_T &= \frac{x_P}{v}\,, \\ \Delta t_u &= \frac{x_P}{u-v}\,, \qquad u - v > 0\,, \end{aligned}\right\} \tag{189}$$

and from the velocity u after the change of reference systems, the observer B_o calculates according to our Eq. (121) the hand setting of brother A_o's clock U_o^A as the coefficient of measure $\tilde{t}_{S(u)} = t_T + \Delta t_u \sqrt{1 - u^2/c_o^2}$ and thus

$$\tilde{t}_{S(u)} = \frac{x_P}{v}\left(1 + \frac{v}{u-v}\sqrt{1 - \frac{u^2}{c_o^2}}\right)\,. \tag{190}$$

On the other hand, due to the fact that twin A constantly moves with the velocity v, the observer B_o calculates according the hand setting of brother A's clock U^A as the coefficient of measure $t'_{S(u)} = (t_T + \Delta t_u)\sqrt{1 - v^2/c_o^2}$ and thus

$$t'_{S(u)} = \left(\frac{x_P}{v} + \frac{x_P}{u-v}\right)\sqrt{1 - \frac{v^2}{c_o^2}} = \frac{x_P}{v}\left(1 + \frac{v}{u-v}\right)\sqrt{1 - \frac{v^2}{c_o^2}}\,,$$

$$t'_{S(u)} = \frac{x_P}{v} \frac{u}{(u-v)} \sqrt{1 - \frac{v^2}{c_o^2}} \,. \tag{191}$$

We will now just simply check that, according to (190) and (191) for an arbitrary v and an arbitrary u, with $c_o > u > v$, brother A_o's clock U_o^A does in fact go behind that of twin A's clock U^A, thus

$$\frac{\tilde{t}_{s(u)}}{t'_{s(u)}} < 1 \quad \text{for arbitrary} \;\; v < u < c_o \,.$$

The proof lies in the inequality between the geometric and the arithmetic means for the velocities u and v, i.e. $\sqrt{u\,v} < \frac{1}{2}(u+v)$. Let us simplify our conclusion even more and keep in mind that $0 < u - v$,

$$\begin{aligned}
&0 < (u-v)^2 \,, \\
&2uv < (u^2+v^2) \,, \\
&2c_o^2 uv < c_o^2(u^2+v^2) \,, \\
&-u^2c_o^2 - v^2c_o^2 < -2c_o^2 uv \,, \\
&c_o^4 - u^2c_o^2 - v^2c_o^2 + u^2v^2 < c_o^4 - 2c_o^2 uv + u^2v^2 \,, \\
&(c_o^2 - v^2)(c_o^2 - u^2) < (c_o^2 - uv)^2 \,, \\
&\sqrt{c_o^2 - v^2}\sqrt{c_o^2 - u^2} < c^2 - uv \,, \\
&2uv\sqrt{1 - \frac{v^2}{c_o^2}}\sqrt{1 - \frac{u^2}{c_o^2}} < 2uv - \frac{2u^2v^2}{c_o^2} \,, \\
&-2uv < \frac{2u^2v^2}{c_o^2} - 2uv\sqrt{1 - \frac{v^2}{c_o^2}}\sqrt{1 - \frac{u^2}{c_o^2}} \,, \\
&u^2 - 2uv + v^2 < u^2 - \frac{u^2v^2}{c_o^2} + v^2 - \frac{u^2v^2}{c_o^2} - 2uv\sqrt{1 - \frac{v^2}{c_o^2}}\sqrt{1 - \frac{u^2}{c_o^2}} \,, \\
&u - v < u\sqrt{1 - \frac{v^2}{c_o^2}} - v\sqrt{1 - \frac{u^2}{c_o^2}} \,, \\
&u - v + v\sqrt{1 - \frac{u^2}{c_o^2}} < u\sqrt{1 - \frac{v^2}{c_o^2}} \,, \quad \frac{u - v + v\sqrt{1 - \frac{u^2}{c_o^2}}}{u\sqrt{1 - \frac{v^2}{c_o^2}}} < 1 \,,
\end{aligned}$$

and thus, as claimed, the so-called twin inequality, cf. GÜNTHER and MÜLLER [], Chap. 8

$$\frac{\tilde{t}_{S(u)}}{t'_{S(u)}} = \frac{1 + \dfrac{v}{u-v}\sqrt{1 - \dfrac{u^2}{c_o^2}}}{\dfrac{u}{u-v}\sqrt{1 - \dfrac{v^2}{c_o^2}}} < 1 \,. \tag{192}$$

So much to the observations made by B_o in the reference system Σ_o.

From twin A's point of view, who is in the reference system Σ', it is also clear that brother A_o has to be younger than A when they meet, because only A_o moved with respect to Σ'. With respect to the inequality, all of the calculations made by observer B_o in Σ_o correspond with the expectations of twin A in Σ'. However, if we numerically compare the statement (192) made by the observer B_o with the measurements made by the twin A in Σ' moving with an arbitrary velocity $u > v$, as we did in the case of the super-train, a new problem arises. This problem even arises if we observe this whole process from the viewpoint of the brother A_o, which we will save up for the end.

Here, the problem has to do with the composition of velocities. We will firstly have to deal with this. The main essence here is to differentiate between two completely different situations.

1. We observe two objects K and L in a reference system Σ_o at the positions x_1 and x that move, according to $x_1 = x_1(t)$ and $x = x(t)$, with the velocities $v = \mathrm{d}x_1/\mathrm{d}t$ and $u = \mathrm{d}x/\mathrm{d}t$. The relative velocity w of both objects in Σ_o is then *per definition* $w = u - v$. In other words, observed from Σ_o, object L moves towards object K with the velocity w. This relative velocity is nothing else, but the change in time of a difference in coordinates. We have used this feature many times in Chap. 12; see Eqs. (130), (131) and (131a). A sound signal approaches an object moving with the velocity w in the same direction with the velocity $c_T - v$. Velocities referring to one and the same reference system are simply per definition added together; see Fig. 17.4,

$$u = w + v\,. \qquad \text{Addition of velocities in a single reference system} \qquad (193)$$

Here, we wish to state that the quantity w is not the velocity of an object, as u and v are, but only represents a mathematical quantity that can even be larger than c_o. For $u = \frac{3}{4}\,c_o$ and $v = -\frac{3}{4}\,c_o$, one receives for example $w = u - v = 1,5\,c_o$.

2. The following situation must be strictly differentiated from the above situation. An object K has, measured from the reference system Σ_o, once again the velocity v that we assume is constant. This object K can by pictured as a complete train defining

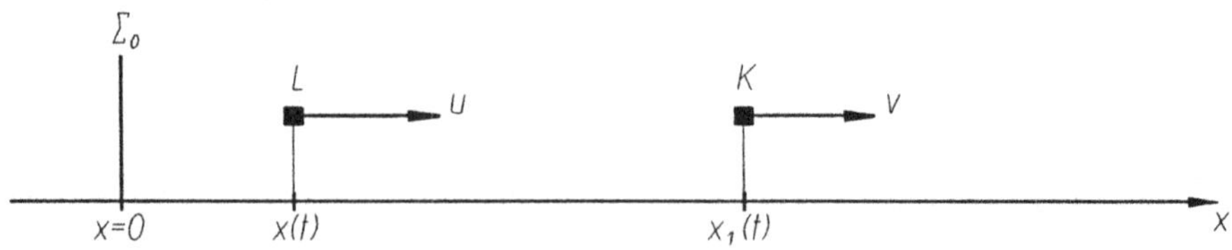

Fig. 17.4 Objects L and K have the velocities $u = \mathrm{d}x/\mathrm{d}t$ and $v = \mathrm{d}x_1/\mathrm{d}t$, respectively, in the reference system Σ_o. In Σ_o, object L approaches object K with the relative velocity $w = \frac{\mathrm{d}}{\mathrm{d}t}\,[x(t) - x_1(t)] = u - v$. In the figure, we have chosen $v = 0,8\,c_o$ and $u = 0,9\,c_o$. Therefore, object L approaches object K with the relative velocity of $w = 0,1\,c_o$

its own reference system Σ'. The space and time coordinates measured from out of this reference system are once again defined as x' and t'. We now measure the motion of a second object in Σ', moving through the positions $x' = x'(t')$ (e.g. the conductor moving through the train). In Σ', object L has the velocity $u' = \mathrm{d}x'/\mathrm{d}t'$. We now ask for the velocity $u = \mathrm{d}x/\mathrm{d}t$ that is measured for the same object L (the conductor moving through the train) from out of the reference system Σ_o. Using Eq. (151a), we discover

$$\frac{\Delta x}{\Delta t} = \frac{\left(\Delta x' + v\,\Delta t'\right)/\sqrt{1 - v^2/c_o^2}}{\left(\Delta t' + v\,\Delta x'/c_o^2\right)/\sqrt{1 - v^2/c_o^2}} = \frac{\Delta x' + v\,\Delta t'}{\Delta t' + v\,\Delta x'/c_o^2}\,,$$

and thus with $u' = \Delta x'/\Delta t'$

$$u = \frac{u' + v}{1 + u'\,v/c_o^2}\,. \qquad \text{EINSTEIN's composition of velocities} \qquad (194)$$

Equation (194) is EINSTEIN's famous composition of velocities that we received for the internal observers, with their own measuring-rods and clocks, inside of our crystal,[2] cf. also Fig. 17.5.

We will repeat the contents of this theorem, because these contents are relevant for the comprehension of the twin paradox: If an object K has the velocity v in the reference system Σ_o, and if the velocity u' is determined for a further object L in a reference system Σ' defined by the object K, then the velocity u that we register for the same object L in the reference system Σ_o must be the calculated by the prescription (194). We would only receive the well-known Eq. (193) for the composition of velocities according to $u \approx u' + v$ if we have the case $|u'\,v| \ll c_o^2$. The velocity $u' = \mathrm{d}x'/\mathrm{d}t'$ of the object L in the reference system Σ', defined by the object K, is in this case a close approximation of the relative velocity w determined in Σ_o. In other words, there is another interpretation of Eq. (193), which describes quite another physical situation. Asuume GALILEI transformation (152). In this case, the composition of velocities u' and v measured in the frames Σ' and Σ_o for an object K yields the velocity u of K measured in Σ_o according to

$$u = u' + v\,. \qquad \text{GALILEI's composition of velocities} \qquad (195)$$

Let us turn to EINSTEIN's composition (194) again. The constancy of signal velocity c_o for every reference system, as we derived in Chap. 12, is a special case. Assume $u' = c_o$ in Σ', then u also becomes, with arbitrary velocity v of Σ' with respect to Σ_o according to (194) $u = \frac{c_o + v}{1 + c_o\,v/c_o^2} = c_o$. Hence, we already knew this case of the

[2]Notice that no condition is assumed for the velocity u' in deriving Eq. (194). Thus, (194) also applies to $c_o < |u'|$. However, in that case the 'object' with velocity u' cannot be identified with a reference system.

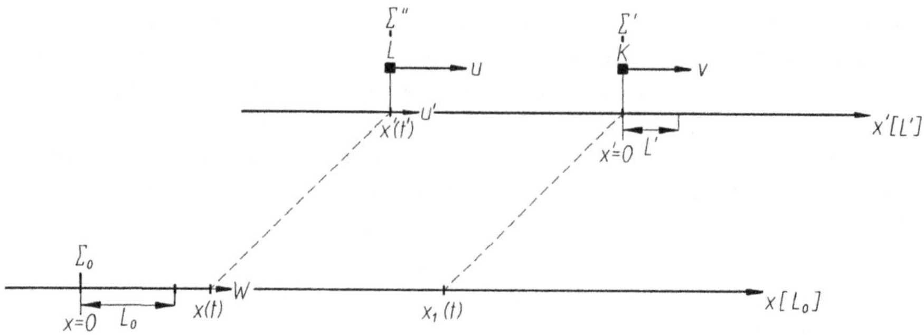

Fig. 17.5 Object K moving with the velocity $v < c_o$ with respect to Σ_o is our reference system Σ'. The observer sitting on object K locates an object L at the positions $x' = x'(t')$, which approaches him with the velocity $u' = \mathrm{d}x'/\mathrm{d}t'$. The object L, which can be identified as reference system Σ'' (if $u' < c_o$), has the velocity $u = \mathrm{d}x/\mathrm{d}t$ in the reference system Σ_o, whereas we measure in Σ_o the velocity $v = \mathrm{d}x_1/\mathrm{d}t$ for the object K (the reference system Σ'). The observer in Σ_o registers that L approaches the object K with the relative velocity $w = u - v$. Generally, this velocity w is completely different to the velocity u', with which, according to the observer in Σ' the object L approaches the object K. We once again calculate using $v = 0,8\,c_o$. We also assume, as in Fig. 17.4, that the object L has a velocity of $u = 0,9\,c_o$ in Σ_o, so that this object, when observed from Σ_o, once again approaches object K with the relative velocity $w = u - v = 0,1\,c_o$. On the other hand, one calculates for the velocity u' according to (194) $u' = \frac{u-v}{1-u\,v/c_o^2} = \frac{0,9\,c_o-0,8\,c_o}{1-0,9\,c_o\cdot 0,8\,c_o/c_o^2} = 0,36\,c_o$. The observer sitting on object K therefore registers that the velocity u', with which the object L approaches him, is 3,6 times that of the relative velocity w observed from Σ_o. Put in other words, the point $x'(t)$ on the x'-axis approaches the point $x' = 0$ with a velocity of $\mathrm{d}x'/\mathrm{d}t' = 0,36\,c_o$; and the point $x(t)$ on the x-axis approaches the point $x_1(t)$ on the x-axis with the velocity $\mathrm{d}x/\mathrm{d}t = 0,1\,c_o$. It is necessary to keep in mind that the observers use their own measuring-rods and clocks that they have with them for their measurements. Dotted lines once again belong to one and the same event

composition of velocities and were thus able to apply this to our thoughts concerning the super-train in the twin paradox.

We also wish to point towards a hasty and false conclusion made by the composition theorem, namely that the signal velocity c_o constitutes the *upper* limit of every 'moving object'. Correct would be to say that an object possessing the velocity $|v| < c_o$ in some reference system would also fulfil this condition in any other reference system. The composition of two velocities $|u'| < c_o$ and $|v| < c_o$ would according to (194) always result in a velocity $|u| < c_o$. Our reference systems are 'objects' with velocities $|V| < c_o$. If we, however, only have *one single* 'moving object' with $|u'| > c_o$ (and taking into account the condition $|V| < c_o$ for reference systems), we would receive for the composed velocity u, according to (194) $u = \frac{u'+V}{1+u'\,V/c_o^2}$, with the result $|u| > c_o$ as one can easily check. Thus, if $|u| > c_o$ applies for any 'object' in a single reference system, then this property is also valid in all other reference systems. For these 'moving objects', the signal velocity c_o appears to be a *lower* limit. The existence of such objects with $|u| > c_o$ causes a fundamental problem to arise, namely that of the breaking of causality. The existence of such objects cannot be completely put out of question by using the principles of Special Relativity. We will deal with the causality problem connected to this in Chaps. 20 and 28–30.

In order to have something specific in front of us for our following considerations of the twin paradox, and in order to keep the equations as simple as possible, we will now proceed to analyse the following case from the viewpoints of all three observers. These would be the observer B_o in Σ_o with the coordinates (x, t), twin A in Σ' with the coordinates (x', t') and finally brother A_o, who starts in the reference system Σ_o and then moves to the reference system Σ'', where the coordinates (x'', t'') are ascertained. The hand settings of brother A_o's personal clock U_o^A will be especially marked using a tilde, $\tilde{t}$. The personal clocks U^A and U_o^A of both brothers have the same setting, zero, at their common point of departure, the common origin of coordinates, our event O. They both agree on instigating a neutral observer, B_o, who observes the complete process from the reference system Σ_o. Twin A finds himself in the reference system Σ' and thus moves with the velocity v with respect to the reference system Σ_o, where his brother A_o is momentarily positioned at $x = 0$. Event T then occurs, as illustrated in Fig. 17.3. Brother A_o changes his reference system to Σ'', which moves with the velocity u with respect to Σ_o. Furthermore, Σ'' possesses the velocity u' with respect to Σ'. This situation is illustrated in Fig. 17.5. We now wish to conditionally determine the velocity u by stating that twin A in his reference system Σ' finds out that: 'Brother A_o is approaching me with the velocity $u' = v$'. Now, applying the composition (194), we see for the velocity $u = \frac{v+v}{1+v\cdot v/c_o^2}$ the expression

$$u = \frac{2v\,c_o^2}{c_o^2 + v^2}\ . \tag{196}$$

Notice that $v < u$.

The following expressions of this velocity u will be needed for later use,

$$\left.\begin{aligned} \sqrt{1-\frac{u^2}{c_o^2}} &= \frac{c_o^2 - v^2}{c_o^2 + v^2}\,, \\ 1-\sqrt{1-\frac{u^2}{c_o^2}} &= \frac{2v^2}{c_o^2 + v^2}\,, \\ u - v &= \frac{c_o^2 - v^2}{c_o^2 + v^2}\, v\,. \end{aligned}\right\} \tag{197}$$

We now have the situation as described by Eqs. (189)–(191) for an arbitrary u with $0 < v < u < c_o$ in front of us. We now choose the velocity $u = \frac{2v\,c_o^2}{c_o^2+v^2}$ for the reference system Σ'' according to (196) and define, in this situation, the event of reunion as $S(v)$ and the time measured in Σ_o that brother A_o spent in Σ'' as Δt_v. By applying Eq. (197), we find for both of the times measured in Σ_o,

$$\left.\begin{aligned} t_T &= \frac{x_P}{v}\,, \\ \Delta t_v &= \frac{x_P}{v}\,\frac{c_o^2 + v^2}{c_o^2 - v^2}\,. \end{aligned}\right\} \tag{198}$$

We receive from the sum of both times the total time measured by observer B_o, $t_{S(v)}$, the time from the departure of the twins, event O, until up to the point of reunion, event $S(v)$,

$$t_{S(v)} = \frac{x_P}{v} + \frac{x_P}{v}\frac{c_o^2 + v^2}{c_o^2 - v^2} = \frac{x_P}{v}\left(1 + \frac{c_o^2 + v^2}{c_o^2 - v^2}\right) = \frac{x_P}{v}\frac{2c_o^2}{c_o^2 - v^2},$$

$$t_{S(v)} = \frac{x_P}{v}\frac{2c_o^2}{c_o^2 - v^2} = \frac{2x_P}{v}\frac{1}{1 - v^2/c_o^2}. \qquad \text{Total time measured by observer } B_o \qquad (199)$$

Twin A moved with the velocity v throughout this whole process. Thus, instead of (191), B_o calculates the setting $t'_{S(v)}$ of A's clock U^A as

$$t'_{S(v)} = t_{S(v)}\sqrt{1 - \frac{v^2}{c_o^2}},$$

$$B_o : t'_{S(v)} = \frac{2x_P}{v}\frac{1}{\sqrt{1 - v^2/c_o^2}}. \qquad \text{Hand setting of the clock } U^A \text{ at the point of reunion} \qquad (200)$$

In order to determine the setting $\tilde{t}_{S(v)}$ of brother A_o's clock U_o^A, observer B_o has only to insert Eq. (197) into (190), so that

$$\tilde{t}_{S(v)} = \frac{x_P}{v} + \frac{x_P}{v}\frac{c_o^2 + v^2}{c_o^2 - v^2}\frac{c_o^2 - v^2}{c_o^2 + v^2},$$

$$B_o : \tilde{t}_{S(v)} = \frac{2x_P}{v}. \qquad \text{Hand setting of the clock } U_o^A \text{ at the point of reunion} \qquad (201)$$

In accord with (192), we find $\tilde{t}_{S(v)} < t'_{S(v)}$ confirmed, namely

$$\frac{\tilde{t}_{S(v)}}{t'_{S(v)}} = \sqrt{1 - \frac{v^2}{c_o^2}} < 1. \qquad (202)$$

Twin A analyses the procedure from his reference system Σ' as follows: Up to event T, his brother A_o moves away from him with the velocity $-v$, from then on he moves back towards him with the velocity $+v$. For event T we found, according to (188), cf. also with Fig. 17.3, the space coordinate $x'_T = \frac{-x_P}{v\sqrt{1-v^2/c_o^2}}$ and the time coordinate $t'_T = \frac{x_P}{\sqrt{1-v^2/c_o^2}}$. Taking the time dilatation of moving clocks into consideration, the hand of the clock U_o^A moves forward on the first half of the journey by $t'_T\sqrt{1 - v^2/c_o^2} = x_P/v$ and on the second part of the journey it moves forward by the same amount, because the distance covered is the same, the only difference being the reverse velocity. Twin A therefore discovers that, at the reunion at $x' = 0$ in Σ', the setting $\tilde{t}_{S(v)}$ of his twin's clock U_o^A during event $S(v)$ has the same coefficient of measure as determined by the observer B_o in Σ_o according to (189),

$$A \,:\, \tilde{t}_{S(v)} = \frac{2x_P}{v} \,, \qquad \text{Hand setting of the clock } U_o^A \text{ at the point of reunion} \tag{203}$$

and thus for his clock U^A,

$$A \,:\, t'_{S(v)} = \frac{2x_P}{v} \frac{1}{\sqrt{1 - v^2/c_o^2}} \,, \qquad \text{Hand setting of the clock } U^A \text{ at the point of reunion} \tag{204}$$

which once again corresponds to the observations made by B_o according to (200).

This means that twin A in Σ' and the observer B_o in Σ_o both come to the same conclusion. At the end of the journey, brother A_o is without doubt the younger. Brother A_o is the one who changed his velocity. He firstly watched his twin A move away with the velocity v and then rushed after him with the higher velocity $u > v$. Mathematically, his clock simply went behind, because of time dilatation and the inequality between the geometric and arithmetic means, as we have already shown deriving (192). This should have cleared up the situation.

However, brother A_o feels that he has been cheated. He does not want to accept the above calculations and presents his own: 'On the first leg of the journey twin A had the velocity v with respect to me. When I stepped on to the train my clock showed $t_T = x_P/v$. His clock should then have had the setting $\frac{x_P}{v}\sqrt{1 - v^2/c_o^2}$. When I sat in the train, twin A once again had a velocity of the amount $|v|$ with respect to me. Thus, if my journey time is once again x_P/v, as stated by twin A, then again, his clock can only have moved forward by $\frac{x_P}{v}\sqrt{1 - v^2/c_o^2}$. Therefore, at the end of the journey my clock would show $2\,x_P/v$ and his $\frac{2\,x_P}{v}\sqrt{1 - v^2/c_o^2}$. This proves that he is the younger and not I, and that his claim to the farm has no face value'.

What should the calculation from brother A_o's position really look like? Here, our own thought processes set up shrewd traps, and we therefore have to proceed with utmost caution.

Brother A_o's clock shows the hand setting of $\tilde{t}_T = t_T = x_P/v$ exactly then when he jumps on to the train in the reference system Σ''. At exactly this Σ_o-time x_P/v, twin A finds himself with his clock U^A at x_P of the reference system Σ_o. This was defined as our event P; see (178) and (179) and also Fig. 17.3 and the upper plot of Fig. 17.2. Twin A's clock U^A shows the hand setting of $t'_P = \frac{x_P}{v}\sqrt{1 - v^2/c_o^2}$ due to the time dilatation of moving clocks. We can therefore state

$$A_o \,:\, \tilde{t}_T = t_T = \frac{x_P}{v} \,, \qquad \text{Hand setting of the clock } U_o^A \text{ at event } T \tag{205}$$

$$A_o \,:\, t'_P = \frac{x_P}{v}\sqrt{1 - \frac{v^2}{c_o^2}} \,. \qquad \text{Hand setting of the clock } U^A \text{ at event } P \tag{206}$$

What is important here is that the events T and P are *only* simultaneous in Σ_o. Brother A_o can only state, as long as he is in the reference system Σ_o that he knows the setting of twin A's clock U^A at the time he boards the train according to (206). As soon as brother A_o is in the reference system Σ''—and this is where the snare is—his statement is incorrect! The statement 'at the time he boards the train' only means when all clocks in Σ_o show $t = x_P/v$, thus also the Σ_o clock at $x = x_P$, only then does A's clock U^A moving past this exact location with the velocity v shows the setting $\frac{x_P}{v}\sqrt{1 - v^2/c_o^2}$.

Let us remind ourselves that: We now have to determine the rule by which the clocks in Σ'' are started. By this, we have *defined* what it means when we state: The clocks in different positions in Σ'' run synchronically. In Chap. 12, we used our elementary principle of relativity as the rule by which we synchronised clocks. Let us take up again our quotation on p. 119. In 1898, seven years before the discovery of EINSTEIN's Special Theory of Relativity, POINCARÉ [;] wrote: 'Il est difficile de séparer le problème qualitative de la simultanéité du problème quantitative de la mesure du temps; soit qu'on se serve d'un chronomètre, soit qu'on ait à tenir compte d'une vitesse de transmission, comme celle de la lumière, car on ne saurait mesurer une pareille vitesse sans *mesurer* un temps.· · · Nous n'avons pas l'intuition directe de la simultanéité, par plus que selle de l'égalité de deux durées. Si nous croyons avoir cette intuition, c'est une illusion".[3]

After changing to the reference system Σ'', brother A_o has the velocity u with respect to Σ_o and the velocity v with respect to Σ'. We will also position a neutral observer B'' in the reference system Σ'', and we equip him with a clock U_t^T. (The other clocks in the reference system Σ'' as well are distinguished with the index 't ', which refers to the 'train'). The observer B'' should be noticed in the reference system Σ_o at the event T, in other words, B'', who has the velocity u in Σ_o, has the coordinate $x_T = 0$ at time $t_T = x_P/v$ in Σ_o.

We still have one free initial condition. In other words, we can pick out any clock in Σ'' and arbitrarily start it. All other clocks then depend on it. We can also freely choose an initial point for the counting of the space coordinates in Σ''. We wish to compare two cases that have equal rights with one another.

1. We firstly determine that: The static clock in U_t^T in Σ'' takes over the hand setting of brother A_o's clock U_o^A at event T. In other words, the hand setting of observer B'''s clock U_t^T is set to $t''_T = t_T = x_P/v$ during event T. The space coordinate of the event T in Σ'' should be $x''_T = 0$. We therefore choose the first case for the initial condition in Σ'' according to:

$$\Sigma'' \; : \; T\left(x''_T = 0,\; t''_T = \frac{x_P}{v}\right) . \qquad \text{Initial condition in } \Sigma'' \text{ First case} \tag{207}$$

[3]It is difficult to separate the qualitative problem of simultaneity from the quantitative problem of the measurement of time; no matter whether a chronometer is used, or whether account must be taken of a velocity of transmission, as that of light, because such a velocity could not be measured without *measuring* a time.· · ·. We have not a direct intuition of simultaneity, nor of the equality of two durations. If we think we have this intuition, this is an illusion'.

We can insert this initial condition into Fig. 17.3; see Fig. 17.6. We still miss the space coordinate x''_P for the event P. We can determine this as follows: A coordinate difference with the amount x_P determined in Σ_o at one and the same time $t_T = t_P$ is the coefficient of measure of a length moving with the velocity u in Σ_o, therefore a static length in Σ'' with the coefficient of measure x''_P. Due to the LORENTZ contraction (113a), $x''_P = \frac{x_P}{\sqrt{1-u^2/c_o^2}}$, hence with the velocity u according to (197)

$$x''_P = x_P \, \frac{c_o^2 + v^2}{c_o^2 - v^2} \ .$$

The coordinate x''_P is the coefficient of measure of the determined distance between the events T and P in Σ'' using the measuring-rod L'' of Σ'', whereby $L'' = L_o \sqrt{1 - u^2/c_o^2} = L_o \frac{c_o^2+v^2}{c_o^2-v^2}$ because of the LORENTZ contraction (112), as shown in Fig. 17.6.

At x''_P we observe the clock U_t^P resting in Σ''. In order to be able to determine the hand setting of the clock U_t^P at event P, the time coordinate P in Σ'', we need to take into consideration that the events T and P occur simultaneously in Σ_o. In Eq. (142a), see also Figs. 12.6 and 12.7, we managed to calculate with the help of our elementary principle of relativity, the number of scale marks $\Delta t'$ that the hand of the clock at the position $x_o + \Delta x$ in Σ_o moving with the velocity v with respect to Σ_o has to be moved forwards or backwards, as against the clock moving with the velocity v at x_o. We use this in our case where u replaces v, as well as $x_o = 0$ and x_P instead of Δx. We now write $\Delta t'' = t''_P - t''_T$ for $\Delta t'$ (our clock U_t^T in Σ'' already shows the different from zero initial position $t_T = t''_T$), thus

$$\Delta t'' = t''_P - t''_T = \frac{-\Delta x \, u/c_o^2}{\sqrt{1 - u^2/c_o^2}} = \frac{-x_P \, u/c_o^2}{\sqrt{1 - u^2/c_o^2}} \ . \tag{208}$$

With (196) and (197) for u, the following is valid

$$\Delta t'' = -\frac{x_P}{v} \, \frac{2v^2}{c_o^2 - v^2} \ . \tag{209}$$

Difference of the clock's U_t^P and U_t^T hand setting in Σ''

We note that the difference $\Delta t'' = t''_P - t''_T$ of the hand settings of both clocks U_t^P and U_t^T is of course independent of the arbitrary choice of initial condition. The initial position does however determine the absolute hand setting. With $t''_T = x_P/v$ according to (207), we discover for the first case of the initial condition

$$t''_P = \frac{x_P}{v} - \frac{x_P}{v} \, \frac{2v^2}{c_o^2 - v^2} = \frac{x_P}{v} \left(1 - \frac{2v^2}{c_o^2 - v^2} \right) .$$

Summarising this, we discover for the event P in Σ''; see also Fig. 17.6,

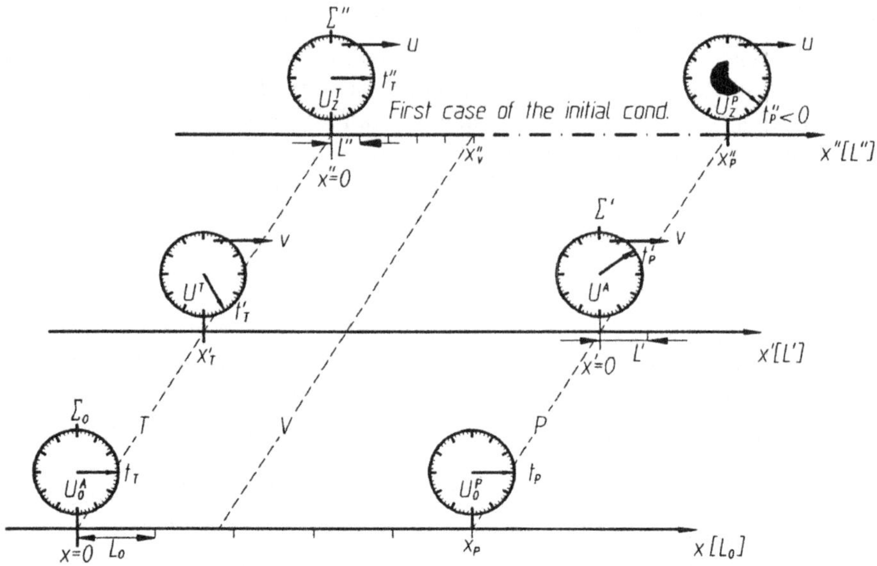

Fig. 17.6 Synchronisation of the clocks in Σ' and Σ'' at the uniform time $t_T = x_P/v$ in Σ_o. We consider the first case according to (207) for the initial condition in Σ''. Brother A_o changes from reference system Σ_o into reference system Σ'' at event T. We once again choose the velocity $v = 0,8\,c_o$, therefore $\gamma = 0,6$ for Σ' with respect to Σ_o. Using (196), we then receive the value $u = \frac{2\cdot 0,8\,c_o^2}{(1+0,64)\,c_o}$ for the velocity of u of Σ'' with respect to Σ_o. According to the initial condition (207) in Σ'', the observer B'''s clock in Σ'', U_t^T takes the same hand setting of brother A_o's clock U_o^A, $t''_T = t_T = x_P/v = 15$ scale parts at event T. Moreover, it is set $x''_T = x_T = 0$. Supplementary to Fig. 17.3, the synchronisation is plotted of both clocks U_t^T and U_t^P stationary in Σ'' at the uniform time $t_T = x_P/v$ in Σ_o. The coordinates in Σ'' for the event P (simultaneous with event T only in Σ_o) are determined in (210). This results in the value $t''_P = \frac{x_P}{v}\,(1 - \frac{2\cdot 0,64\,c_o^2}{c_o^2 - 0,64\,c_o^2}) = -38,3$ scale parts for the hand setting of the clock U_t^P. We take the hand setting $t'_T = 25$ scale parts from Fig. 17.3. Decisive for the paradox is the negative sign. Brother A_o, as long as he remains in the reference system Σ_o, determines according to (206) the value $t'_P = \frac{x_P}{v}\,\gamma$ for the hand setting of twin A's clock U^A for time $t_P = t_T$. However, this hand setting refers to a point of time long before event T in Σ'', namely $t''_P = -38,3$. In order to determine the hand setting of twin A's clock U^A at event T in Σ'', brother A_o must first calculate the position x''_V that twin A's clock U^A occupies directly opposite of a clock stationary in Σ'' showing the time $t''_V = t''_T = x_P/v$. This position is calculated in (211), and the result is $x''_V = x_P$ scale parts. In our representation, we have once again chosen the coefficient of measure $x_P = 5$. On the x''-axis, this would be the point with the distance $x_P\,L''$ from $x'' = 0$. In other words, we have to calculate the synchronisation of the clocks in Σ' for a uniform time in Σ'', let us say $t''_T = x_P/v$. This is once again shown in Fig. 17.8. It is obvious that the paradox arises if one does not observe that the hand of twin A's clock U^A continues moving forward between the events P and V, if one does not therefore observe that the position of the clock U^A at event T in Σ_o also belongs to event P, however in Σ'' to event V. We have marked the length that brother A_o ignored on the x''-axis in Fig. 17.6 using dots. This length is responsible for the occurrence of the paradox during the motion of twin A's clock U^A

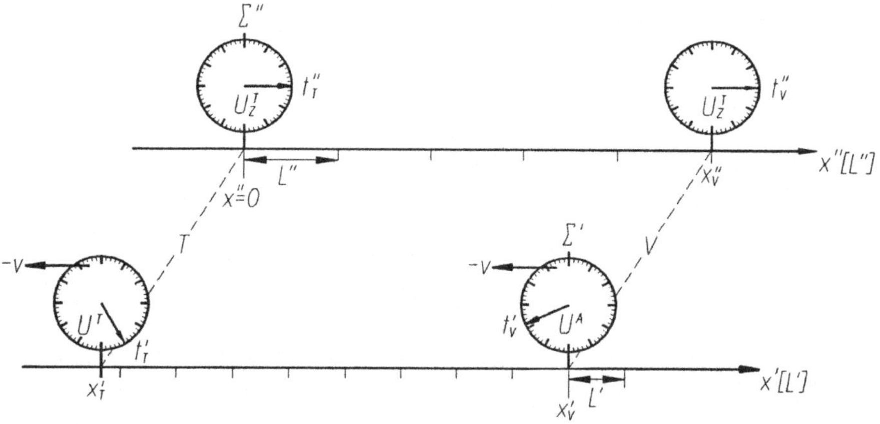

Fig. 17.7 Synchronisation of the clocks in Σ' and Σ'' at the uniform time $t_T = x_P/v$ in Σ_o, for the second case of our initial condition according to (219). The difference between the situation here and in Fig. 17.6 for the first case of the initial condition is that all the clocks in Σ'' were moved forward by $-\Delta t'' = \frac{x_P}{v}\frac{2v^2}{c_o^2-v^2}$ and that the value $x_P'' = x_P\frac{c_o^2+v^2}{c_o^2-v^2}$ was subtracted from all space coordinates in Σ''. We once again calculate using $v = 0,8\,c_o$, $\gamma = 0,6$ and gauge all clocks so that x_P/v shows 15 scale parts. Thus, the hand of the clock U_t^P is at $t_P'' = 15$ scale parts, whilst the hand of the clock U_t^T is at $t_T'' = \frac{x_P}{v}\frac{c_o^2+v^2}{c_o^2-v^2} = 15\,\frac{1+0,64}{1-0,64} = 68,3$ scale parts according to (220). The hand of this clock has therefore completed one whole cycle. We show this on the clock U_t^T by adding a **1** in the display. All other specifications are identical to those in Fig. 17.6

$$\Sigma'' : \quad P\left(x_P'' = x_P\,\frac{c_o^2+v^2}{c_o^2-v^2}\,,\; t_P'' = \frac{x_P}{v}\left(1-\frac{2v^2}{c_o^2-v^2}\right)\right). \qquad \text{First case} \tag{210}$$

The time display $\tilde{t}$ of brother A_o's clock U_o^A is now identical to the display of the clock U_t^T, because of the initial condition (207). In order to be able to determine the journey time $\tilde{t}_v$ from the point of entering the train to the point of reunion from brother A_o's view, we must determine the position of twin A in Σ'' at time $t_T'' = x_P/v$. We will call this event V. Twin A has the velocity $-v$ in Σ''. We know his position x_P'' at time t_P'' according to (208). Hence, at time $t_T'' = x_P/v$, he is located at $x_V'' = x_P'' - v\,(t_T'' - t_P'') = x_P'' - v\,\frac{x_P}{v}\frac{2v^2}{c_o^2-v^2}$ in Σ'' and thus

$$x_V'' = x_P\left(\frac{c_o^2+v^2}{c_o^2-v^2} - \frac{2v^2}{c_o^2-v^2}\right).$$

We therefore receive, for the coefficient of measure x_V'' of the position of twin A in Σ'', the value

$$x_V'' = x_P \qquad \begin{array}{l}\text{Space coordinate of twin } A \text{ in } \Sigma''\\ \text{at the point of time } t_T'' = x_P/v\\ \text{First case}\end{array} \tag{211}$$

or, respectively, for the coordinates of event V

$$\Sigma'' : \quad V\left(x''_V = x_P,\ t''_V = \frac{x_P}{v}\right). \qquad \text{First case} \qquad (212)$$

Due to the velocity $-v$ of twin A with respect to brother A_o, the hand of brother A_o's clock U_o^A moves forward by the value $\Delta\tilde{t}_v$ when he steps on to the train,

$$\Delta\tilde{t}_v = \frac{x_P}{v}. \qquad \text{At the clock } U_o^A \text{ passed time up to the point of reunion} \qquad (213)$$

This results, according to brother A_o, together with (205) in a hand setting $\tilde{t}_{S(v)} = \tilde{t}_T + \Delta\tilde{t}_v$ for U_o^A at event $S(v)$, the reunion of the brothers, thus,

$$A_o : \tilde{t}_{S(v)} = 2\,\frac{x_P}{v}, \qquad \text{Hand setting of the clock } U_o^A \text{ at the point of reunion} \qquad (214)$$

corresponding to the results of not only observer B_o according to (201), but also twin A according to (203).

How does brother A_o evaluate the passed time of twin A's clock U^A?

Due to time dilatation, the hand of twin A's clock U^A moves forward by $\Delta t'_v$ during time $\Delta\tilde{t}_v = x_P/v$ according to

$$\Delta t'_v = \frac{x_P}{v}\sqrt{1-\frac{v^2}{c_o^2}}\,. \qquad (215)$$

This is therefore the duration of brother A_o with his clock U_o^A in the reference system Σ'' as measured from the reference system Σ'.

What is however the position of clock's U^A hand at the beginning of this part of the journey, in other words at event V which has the coordinates $x''_V = x_P$ and $t''_V = x_P/v$ in Σ''? The paradox arises if one simply uses the statement (206) for this hand setting that brother A_o made before he jumped on to the train. This statement however is incorrect, because the events P and V must not be confused! We need the hand setting of the clock U^A at event V! If we used the hand setting of U^A at event P, according to (206), then we would, together with $\Delta t'_v$, receive a journey time of $\frac{2x_P}{v}\sqrt{1-v^2/c_o^2}$, resulting in the fact that twin A would be the younger. This calculation is not allowable, as we have now seen, and the also result contradicts the observations made by twin A on his own clock, according to (204), as well as those made by observer B_o, according to (200).

In Σ'', between the events P and T (event T in Σ'' being simultaneous with event V) the following time passes, according to (209)

$$t''_{PT} := t''_T - t''_P = t''_V - t''_P := t''_{PV} = -\Delta t'' = \frac{x_P}{v}\,\frac{2v^2}{c_o^2 - v^2}\,. \qquad (216)$$

Due to time dilatation of the clock U^A moving with the velocity v with respect to Σ'', the hand of the clock moves forward by $t'_{PV} = t''_{PV}\sqrt{1 - v^2/c_o^2}$, so that they take up the position $t'_V = t'_P + t'_{PV}$ at event V, thus with t'_P, according to (206),

$$
\begin{aligned}
t'_V &= \frac{x_P}{v}\sqrt{1 - \frac{v^2}{c_o^2}} + \frac{x_P}{v}\frac{2v^2}{c_o^2 - v^2}\sqrt{1 - \frac{v^2}{c_o^2}} = \frac{x_P}{v}\sqrt{1 - \frac{v^2}{c_o^2}}\left(1 + \frac{2v^2}{c_o^2 - v^2}\right) ,\\
&= \frac{x_P}{v}\sqrt{1 - \frac{v^2}{c_o^2}\frac{c_o^2 + v^2}{c_o^2 - v^2}} = \frac{x_P}{v}\sqrt{1 - \frac{v^2}{c_o^2}\frac{1 + v^2/c_o^2}{1 - v^2/c_o^2}} = \frac{x_P}{v}\frac{1 + v^2/c_o^2}{\sqrt{1 - v^2/c_o^2}} .
\end{aligned}
$$

This is therefore the hand setting of twin A's clock U^A at event T in Σ'', as registered by brother A_o after transferring into this reference system; see also Fig. 17.8,

$$A_o \,:\, t'_V = \frac{x_P}{v}\frac{1 + v^2/c_o^2}{\sqrt{1 - v^2/c_o^2}} . \qquad \text{Hand setting of the clock } U^A \text{ at the event } T \text{ in } \Sigma'' \tag{217}$$

Up to the reunion of the brothers at event $S(v)$, the hand of brother A_o's clock U_o^A moved forward, according to (213) by $\Delta\tilde{t}_v = x_P/v$ scale marks. The hand of the clock U^A therefore only moved forward, due to time dilatation, by $\Delta t'_v = \Delta\tilde{t}_v\sqrt{1 - v^2/c_o^2}$ and takes up the position at event $S(v)$

$$
\begin{aligned}
t'_{S(v)} &= \frac{x_P}{v}\frac{1 + v^2/c_o^2}{\sqrt{1 - v^2/c_o^2}} + \frac{x_P}{v}\sqrt{1 - \frac{v^2}{c_o^2}}\\
&= \frac{x_P}{v}\frac{1}{\sqrt{1 - v^2/c_o^2}}\left(1 + \frac{v^2}{c_o^2} + \sqrt{1 - \frac{v^2}{c_o^2}}\sqrt{1 - \frac{v^2}{c_o^2}}\right) = \frac{x_P}{v}\frac{2}{\sqrt{1 - v^2/c_o^2}} .
\end{aligned}
$$

Brother A_o in fact observes

$$A_o \,:\, t'_{S(v)} = \frac{2x_P/v}{\sqrt{1 - v^2/c_o^2}} , \qquad \text{Hand setting of the clock } U^A \text{ at the point of reunion} \tag{218}$$

which is completely in accordance with the statement (204) of twin A, as well as the statement (200) made by observer B_o. Only the hand of brother A_o's personal clock U_o^A goes behind that of twin A's personal clock U^A, just as we awaited and expected.

The paradox is there clarified. We now also want to see exactly where the trap is set for a second case of the initial condition of space and time measurement in the reference system Σ''. In order to enable brother A_o, even after he has changed reference systems, to measure the passed time on twin A's clock U^A, we choose for this second case the following initial condition:

2. The clock U_t^P resting in Σ'' takes over the hand setting of the clock U_o^A at event P. This means that the hand of the clock U_t^P is set to $t''_P = t_P = x_P/v$ at event P. We also set the initial point for the counting of the space coordinates in Σ'' to $x''_P = 0$;

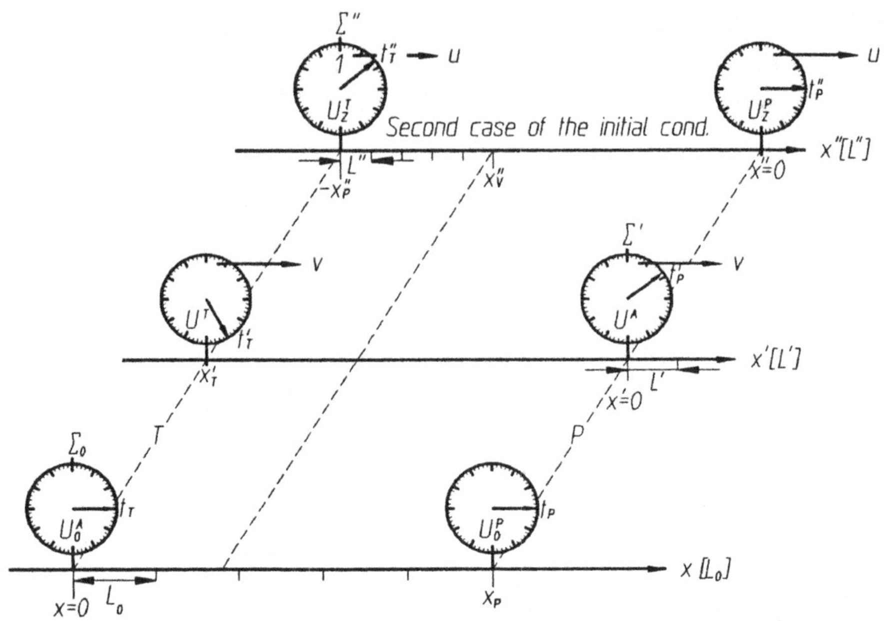

Fig. 17.8 Synchronisation of the clocks in the reference system Σ' from the viewpoint of an observer stationary in Σ'' at the uniform time $t''_T = t''_V = x_P/v$ for both events T and V. The reference system Σ' has the velocity $-v$ with respect to Σ''. We once again calculate using $v = 0,8\,c_o$, so that $\gamma = 0,6$. The clocks in the reference system Σ'' read $t''_T = x_P/v = 15$ scale parts. Due to the relativity of the LORENTZ contraction, the observer stationary in Σ'' notices that the measuring-rod L' is shorter than L'' according to $L' = \gamma\, L''$; see also Fig.15.1. (We have chosen the length L_o from Fig. 17.6 for the arbitrary length L'' in our picture). We adopt the hand setting of the clock U^T from Fig. 17.3 according to $t'_T = t_T/\gamma = 15/0,6 = 25$ scale parts. In (211), we found $x''_V = x_P$. We apply Eq. (142a) for the synchronisation of two clocks in the reference system Σ'' by replacing Δx with $x''_V = x_P$ and v with $-v$ and receive the coefficient of measure $\Delta t' := t'_V - t'_T$ which we must add to t'_T in order to receive the hand setting t'_V of twin A's clock U^A. With $\Delta t' = \frac{x_P\, v/c_o^2}{\gamma} = \frac{x_P}{v}\,\frac{v^2/c_o^2}{\gamma}$, $t'_V = \frac{x_P}{v}\,\frac{1+v^2/c_o^2}{\gamma}$ follows as we have already seen using a different approach in (217). Using this in our example results in the hand setting for the clock U^A at event V of $t'_V = 15\,\frac{1+0,64}{0,6} = 41$ scale parts. We still explain the location of the point x'_V. The coordinate x''_V is the coefficient of measure of a moving length in Σ'', whose end points are defined in Σ' by x'_T and x'_V, so that $\Delta x'_V = x'_V - x'_T = x''_V/\gamma$ and thus $x''_V\, L'' = \Delta x'_V\, L'$ as plotted in the picture

see Fig. 17.7. We therefore consider for our free initial condition in Σ''

$$\Sigma'' : \quad P\left(x''_P = 0,\; t''_P = \frac{x_P}{v}\right) . \qquad \text{Initial condition in } \Sigma'' \text{ Second case} \tag{219}$$

In this case, brother A_o can rightly state after exchanging reference systems that he knows the hand setting of twin A's clock U^A as measured at event P in the reference system Σ''. This reading concurs per definition $t'_P = \frac{x_P}{v}\sqrt{1 - v^2/c_o^2}$. However, this is not the hand setting of observer B'''s clock U_t^T in Σ'', to whom he has just moved to and according to whose clock he has just set his. We will come to this in a moment.

All differences in the coordinates for space and time measurements in Σ'' remain unchanged, if the initial point is changed. The difference between the hand settings $\Delta t'' = t''_P - t''_T = -\frac{x_P}{v}\frac{2v^2}{c_o^2 - v^2}$, according to (209), also remains unchanged, just as the difference between the space coordinates of the events T and P in Σ''.

For $\Delta x''$ we have once again according to (210) $\Delta x'' = x''_P - x''_T = x_P\,\frac{c_o^2+v^2}{c_o^2-v^2}$. We have displaced the initial point for counting the coordinates in Σ'', according to (219) by x''_P to the right. For event T, we therefore receive in the second case the coordinates in Σ'',

$$\Sigma'' : \quad T\left(x''_T = x_P\,\frac{c_o^2+v^2}{c_o^2-v^2},\; t''_T = \frac{x_P}{v}\,\frac{c_o^2+v^2}{c_o^2-v^2}\right). \qquad \text{Second case} \tag{220}$$

And how does this appear in the case concerning the twins and their clocks? Let us start with twin A. His clock U^A moves towards the observer B'' (where brother A_o is also situated) with the velocity $-v$. Up to the reunion, event $S(v)$, the time $t''_{PS} = \Delta x''/v$ passes between the events P and $S(v)$ in Σ'', thus

$$t''_{PS} = \frac{x_P}{v}\,\frac{c_o^2+v^2}{c_o^2-v^2}. \qquad \text{Period of time between } P \text{ and } S(v) \text{ in } \Sigma'' \tag{221}$$

The clock U^A of twin A moving with the velocity $-v$ with respect to Σ'' underlies time dilatation. Its hand only moves forward by $t'_{PS} = t''_{PS}\sqrt{1 - v^2/c_o^2}$ and thus takes up the position at event $S(v)$ of

$$\begin{aligned} t'_{S(v)} = t'_P + t'_{PS} &= \frac{x_P}{v}\sqrt{1 - \frac{v^2}{c_o^2}} + \frac{x_P}{v}\,\frac{c_o^2+v^2}{c_o^2-v^2}\sqrt{1 - \frac{v^2}{c_o^2}} \\ &= \frac{x_P}{v}\sqrt{1 - \frac{v^2}{c_o^2}}\left(1 + \frac{c_o^2+v^2}{c_o^2-v^2}\right) = \frac{x_P}{v}\,\frac{2\sqrt{1 - v^2/c_o^2}}{1 - v^2/c_o^2}\,, \end{aligned}$$

and thus brother A_o discovers, as in the first case, according to (218)

$$A_o : t'_{S(v)} = \frac{2x_P/v}{\sqrt{1 - v^2/c_o^2}}. \qquad \text{Hand setting of the clock } U^A \text{ at the point of reunion} \tag{218}$$

Here, we also note the coordinates of the event $S(v)$ in the reference system Σ',

$$\Sigma' : \quad S(v)\left(x'_{S(v)} = 0, \quad t'_{S(v)} = \frac{2x_P/v}{\sqrt{1 - v^2/c_o^2}}\right). \tag{222}$$

Let us now consider brother A_o's clock U_o^A.

The paradox occurs, when brother A_o states, because he calculated this hand setting (219) for twin A, that his clock shows the hand setting $\breve{t}_{S(v)} = t_P + t'_{PS}$, namely the time $t_P = x_P/v$ during his stay in the reference system Σ_o, increased by the above calculated time in Σ'', namely $t''_{PS} = \frac{x_P}{v}\frac{c_o^2+v^2}{c_o^2-v^2}$, so that

$$\breve{t}_{S(v)} = \frac{x_P}{v} + \frac{x_P}{v}\frac{c_o^2+v^2}{c_o^2-v^2} = \frac{x_P}{v}\left(1+\frac{c_o^2+v^2}{c_o^2-v^2}\right) = \frac{x_P}{v}\frac{2c_o^2}{c_o^2-v^2},$$

thus

$$\breve{t}_{S(v)} = \frac{2x_P/v}{1-v^2/c_o^2} = \frac{t'_{S(v)}}{\sqrt{1-v^2/c_o^2}}. \tag{223}$$

According to this equation, brother A_o would obviously be the elder, since it is without doubt $\breve{t}_{S(v)} > t'_{s(v)}$. And brother A_o now states that the hand of twin A's personal clock U^A goes behind the one of his clock U_o^A, which everybody can easily check on. Thus he, brother A_o, is the elder, and the farm belongs to him and not vice versa. This is however correct if the hands of brother A_o's clock U_o^A moved forward by $\breve{t}_{S(v)}$ during the whole journey completely by itself, without the help of anybody. And here, brother A_o knows that he is not telling the complete truth. Correct is that his clock in the reference system Σ_o shows how much time has passed, $t_P = x_P/v$. In the reference system Σ_o, and only there, the events T and P are simultaneous. In other words, during the event of transferring reference systems T, thus shortly before and shortly afterwards, brother A_o's clock U_o^A has the setting $t_T = t_P = x_P/v$. He therefore arrives in the reference system Σ'' with this hand setting. And then he has brought the hand setting of his clock U_o^A, according to (220) to exactly the same hand setting t''_T of the clock U_t^T resting in Σ'', in other words, he has secretly moved the hand of his clock forward by the value $|\Delta t''|$. This is the value calculated in (209) that the clock U_t^T has to be moved forward with respect to U_t^P, so that both clocks run synchronically in Σ''. The events T and P are only simultaneous in Σ_o; see Fig. 17.6 and Fig. 17.7. The hand of brother A_o's clock U_o^A has therefore in fact not moved forward by the value $-\Delta t'' = \frac{x_P}{v}\frac{2v^2}{c_o^2-v^2}$, and he has therefore not aged by the value $|\Delta t''|$. Brother A_o's duration of stay in the reference system Σ'' is not the time t''_{PS}. From the viewpoint of the reference system Σ'', brother A_o had not just changed reference system at event P. This occurred at event T. And these two events show, according to (216), the difference in time $t''_{PT} = -\Delta t''$ in Σ''. Brother A_o's duration of stay in Σ'' is thus only

$$\Delta\tilde{t}_v \equiv \Delta t''_v = t''_{PS} - t''_{PT} = t''_{PS} + \Delta t'' = \frac{x_P}{v}\frac{c_o^2+v^2}{c_o^2-v^2} - \frac{x_P}{v}\frac{2v^2}{c_o^2-v^2} = \frac{x_P}{v},$$

as we already discovered in (213). This equation also directly results from time dilatation of brother A_o's clock U_o^A moving with the velocity $u = \frac{2v\,c_o^2}{c_o^2-v^2}$ in Σ'' with respect to Σ_o, for which we calculated, according to (198) the duration of stay in Σ_o according to $\Delta t_v = \frac{x_P}{v} \frac{c_o^2+v^2}{c_o^2-v^2}$. Applying (197), it follows immediately $\Delta \tilde{t}_v = \Delta t_v \cdot \sqrt{1-u^2/c_o^2}$.

It thus remains as it was, during the complete journey the hand of brother A_o's, the one who rushes back towards twin A, clock U_o^A only moves forwards by $\tilde{t}_{S(v)} = 2x_P/v$,

$$A_o : \tilde{t}_{S(v)} = 2\,\frac{x_P}{v}\,. \qquad \text{Hand setting of the clock } U_o^A \text{ at the point of reunion} \tag{214}$$

It therefore goes behind twin A's clock U^A by the factor $\sqrt{1-v^2/c_o^2}$, which shows $t'_{S(v)} = \frac{2x_P/v}{\sqrt{1-v^2/c_o^2}}$ at the reunion.

The returning brother, or the brother who rushes after the other respectively is the younger.

Brother A_o had manipulated his clock. He has been unmasked. The paradox has been solved; see Fig. 17.9.

The geometric versified reader will anticipate that these connections virtually demand for geometric description. For this, we especially refer the reader to LIEBSCHER's [] presentation of Special Relativity.

It is fascinating to imagine how all these processes occur inside of a crystal, how the oscillating breathers move through the crystal with varying velocities and frequencies, how various breather groups with the same velocity can be formed into a reference system, how these are synchronised with respect to one another, such that the time displayed in one system behaves exactly as shown in Fig.12.7, to that of another breather system moving through the crystal with a different velocity. The end result is that every single phase of the twin paradox can be illustrated using an oscillating breather moving through the crystal.[4]

We outside experimenters have a privileged position with respect to that of the internal observers, because we can determine the motion state of the crystal as a whole by using our outside simultaneity, as defined by the speed of light. Although our 'outside clocks' and 'outside measuring-rods', the clocks and measuring-rods that every physicist uses in his measurements, underlie the laws of time dilatation and length contraction, we use the speed of light c_L, whereas the internal observer uses the critical velocity c_o. As explained in Chap. 11, the physically identical clocks, which only move relatively with respect to one another, used by the internal observers inside of the crystal are from the point of view of the outside observers not physically identical clocks. Therefore, for the outside observer, the whole complicated and strenuous internal procedure of synchronising the clocks seems to be most strange and at first unnecessary. However, the internal time processes registered by the internal

[4]Notice that all formulas for the coordinates of the various events in different frames Σ_o, Σ', Σ'', ... directly can be controlled with the help of LORENTZ transformation (151).

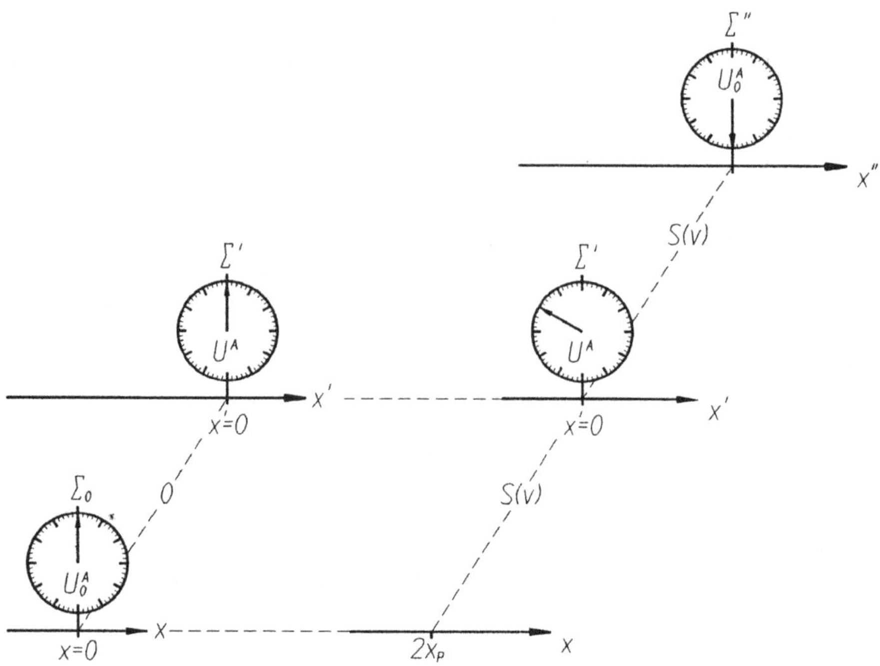

Fig. 17.9 End of the story of the twins. The twins meet at event $S(v)$. There departure occurred at event O. We once again calculate using $v = 0,8\,c_o$, $\gamma = 0,6$ and gauge all clocks according to $x_P/v = 15$ scale parts. The reference system Σ' has the velocity v with respect to Σ_o, and the reference system Σ'' has the velocity v with respect to Σ'. The hand of twin A's personal clock U^A reads according to (222), as we discovered many times over, $t'_{S(v)} = 2 \cdot 15/0,6 = 50$ scale parts at event $S(v)$, whilst the hand of brother A_o's personal clock U_o^A goes behind, as we have also discovered many times over, by the factor of γ and only reads, according to (214) $\tilde{t}_{S(v)} = 30$ scale parts, assuming of course that brother A_o did not secretly reset his clock (as he did in our story when he changed reference systems)

observer are alone responsible for the dynamic processes in the interior of a crystal. The energy E of a kink q^I moving with the velocity v,

$$q^I = q^I(x,t) = \frac{2\pi}{a} \arctan \exp\left[\frac{\pi(x - vt)}{L_o\sqrt{1 - v^2/c_o^2}}\right],$$

has as an example nothing to do with the speed of light c_L, however this energy increases with the approximation of the parameter v to the critical velocity c_o of the sine-Gordon equation. We will show for this energy in Chap. 23, Eq. (297) on p. 251 that $E = \frac{E_o}{\sqrt{1-v^2/c_o^2}}$, where E_o represents the energy of the kink at rest; see also Seeger [].

From the dependence of an object's energy to its velocity, where for our 'outside objects' the speed of light is valid, Einstein [;], three months after his famous paper founding the Special Theory of Relativity, came to the conclusion of the inertia of energy, a result known to virtually everyone as the famous equation $E = m\,c^2$ (cf. also Chap. 3). We will discuss the connected processes, including those less spectacular processes in our solid objects, based on the foundations of our general equation of motion for dislocations (79) in Chap. 23.

The simple reason that the effects, being plastic deformations, resulting from the energy mass equivalence inside of a solid are so relatively harmless, is that the sound velocity, or our critical velocity c_o of the sine-GORDON equation is, at least with respect to the accessible material atomic lattices, minute in comparison with the speed of light c_L. However, under different physical conditions there may be other types of atomic lattices—and such atomic lattices are indeed favoured by astrophysicists—where the difference between the critical velocity c_o and the speed of light c_L is only very small, therefore, taking EINSTEIN's Special Theory of Relativity into consideration, $c_o < c_L$, but $c_o \approx c_L$. In such an atomic lattice, but not in our everyday earthly crystals, where the whole processes are harmless, the mutual destruction of two dislocations with opposite signs would lead to an inferno, which when compared to nuclear fusion would make the latter look like a little blaze. The enormous dimension of such a process simply lies in the explanation that dislocations traverse a very large part of the crystal, in other words their space, whereas our elementary particles are confined to space points. A dislocation mass therefore exceeds the masses of the point particles by many powers of ten.

We should be happy that the sound velocity is so small. On the other hand, it would be attractive to consider a crystal with a critical velocity only slightly less that of speed of light. Here, we must of course admit that when considering such possibilities, we overstep the limits of our self-restricted calculations and now only argue qualitatively, whereby we inevitably accept incalculabilities and uncertainties concerning the conclusions. For atoms moving inside of the lattice with a velocity comparable to that of light c_L, their inertia m would no longer be a constant, a constant that we took as granted, but according to Special Relativity (see Chap. 3) would be dependent on its own velocity. Our accepted assumptions (68) of linear theory of elasticity could no longer be upheld. However, even in this case we would qualitatively await such states of excitation of a lattice which would be very similar to the solutions of the wave equation and of the sine-GORDON equation discussed by us, which originally brought us into a position where we were able to discuss the sound velocity c_T and the critical velocity c_o for plastic deformations, and subsequently comparing these to the speed of light c_L. If we however assume a critical velocity $c_o \approx c_L$, $c_o < c_L$, then the clocks of the internal observer moving with the velocity v inside of the crystal, when observed from outside, *almost* go behind by the speed of light, thus they behave *almost* like the outside observer's clocks, as Special Relativity demands. Yet those breather clocks moving inside of the crystal and those resting relative to the crystal are, when observed from outside, physically completely different clocks, and their difference in pace therefore cannot result in a conclusion being made by an outside observer, on the behaviour of a moving clock since outside Special Relativity is not applicable in this case, as discussed in Chap. 11. The numerical value of the sound velocity is for this set of circumstances of no importance.

However, what about the dynamic consequences? Here, the numerical value of the sound velocity (or the equivalent critical velocity c_o) figures in the energy equation as a square factor; see Chap. 23. The sound velocity in ice for example is around $3 \cdot 10^5$ m/s, and is therefore smaller than that of light by the factor 10^5. For the corresponding energy, a factor of 10^{10} arises. Whilst the destruction of dislocations is normally an energetically harmless procedure (at the most an ice crystal would melt), including the factor 10^{10} would lead to a far more infernic reaction.

Chapter 18
The DOPPLER Effect

In every advanced course on physics the DOPPLER, effect is usually the standard example for demonstrating the fundamental physical differences between the propagation of elastic waves through a mechanical medium and the propagation of electromagnetic waves through our physical space. Compared to this MACH's quotation in front of the preface, 'Light somewhat like sound. Sound somewhat like light.', looks like a relict from the last century. And yet we will illustrate, with the help of the DOPPLER effect, in what sense this MACHian thesis proves true without actually coming into conflict with our common knowledge of the DOPPLER effect, as long as we agree to incorporate the means with which we gained our knowledge into our considerations. This is a concession that we owe the philosopher ERNST MACH.

What is the DOPPLER effect?—A fire engine rushes towards us. At the moment when the vehicle passes by, the pitch of the signal horn falls to lower frequencies. If we question the driver of this fire engine, he would assure us that the signal did not change at all.

When experimenting with light, we can identify the chemical nature of a light source by spectrally splitting light using a prism, for example, the two yellow lines lying together very close that are unmistakably emitted by excited sodium atoms. By measuring the wave length λ of this light, we automatically determine its frequency due to $c_L = \lambda\,\nu$. For the yellow sodium line, we get $\lambda_{Na} = 5890\,\text{Å} = 58,9 \times 10^{-8}$ m and thus $\nu_{Na} = 5,09 \times 10^{14}\,\text{Hz}$. The light we receive from stars far away, from stars in other galaxies we notice that all spectral lines are displaced into the long-wave red area. Due to the mutual positions of the lines, all light sources, e.g. the excited sodium atoms, can be exactly identified; however, the absolute frequencies that we measure are all slightly lower. The explanation for this lies in the enormous velocities with which the galaxies and thus the stars in them move away from us.[1] The diminished frequency of the light we receive is analogous to the deepening tone of the fire engine's horn moving away from us. Such a change in the frequency of

[1]The law of the expansion of the universe discovered in 1929 by E.P. HUBBLE states that this recession velocity increases proportionally to the distance between the star systems.

H. Günther, *Elementary Approach to Special Relativity*,
https://doi.org/10.1007/978-981-15-3168-2_18

light can be illustrated in a laboratory. If we heat up a gas of sodium atoms, for example, the radiating atoms receive an increase in thermal velocity, with which the atoms alternatively move towards and away from us. If they move towards us, the frequency of the light we receive increases; if the atoms move away, the frequency decreases just as with the fire engine. Due to the rapid changing velocities, we cannot determine the singular changing frequencies, we can only determine the sum of the changes—as if we had a squadron of fire engines on an airfield moving back and forth with varying velocities, whose audible signals we register. We therefore register the sum of all sodium lines displaced to the left or to the right, in other words a thick line which becomes even thicker the more a gas is heated up.

This is the DOPPLER effect: An emitter sends out a harmonic wave, in other words, a sine-shaped oscillation of a certain constant ν_o frequency determined by the construction of the emitter. The experimenter, who is resting with respect to this emitter, therefore measures just this frequency ν_o. If however the observer and the emitter move towards one another, then the observer registers a higher frequency; if both of them move away from each other, then the observer register a lower frequency.

The DOPPLER effect occurs independent of the nature of the waves that are sent out and observed. It is of no difference if the waves are sound waves in water or in air or even light waves. We can always register the DOPPLER effect if the emitter and observer move relatively to one another. Do the same equations apply to all these DOPPLER effects? Aren't sound and light waves something completely different? Doesn't the former just cause a mechanical medium to oscillate, and these oscillations then propagates towards us, whereas we receive light as factual messengers from distant stars?

Let us begin with the DOPPLER effect as it was originally discovered by CHRISTIAN DOPPLER in 1842, who has already completely described it. It is remarkable that C. DOPPLER originally predicted this effect for light. Not until three years later was this effect tested in acoustics. (The fast fire engines that today incite us to contemplate on the change of frequency did not exist in those days). Let us therefore begin with a description, a description that could have been made in 1842, and a description that can be found in physics course books today. We will start with the less complicated acoustic DOPPLER effect out of grounds of simplicity.

In order to be able to measure something, we need three things. Firstly the source, let us say, a standard emitter S. This emitter, a tuning fork, is able, due to its construction regulations, to oscillate. Its two arms are able to oscillate 440 times per second when knocked. We now also need a medium that is also in a position to oscillate in response to the oscillating arms of the tuning fork. In other words, the tuning fork excites a wave in the medium, which then moves through the medium with the characteristic wave velocity for exactly this medium. A medium that can oscillate with any arbitrary frequency would be the best choice, so that the medium itself would not be the cause for any frequency distorsions. In Chap. 4, we described that the mechanical continuum, the limiting case of the linear chain, is just such a medium. Here, waves of any frequency can propagate dispersion free, in other words with a velocity independent of frequency, for example with the transversal sound velocity c_T. Our mechanical continuum is therefore ideal for our purposes.

The third and last point that needs to be fulfilled is that we need an observer with an instrument, a sensitive tuning fork that oscillates in response to the oscillations of the medium. For example, if one places a tuning fork on a piano, the fork begins to oscillate. This is the principle of a resonator. Now every resonator can do more than oscillate at a single frequency, but the oscillations made at its eigen frequency are the most strongest. Due to the fact that the observer cannot know the frequency of the arriving oscillation, he sets up a large number of resonators. The resonator with its eigen frequency that oscillates the most under the impression of the arriving wave therefore gives us the frequency of the incoming wave.

We will describe this experiment schematically. The membrane of the standard emitter oscillates harmonically with a period of oscillation T_o, i.e. with frequency $\nu_o = 1/T_o$. The observer S sitting on top of the emitter will therefore register that his membrane strikes the surrounding medium Δn_o times during the time Δt_o,

$$\nu_o = \frac{1}{T_o} = \frac{\Delta n_o}{\Delta t_o}. \qquad \text{Frequency of the emitter } S \qquad (224)$$

This produces a wave of the frequency $\tilde{\nu}$ in the medium. This waves moves through the medium with the velocity c_T,

$$\tilde{\nu}_o. \qquad \text{Frequency of the wave produced in the medium} \qquad (225)$$

The observer O sitting on the receiver registers an arriving frequency ν, in other words a harmonic wave with a period of oscillation T, by counting the number Δn of wave crests that arrive during the time Δt,

$$\nu = \frac{1}{T} = \frac{\Delta n}{\Delta t}. \qquad \text{Frequency registered by observer} \qquad (226)$$

The comparison case that can be seen as a test for the quality of the transmitting medium will consist in the demand that both emitter and observer are stationary with respect to the medium. We will also only allow such media to be included, for which all three frequencies coincide. Taking our considerations from Chap. 4 into mind, this would apply to the mechanical continuum, which we receive as a limiting case of the linear chain, see Fig. 18.1,

$$\nu_o = \tilde{\nu} = \nu. \qquad \text{The static comparison case} \qquad (227)$$

In all, we will observe three cases.

(I) The emitter moves towards the observer with the velocity v, see Fig. 18.2. From observer O's point of view, this would look like the following: The first wave crest was emitted when the emitter S was located at x_o. This wave crest moves with

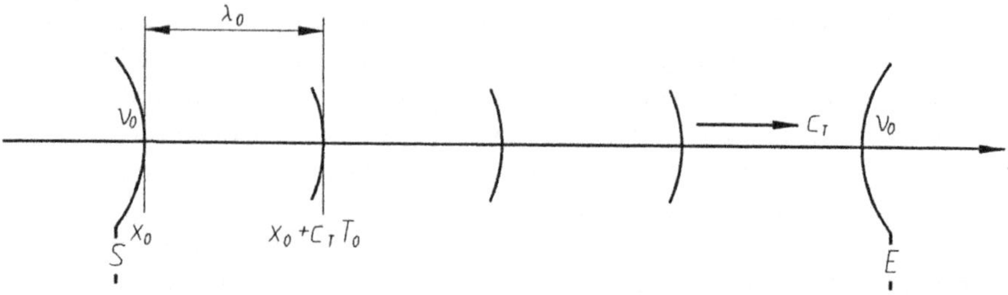

Fig. 18.1 The emitter S and the observer O are at rest in the medium. The oscillation of the frequency $\nu_o = 1/T_o$ sent out by the emitter causes a wave of the length λ_o in the medium, which moves towards the observer with the sound velocity $c_T = \lambda_o \times \nu_o$ who in turn finds out the emitter's frequency ν_o

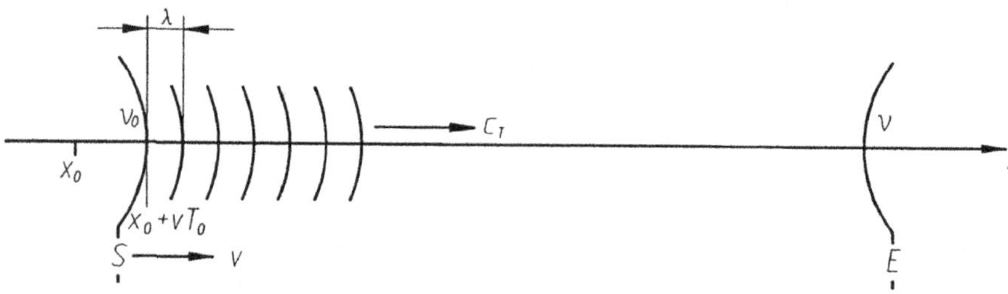

Fig. 18.2 Moving emitter. The emitter S once again oscillates with the frequency $\nu_o = 1/T_o$; however, it now also moves with the velocity v towards the static observer O. Here, according to (228) a wave of the length $\lambda = (c_T - v)\,T_o$ is produced. We will take as an example $v = 0,8\,c_T$ and get $\lambda = 0,2\,\lambda_o$. The frequency registered by the observer is then, according to (229) $\nu = \nu_o \frac{1}{1-v/c_T}$, and in our example $\nu = \frac{1}{1-0,8}\,\nu_o = 5\,\nu_o$

the velocity c_T towards him. After time $T_o = 1/\nu_o$, the emitter S is located, due to its own velocity v at $x_1 = x_o + v\,T_o$. According to the preparations made (respectively according to the construction of the emitter with the eigen frequency ν_o), emitter S emits a second wave crest. The first wave crest is, at this exact time point, located at $x_2 = x_o + c_T\,T_o$, so that the distance between it and the next wave crest is λ with

$$\lambda = x_2 - x_1 = x_o + c_T\,T_o - (x_o + v\,T_o) = (c_T - v)\,T_o\,. \tag{228}$$

Both wave crests move towards the observer with the velocity c_T. When the first wave crest arrives, it will take the second wave crest a certain amount of time to arrive, $T = \lambda/c_T$. We get a value ν for the frequency the observer determined according to,

$$\nu = \frac{1}{T} = \frac{c_T}{\lambda},$$

therefore,

$$\nu = \nu_o \frac{1}{1 - v/c_T}. \qquad \text{DOPPLER effect, Moving emitter} \qquad (229)$$

This is the formula for the DOPPLER increase of frequency if the observer is at rest and the emitter moves towards the observer with a velocity v. According to our definition on the sign of velocity (see Fig. 18.2), v should be used in Eq. (229) positively if the emitter moves towards the observer. The moment it passes by the observer we would have to change the sign of the velocity v in Eq. (229) to $-v$. The emitter then moves away from the observer with the velocity v, and the later registers a lowering of the frequency.

For the case where the velocity v is very small compared to the sound velocity c_T, thus $v/c_T \ll 1$, we can replace Eq. (229) according to $\frac{1}{1+x} \approx 1 - x$ for $x \ll 1$ with

$$\nu = \nu_o \left(1 + \frac{v}{c_T}\right). \qquad \text{Moving emitter, Low velocity,} v \ll c_T \qquad (229a)$$

Equation (229a) can be applied very easily for our opening case with the fire engines. The velocity of the fire engines is nowhere near that of sound, and we determine for the jump in frequency that we register the value $\Delta\nu = 2\,\nu_o\, v/c_T$. The fireman emits a signal tone of the frequency ν_o. Whilst he moves towards us, we register a higher frequency according to (229a), and when the fire engine drives past us and moves away, we register a lowered frequency, thus the factor 2.[2]

(II) We now let the emitter be at rest and the observer moves, with its receiving apparatus[3] towards the emitter with the velocity v. Due to the fact that our observer is to the right of the emitter, v is used with the opposite sign. The velocity is counted greater than zero when the observer approaches the emitter, see Fig. 18.3. The latter once again emits its eigen frequency ν_o, which, because of the emitter's condition of rest in the medium, transmits to the medium itself. Therefore, a harmonic wave of the frequency ν_o and the wavelength $\lambda_o = c_T/\nu_o$ moves through the medium, see Fig. 18.3. The observer O is at the position $x_o + \lambda_o = x_o + c_T/\nu_o$, when the first wave crest arrives. The next arriving wave crest then finds itself at x_o. If it arrives after time T, then O measured a frequency of $\nu = 1/T$. This wave crest now moves

[2]With a sound velocity of $c_T = 330\,\text{ms}^{-1}$ and a velocity for the fire engine of $v = 80\,\text{kph} = 80 \times \frac{1000\,\text{m}}{60\times 60\,\text{s}} = 22,2\,\text{ms}^{-1}$ (so that our approximation condition $v/c_T = \frac{330}{22,2} = 0,067 \ll 1$ is fulfilled) we get if we have a signal tone of the frequency $\nu_o = 440\,\text{Hz}$, a change in the frequency of $\Delta\nu = 54\,\text{Hz}$ the moment the fire engine rushes past us, in other words the signal tone frequency suddenly falls from around 467 Hz to around 413 Hz, which cannot be overheard.

[3]The human ear is an excellent receiver for sounds in the frequency area between 50 Hz and 16,000 Hz, and readily responds to changes of frequency in this area. Our eye cannot copy this out of purely physiological reasons. The eye cannot distinguish between pure spectral colours, which are relevant for the DOPPLER effect, and mixed colours. It is therefore not alone the velocity of light that is responsible for there not being a simple view of an optical DOPPLER effect.

towards O with the velocity c_T whilst O moves towards the wave crest with the velocity v. Thus, the distance λ_o is bridged by a total combined velocity of $c_T + v$ and the time needed for this is

$$T = \frac{\lambda_o}{c_T + v} = \frac{c_T}{\nu_o} \frac{1}{c_T + v} = \frac{1}{\nu_o} \frac{1}{1 + v/c_T}.$$

The moving observer registers a frequency of $\nu = 1/T$ according to

$$\nu = \nu_o \left(1 + \frac{v}{c_T}\right). \qquad \text{DOPPLER effect, Moving observer} \qquad (230)$$

We count $v > 0$ for an approach between emitter and observer.

The DOPPLER effect for a moving observer (230) only coincides with the approximation formula (229a) of the DOPPLER effect for a moving emitter. In such a case that velocities v are far smaller than sound velocity c_T, and only in such cases, is it of no difference if either the emitter or the observer move. However, the principle difference between Eqs. (229) and (230) for the DOPPLER effect of a moving emitter or observer respectively has a profound consequence that we now wish to discuss.

Up to now, we have taken it as granted that we could determine the velocity v of the emitter or the observer with respect to the prescribed motionless state of the medium in which the wave propagation took place. We therefore had to know previously which is the state of rest for the medium. We will now drop this assumption.

We will now only assume that the emitter (per construction) produces its eigen frequency and that the emitter moves towards the observer with the velocity v. The velocity v is therefore the relative velocity between emitter and observer that we note as $v = v_1 + v_2$. Here $v_1 = q$ is the unknown, absolute emitter velocity with respect to the medium, and v_2 once again the calculated, with opposite sign, absolute velocity of the observer with respect to the medium, whereby the observer is positioned to the right-hand side of the emitter (see Fig. 18.3). What frequency ν does the observer measure in dependence of the unknown velocity v_1? First of all the emitter produces, because of its absolute velocity with respect to the medium according to (229), a

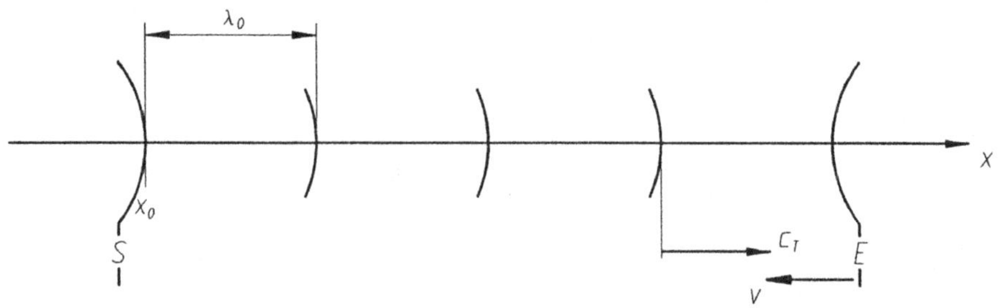

Fig. 18.3 Moving observer. The emitter stationary in the medium produces a wave of the frequency ν_o. The observer moves towards this wave with the velocity v and registers according to (230) a frequency of $\nu = \nu_o (1 + v/c_T)$. We once again choose $v = 0,8\, c_T$ and thus get $\nu = \nu_o (1 + 0,8) = 1,8\, \nu_o$. We count $v > 0$ for an approach between emitter and observer

wave of the frequency ν_1 in this medium,

$$\nu_1 = \nu_o \frac{1}{1 - v_1/c_T} = \frac{\nu_o c_T}{c_T - v_1}$$

that has the wavelength λ_1,

$$\lambda_1 = \frac{c_T}{\nu_1} = \frac{c_T - v_1}{\nu_o}.$$

As above, the observer O covers this distance, because of its absolute velocity v_2 with respect to the medium, with the velocity $c_T + v_2$. The observer therefore measures a frequency ν according to

$$\nu = \frac{c_T + v_2}{\lambda_1} = \nu_o \frac{c_T + v_2}{c_T - v_1} = \nu_o \frac{c_T + v - v_1}{c_T - v_1}$$

thus,

$$\nu = \nu_o \left(1 + \frac{v}{c_T - v_1}\right). \tag{231}$$

The observer thus measures the frequency (231) if the emitter has the unknown velocity v_1 with respect to the medium (and the observer the unknown velocity v_2).

Upholding the relative velocity v between emitter and observer, the emitter and the observer should swap positions. We could also imagine that both emitter and observer possess identical transmitting installations as well as receiving systems. They would then only have to swap functions; the former receiver emits and the former emitter receives. This results in the emitter being to the right-hand side of the receiver. For the relative velocity $v = v_1 + v_2$ is still valid. However, now $v_1 = q$ is the absolute velocity of the observer with respect to the medium, and the emitter has with respect to the medium the absolute velocity v_2, which we once again count positive, if the emitter moves towards the observer. In order to measure the receiver's frequency $\bar{\nu}$, the above calculation is reused, with v_1 and v_2 in swapped positions with an unchanged relative velocity $v = v_1 + v_2$, thus

$$\bar{\nu} = \nu_o \left(1 + \frac{v}{c_T - v_2}\right), \tag{231a}$$

thus

$$\bar{\nu} = \nu_o \left(1 + \frac{v}{c_T + v_1 - v}\right). \tag{231b}$$

The observer thus measures the frequency (231b) if the emitter has the unknown velocity v_2 with respect to the medium (and the observer the unknown velocity v_1).

We add up both frequencies (231) and (231b), write the variable q for the unknown velocity v_1 and look in the developed function $f(q)$,

$$f(q) = \nu + \bar{\nu} = 2\nu_o + \nu_o \left(\frac{v}{c_T - q} + \frac{v}{c_T + q - v}\right),$$

$$f(q) = 2\nu_o + v\nu_o \frac{2c_T - v}{c_T^2 - q^2 + vq - vc_T},$$

for an extreme value. From

$$f'(q) = v\,\nu_o\,(2c_T - v)\,\frac{2\,q - v}{(c_T^2 - q^2 + vq - vc_T)^2} = 0$$

we determine

$$q = q_o = \frac{v}{2}.$$

From

$$f''(q) = v\,\nu_o\,(2c_T - v)\,\frac{2(c_T^2 - q^2 + vq - vc_T) - (2q - v)(-2q + v)}{(c_T^2 - q^2 + vq - vc_T)^3}$$

follows

$$f''(q_o) = f''(\frac{v}{2}) = v\,\nu_o\,(2c_T - v)\,\frac{2}{(c_T - v/2)^6} > 0.$$

We find: The function $f(q)$ has at $q = q_o = v/2$ a minimum value. This means that the sum of the frequencies $\nu + \bar{\nu}$ measured by the observer attains the lowest value if, with a constantly relative velocity v between emitter and observer, the absolute velocity v_1 of the emitter is with respect to the medium $v_1 = v/2$. Therefore, the observer also has, with respect to the medium, the velocity $v_2 = v/2$. By patiently measuring the DOPPLER frequencies, we are in a position of determining our absolute velocity with respect to the transport medium of the waves. This is the first characteristic of the acoustic DOPPLER effect:

> Using the DOPPLER effect, the absolute velocity of the receiver with respect to the transport medium of waves can be determined.

We will now consider a third situation that at first seems to produce a trivial result.

(III) The emitter should once again produce its eigen frequency, should however now not move directly towards the receiver, but should move past the receiver at a large distance. This is the question of the so-called transversal DOPPLER effect that applies to waves excited perpendicular to the emitter's direction of motion, and that are observed in this direction, see Fig. 18.4. What frequency ν does the observer measure for the waves that the emitter produces in the moment of nearest approximation R_o? Our assumption 'large distance' R means that the change of distance can simply be ignored at the moment of nearest approximation during the period of one natural oscillation. One can make this elementary geometrically clear what this would look like in an equation. The emitter S moving with the velocity v moved during the period $T_o = 1/\nu_o$ of one natural oscillation by v/ν_o see Fig. 18.5.

Now with $v/\nu_o \ll R_o$ and the approximate formula $\sqrt{1+x} \approx 1 + x/2$ for $x \ll 1$ it is

$$L = \sqrt{R_o^2 + \frac{v^2}{\nu_o^2}} = R_o \sqrt{1 + \frac{v^2}{R_o^2 \nu_o^2}} \approx R_o \left(1 + \frac{v^2}{2R_o^2\nu_o^2}\right) = R_o + \frac{v^2}{2R_o^2\nu_o^2}.$$

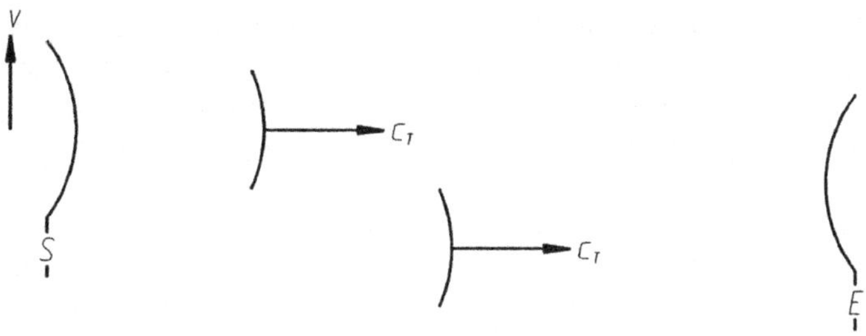

Fig. 18.4 Experimental construction for the transversal DOPPLER effect. The observer O rests in the medium. The emitter S moves with a velocity v perpendicular to the direction of the waves the emitter sends out

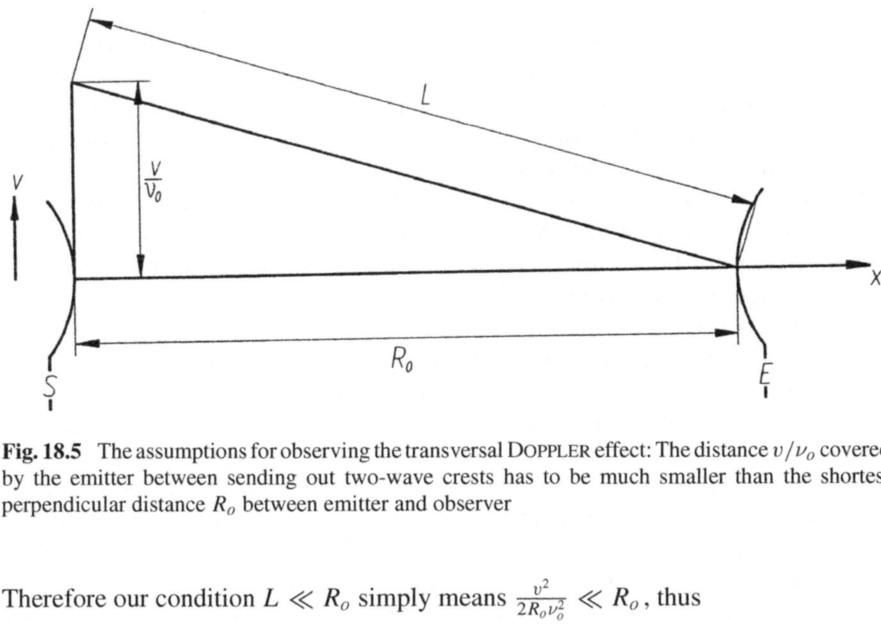

Fig. 18.5 The assumptions for observing the transversal DOPPLER effect: The distance v/ν_o covered by the emitter between sending out two-wave crests has to be much smaller than the shortest perpendicular distance R_o between emitter and observer

Therefore our condition $L \ll R_o$ simply means $\frac{v^2}{2R_o\nu_o^2} \ll R_o$, thus

$$\frac{v^2}{c_T^2}\frac{\lambda_o^2}{2R_o} \ll R_o.$$

We therefore find the equation $v/c_T \ll R_o\sqrt{2}/\lambda_o$ for the condition 'large distance', or simply

$$\frac{v}{c_T} \ll \frac{R_o}{\lambda_o}. \tag{232}$$

In the limit of an 'infinitely far away' emitter, this condition would always be fulfilled.

The second wave crest, sent on to the first wave crest after time T_o, moves with the same velocity c_T as the first, and must according to (232) have the same distance R_o to cover to the observer. Thus, the second wave crest arrives exactly T_o later at the observer, so that the observer determines the same frequency ν_o as the emitter sent out,

$$\nu = \nu_o. \qquad \text{Moving emitter, Transversal observation} \tag{233}$$

This is our second characteristic for the acoustic DOPPLER effect:

There is no transversal acoustic DOPPLER shift.

All of this has been known for over 150 years. However, that what is going to happen next is new and exciting. We once again move into the world of our infinite crystal, into a crystal where our internal observers are operating with their natural measuring-rods and clocks (and where these observers perhaps philosophically discuss the results of their observations). We think about how the DOPPLER effect could be determined and measured using only the natural measuring-rods and clocks that the internal observers have at their disposal (see GÜNTHER [34]). Here, we once again assume that the critical velocity c_o of the sine-GORDON equation coincides with the transversal sound velocity c_T. In fact, the whole experimental situation is exactly as we have previously described and examined.

An internal observer S sits on an emitter, a standard emitter that produces per construction an oscillation of the frequency ν_o. This observer S determines this frequency by counting the number Δn_o of oscillations produced by the emitter during a certain time interval Δt_o in the same manner as described by us in (224). However, here we have to be a little more careful. Observer S's internal standard emitter is per definition situated in its own reference system. Therefore, time interval Δt_o is a numerical date that is only valid for this special reference system. We will have to take this into account. The same applies to observer O on his receiving station. He determines the frequency from the number Δn of arriving wave crests during a certain time interval Δt. Here Δt is a statement of time that only makes sense in the reference system in which observer O as well as his receiver are at rest.

To be consistent, we will only take those standard emitters into consideration where the internal observer S measures one and the same frequency ν_o, irrespective of the reference system in which he and his standard emitter are at rest. Exactly this is what is meant when we make the demand: A standard emitter shall produce per construction an oscillation of the frequency ν_o. Here, one can, for example, think of breather-similar oscillation systems or other oscillating dislocation configurations. This (only meaningful) definition of a standard emitter of constant frequency has one important consequence. An internal observer's standard emitter is not a standard emitter as defined by an outside observer. An oscillating breather (the internal observer's standard emitter) is not a standard emitter for the outside observer. For the outside observer, the moving breather is with respect to the crystal of a different oscillationary construction than the breather being at rest with respect to the crystal, as we have already discussed in Chap. 11. If we ignore these facts, then we slip into the paradox described in Chap. 11.

In order to describe the experimental situation, we choose an arbitrary reference system Σ_o (after which this reference system is our norm) in which emitter and observer are alternatively situated and examine the above-described cases in the same sequence.

(I) Observer O rests in the reference system Σ_o in which we designate the space and time coordinates as x and t. Emitter S moves towards the observer with the positive velocity v. An observer S on top of the standard emitter S is at rest in a reference system Σ', which possesses the velocity $-v$ when observed from Σ_o. The space and time coordinates in Σ' are designated as x' and t', see Fig. 18.6. Observer S controls the frequency ν_o of his emitter in his reference system Σ'. He therefore counts the number Δn_o of oscillations during the time interval $\Delta t'$ shown by his clock and finds $\nu_o = \Delta n_o / \Delta t'$, or a period of oscillation T_o for the membrane according to

$$T_o = \frac{1}{\nu_o} = \frac{\Delta t'}{\Delta n_o}. \tag{234}$$

An internal observer S on emitter S sees that he produces a new wave crest in space every T_o seconds (or in any other time unit, for example lattice seconds, see Eq. (91)). However, observer O resting in Σ_o determines that S's clock in Σ' goes behind according to our Eq. (121) (here, we write c_T in the place of c_o). Therefore, when emitter S's clock reads T_o seconds and the next wave crest is emitted, observer

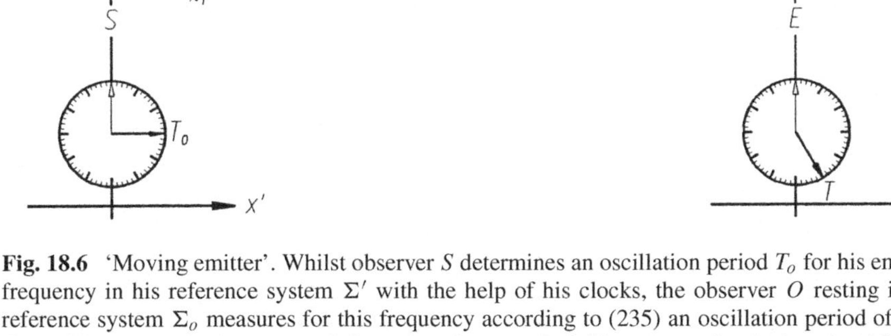

Fig. 18.6 'Moving emitter'. Whilst observer S determines an oscillation period T_o for his emitted frequency in his reference system Σ' with the help of his clocks, the observer O resting in his reference system Σ_o measures for this frequency according to (235) an oscillation period of $T = T_o/\gamma$. We once again choose $v = 0,8\,c_T$, therefore $\gamma = 0,6$. If we gauge the clock so that 15 scale marks are displayed for T_o, then observer O observes 25 scale marks on his clock for the oscillation period $T = 15/0,6 = 25$ scale marks. This distance λ_r of two running wave crests in the medium can be determined, using (236), as $\lambda_r = (c_T - 0,8\,c_T)\,T_o/0,6 = c_T\,T_o\,0,2/0,6 \approx 0,3\,\lambda_o$. Here, λ_o is the wavelength as shown in Figs. 18.1 and 18.3. One should note the difference as compared with the wavelength λ in Fig. 18.2

O's clock has to have been moved forward T seconds according to

$$T = \frac{T_o}{\sqrt{1 - v^2/c_T^2}}. \tag{235}$$

Our further argumentation can be literally taken from the above classic mode of description. The first wave crest produced in space runs towards observer O with the velocity c_T. Due to the fact that emitter S follows the wave crests with the velocity v, observer O calculates the distance λ_r between two successive wave crests moving towards him, using Eq. (228) (only with T in place of T_o),

$$\lambda_r = (c_T - v)\,T, \tag{236}$$

and thus from (234)–(236) for the frequency ν_r that he determines,

$$\nu_r = \frac{c_T}{\lambda_r} = \frac{c_T}{(c_T - v)\,T} = \frac{c_T\sqrt{1-v^2/c_T^2}}{T_o\,(c_T - v)} = \frac{1}{T_o}\,\frac{\sqrt{c_T^2 - v^2}}{c_T - v} = \nu_o\,\frac{\sqrt{(c_T - v)(c_T + v)}}{\sqrt{(c_T - v)^2}},$$

therefore

$$\nu_r = \nu_o\sqrt{\frac{c_T + v}{c_T - v}}. \qquad \text{Internal observer, Moving emitter} \tag{237}$$

This is the equation for the DOPPLER shift for moving emitters as registered by an internal observer in our crystal.

From the viewpoint of an outside observer, this equation is just the result of the fact that the internal observer moving with the velocity v produces a changed frequency $\nu_o\sqrt{1 - v^2/c_T^2}$ according to time dilatation (121). The moving breather oscillates slower. If one inserts this value in place of ν_o in (229), then (237) is in fact the result.

For small relative velocities v between emitter and observer, in other words for $v/c_T \ll 1$ (neglecting terms of higher order in v/c_T), we find with the help of our approximate formulas $\frac{1}{1-x} \approx 1 + x$, $\sqrt{1 + 2x} \approx 1 + x$ for $x \ll 1$,

$$\nu_r = \nu_o\sqrt{\frac{c_T + v}{c_T - v}} = \nu_o\sqrt{\frac{1 + v/c_T}{1 - v/c_T}} \approx \nu_o\sqrt{(1 + \frac{v}{c_T})(1 + \frac{v}{c_T})}$$
$$\approx \nu_o\sqrt{1 + 2v/c_T} \approx \nu_o\,(1 + v/c_T),$$

thus

$$\nu_r = \nu_o\left(1 + \frac{v}{c_T}\right). \qquad \text{Internal observer, Moving emitter, Small velocities, } v \ll c_T \tag{237a}$$

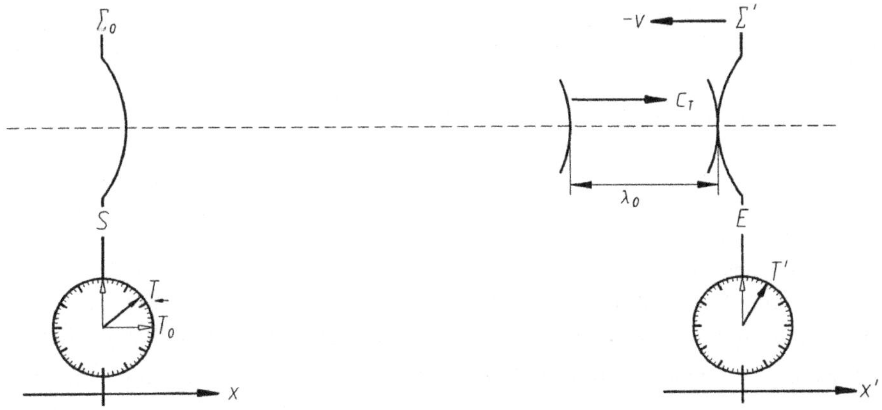

Fig. 18.7 'Moving observer'. The wavelength λ_o, see also Fig. 18.1, is now a difference in coordinates in Σ_o. The received frequency $\nu_r = 1/T'$ is measured by an internal observer O using a receiver resting in Σ'. This measured time T' for the sweeping of two successive wave crests is smaller, due to time dilatation (121), by the factor γ than the time interval $T_\leftarrow$ that the emitter S measures for the sweeping of two successive wave crests by the observer O moving towards the emitter S with the velocity v, $T' = \gamma\ T_\leftarrow$. We once again calculate using $v = 0,8\,c_T$, thus $\gamma = 0,6$, and the emitter's period of oscillation T_o amounts to 15 scale marks on our clock. The observer on the emitter then measures $T_\leftarrow = \frac{\lambda_o}{c_T+v} = \frac{\lambda_o}{(1+0,8)\,c_T} = 0,56\,T_o = 0,56 \times 15 = 8,3$ scale marks, and the observer O on the receiver measures $T' = T_o\,0,6/1,8 = 5$ scale marks

As a comparison of (237a) with (229a) shows, the internal observer registers in this approximatative case the same law for the DOPPLER effect as the classic observer.

(II) Now, the emitter S rests in the reference system Σ_o (so that S now measures the coordinates x and t), and the observer O moves, as seen from Σ_o, with the velocity v towards the emitter. The observer O rests in a reference system Σ', which has the velocity $-v$ when observed from Σ_o. In Σ', the space and time coordinates x' and t' are measured, see Fig. 18.7.

Due to the fact that the emitter S rests in Σ_o, the emitter produces a wave with its eigen frequency $\nu_o = \Delta n_o/\Delta t$ and therefore with a wavelength $\lambda_o = c_T/\nu_o$ that moves towards the observer O. Observed from Σ_o, the observer moves towards the wave crests with the velocity $-v$. Thus, the observer will need the time $T_\leftarrow = \frac{\lambda_o}{c_T+v}$ for the distance between one wave crest to the next (this means for a wavelength λ_o). This time $T_\leftarrow$ is displayed by the clocks in the reference system Σ_o. The observer's clock however goes behind, as we already know, with respect to the other clocks in Σ_o according to our Eq. (121). Thus, as a result, the observer O's clock will measure a shorter time interval T' for the sweeping of the wave crests according to

$$T' = T_\leftarrow \sqrt{1 - \frac{v^2}{c_T^2}}.$$

We once again designate the frequency of the waves emitted by emitter S and registered by observer O as ν_r, thus $\nu_r = 1/T'$ and get

$$\nu_r = \frac{1}{T_{\leftarrow}} \frac{1}{\sqrt{1 - v^2/c_T^2}} = \frac{c_T + v}{\lambda_o} \frac{1}{\sqrt{1 - v^2/c_T^2}} = \frac{\nu_o}{c_T} \frac{c_T + v}{\sqrt{1 - v^2/c_T^2}} = \nu_o \frac{c_T + v}{\sqrt{c_T^2 - v^2}}.$$

We then find as above

$$\nu_r = \nu_o \sqrt{\frac{c_T + v}{c_T - v}}. \qquad \begin{array}{l}\text{Internal observer}\\ \text{Moving observer}\end{array} \qquad (238)$$

The outside observer will explain this internal observer's equation simply by stating: Because *time dilatation* the *standard of frequency* which is reciprocal to the corresponding oscillation period of the moving observer is diminished, so that coefficients of measure for frequency are increased by the factor $1/\sqrt{1 - v^2/c_T^2}$. If we therefore replace the coefficient of measure for frequency ν_o in (238) with $\nu_o/\sqrt{1 - v^2/c_T^2}$ we get the frequency ν according to (229).

The internal observer does not know of the possibilities the outside observer has at his disposal. For the internal observer, we see that:

The equations for the DOPPLER effect for a moving emitter (237) and moving observer (238) are identical!

$$\nu_r = \nu_o \sqrt{\frac{c_T + v}{c_T - v}}. \qquad \begin{array}{l}\text{Internal observer}\\ \text{DOPPLER effect}\end{array} \qquad (239)$$

Apart from the case concerning small velocities, where the observations made by outside observers and internal observers coincide, our internal observers inside of the crystal and outside observers evaluate the DOPPLER effect in completely different ways. According to Eqs. (237) and (238), the internal observer cannot see any difference between a moving emitter or a moving observer. For the internal observer, the DOPPLER effect depends only on the relative velocity v between the emitter and the observer.

The internal observers can check and recheck their DOPPLER shifts as often as they like. They would not find any clue to a motion relative to the crystal lattice. This DOPPLER effect confirms therefore our considerations from Chap. 15. There is absolutely no proof from the internal observer's viewpoint for motion relative to the crystal lattice. The first characteristic of the acoustic DOPPLER effect for an internal observer in contrast to an outside observer would therefore be (as expected after Chap. 15):

No absolute velocity with respect to a wave's transport medium can be determined using the DOPPLER effect.

We will now consider the third experimental arrangement.

(III) Observer O once again finds himself in the reference system Σ_o (with the coordinates x and t), and the emitter S moves at a 'very large distance' R from the observer, for which we once again assume condition (232). According to this, the change in distance between emitter and observer during the period of an natural oscillation can be ignored. The emitter is at rest in a reference system Σ' (with the coordinates x' and t') which has the velocity $+v$ with respect to the reference system Σ_o. The argumentation in this case begins just as it did in the first case. The emitter S determines by counting the number of wave crests, an oscillation period T_o of the emitter's membrane in the reference system Σ', as we have already described in (234),

$$T_o = \frac{1}{\nu_o} = \frac{\Delta t'}{\Delta n_o}. \tag{234}$$

The observer O with his clocks stationary in Σ_o once again finds out that the emitter S's clock goes behind according to Eq. (121). The emitter therefore does not produce a wave crest in space every T_o seconds, but, according to observer O's clocks every T seconds, as we already discovered in Eq. (235),

$$T = \frac{T_o}{\sqrt{1 - v^2/c_T^2}}. \tag{235}$$

These wave crests move towards the observer with the velocity c_T. The second wave crest sent out after time T has to bridge the same distance R_o as the first wave crest and moves with the same velocity c_T. Therefore, the second wave arrives exactly time T after the first crest arrives, which enables the observer to calculate the frequency ν according to

$$\nu = \frac{1}{T} = \frac{\sqrt{1 - v^2/c_T^2}}{T_o}$$

and thus

$$\nu = \nu_o \sqrt{1 - \frac{v^2}{c_T^2}}. \qquad \text{Internal observer, Transversal DOPPLER effect} \tag{240}$$

The internal observer therefore registers, in contrast to the outside observer, for the second characteristic of the acoustic DOPPLER effect:

> There is a transversal acoustic DOPPLER effect.

And now comes the most important thing. We compare our Eqs. (239) and (240) to the corresponding DOPPLER effect equations for light waves, electromagnetic waves generally. We can check up on these in every physics coursebook or monograph on relativity, for example A. P. FRENCH [23], and find:

> The equations formulated by the internal observer inside of a crystal for the acoustic DOPPLER effect are absolutely identical to the complete description of the DOPPLER phenomena for light, if one replaces only the transversal sound velocity c_T with the speed of light c_L.[4]

---

[4]The method, originating from W. VOIGT, of calculating the DOPPLER effect from the relativistic invariance of a harmonic wave's phase (see also the corresponding explanation made by A. PAIS

Just as the internal observers inside of their crystal, we are in no position of defining an absolute motion using the DOPPLER effect for electromagnetic waves (or using any other effect) with respect to our physical vacuum. Furthermore, as can be derived from our crystal, the transversal DOPPLER effect is nothing other than time dilatation of a moving clock converted to frequency. The transversal DOPPLER effect is thus also a test for EINSTEIN's prophetized time dilatation of clocks for our physical reality formulated in his Special Theory of Relativity (for this moment in time we have left the world of the internal observers). In fact, this effect could only be experimentally verified decades after the invention of the Special Theory of Relativity, and EINSTEIN himself observed the verification process of this effect and saw it as the *experimentum crucis* of his theory, as we already stated in Chap. 3, p. 16. It is from today's viewpoint somewhat curious that the first verification experiments made by H. J. IVES [44, 45] and G.J. STILLWELL 1938/39 had as a target the demonstration of the non-existence of the transversal DOPPLER effect, and this 33 years after the founding of the Special Theory of Relativity. Naturally, the experimenters did not achieve the desired result. Compared to the former, the experiments made by G. OTTING [70] aimed at proving this effect and were successful.

We see that the situation of the outside observers concerning light is in fact identical to that of the internal observers of our crystal concerning their experiments with sound. In the light of this and taking all other assumption of our model into account, the MACHian thesis, cf. THIELE [93] holds true:

Light somewhat like sound.
Sound somewhat like light.

We now also understand how this statement can be valid without actually coming into conflict with conventional physics. We have included the measuring instruments, whose usage founds the basis for the physical laws, into our considerations.

We have to remember for the evaluation of the DOPPLER effect, as already stated above, that a standard emitter, an emitter that produces per construction a constant frequency, is for the internal observer in our crystal something different than a standard emitter for the outside observer. The standard emitters constructed by the internal observer are oscillating dislocation configurations that, from the viewpoint of an outside observer, have other construction rules when they move relative to the crystal. Principally, this is not the case for the internal observers, because they can do nothing else but measure any spatial change, or change in time, with the help of such dislocation configurations. If we ignore this point, then the clock paradox described in Chap. 11 arises. Here, we wish to mention: The description of mechanical phenomena of a crystal (such as the DOPPLER effect for example) using the measuring instruments of an internal observer may seem to the reader as artificial.

---

[71]) does in fact speedily lead us to our Eqs. (236)–(240) and is thus exclusively used for electromagnetic waves, see e.g. FRENCH [23]. However, because we aimed at a more detailed and explicit method of illustrating the physical processes, we did without the more mathematically elegant way of deriving the equations for theDOPPLER effect. This method of the relativistic invariance of a harmonic wave's phase together with our measuring-rods and clocks does indeed also deliver a complete description of the acoustic DOPPLER effect.

Nevertheless, this is the only method that accommodates our physical problem. The physical symmetry, here the relativistic symmetry of the DOPPLER effect that we previously only knew for light can only be described using this method. The relevance of such a description using internal measuring instruments will become fully clear when we turn to the examination of dynamic questions, as we will do in Chaps. 22 and 23. In these chapters, we will explicitly show that the properties of inertia and energy of dislocation configurations, which only have physical meaning when brought into relation with a lattice are determined by the critical velocity c_o.

Chapter 19
Aberration

We will in this chapter leave our crystal and turn to the physics of our 'outside' space for which the EINSTEINian Special Theory of Relativity with its specific critical velocity, the speed of light c_L, is valid.

We consider the case of the light emitted from a star S, which in Σ_o is observed as a plane wave propagating along negative y-axis, cf. Figure 19.1. The star is seen, if the rays arriving at $A(t_1)$ and $B(t_1)$ respectively, meet at O'.

Let us now consider an inertial system Σ', which is moving along negative x-axis of Σ_o at a velocity of amount v. A telescope resting in Σ' has the linear dimension $2L$. We ask for the angle α' of inclination, which is measured in Σ' for the telescope if the star is observed there. This angle α' of inclination is defined by the condition that the peripheral rays meet in the middle.[1] Obviously, if the telescope would not be inclined, but positioned parallel to x'-direction of Σ', the star could not be seen, since both rays would not meet at O', cf. Fig. 19.2.

For calculating the angle α' of aberration measured in Σ', we will make use of the postulate (112) of LORENTZ contraction measured in Σ_o, solely,

$$L' = L_o\sqrt{1-\frac{v^2}{c_L^2}}. \tag{112}$$

This is sufficient for the calculation of the exact relativistic angle of aberration. We have the projections,

$$a' = L\cos\alpha' \quad , \quad b' = L\sin\alpha' \; . \tag{241}$$

The telescope is resting in Σ'. Applying the postulate (112) with a' and a for L_o and L' we measure in Σ_o,

[1]This definition also applies to modern very long baseline interferometry.

H. Günther, *Elementary Approach to Special Relativity*,
https://doi.org/10.1007/978-981-15-3168-2_19

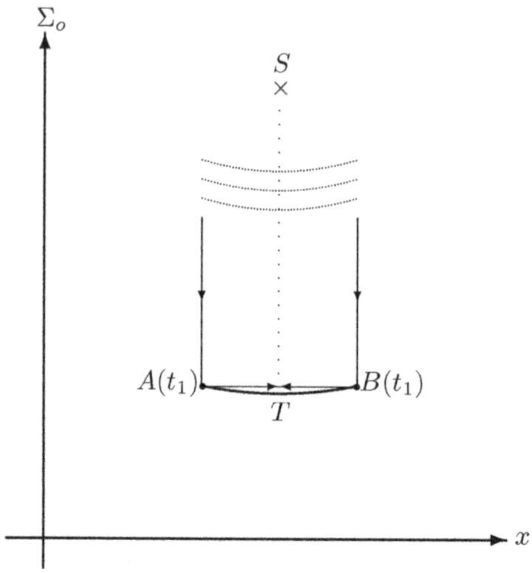

Fig. 19.1 Observation of the star in Σ_o

$$a = a'\gamma = L\gamma\cos\alpha' \ , \quad b = b' = L\sin\alpha' \ . \tag{242}$$

Consider the two rays reaching the left and the right end points $A(t_1)$ and $B(t_1)$ of the telescope at time t_1 in Σ_o. Now again, the star is seen, if these rays meet at O', compare Figs. 19.1 and 19.2. This means, the running time T_r of the right-hand light ray from B to O' must coincide with the running time T_l of the left-hand light ray along the way ACO'.

In order to reach the point O' the right-hand light ray has to overcome the distance $a + vt$ in the negative x-direction (since the telescope is moving with v in this direction) and the distance b in the negative y-direction, as a whole the ray has to overcome the distance $\sqrt{(a + vT_r)^2 + b^2}$.

Using in the following the notations,

$$u = c_L T \ , \quad \beta = \frac{v}{c_L} \ , \quad \gamma = 1 - \beta^2 \ ,$$

we get for the distance cT_r of the way BO'

$$u_r = c_L T_r = \sqrt{(a + \beta u_r)^2 + b^2} \ . \tag{243}$$

The distance of AC is $2b$ and the running time of the left-hand light ray along AC is $2b/c_L$. Now, the telescope approaches the ray in negative x-direction. Hence, for the second part CO' the distance in x-direction is $a - v\Delta t$ with the running time Δt along the way CO'. With $T_l = 2b/c_L + \Delta t$ for the total running time T_l along the way ACO' we get $a - v\Delta t = a - v\big(T_l - 2b/c_L\big)$. The distance of the way CO' along y-direction is b. Finally, we get for the distance $c_L\, T_l$ of the way

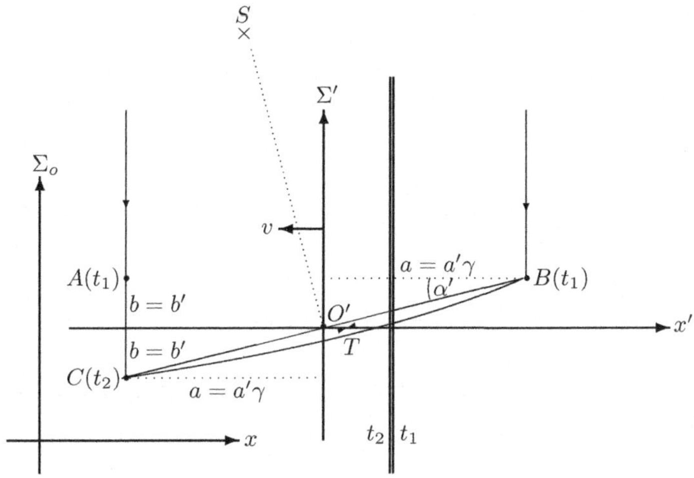

Fig. 19.2 Observation of the star in Σ'

ACO'

$$u_l = c_L T_l = 2b + \sqrt{(a - \beta(u_l - 2b))^2 + b^2} \; . \tag{244}$$

With (242), we get for (243)

$$u_r = \sqrt{(L\gamma cos\alpha' + \beta u_r)^2 + L^2 \sin^2 \alpha'} \; . \tag{245}$$

This is a quadratic equation for u_r with the solution

$$u_r = \frac{L}{\gamma}(\beta \cos \alpha' + 1) \; . \tag{246}$$

The second solution $(L/\gamma)\,(\beta \cos \alpha' - 1)$ has no physical meaning, since it would be a negative distance.

With (242), we get for (244)

$$u_l - 2L \sin \alpha' = \sqrt{\left(L\gamma \cos \alpha' - \beta(u_l - 2L \sin \alpha')\right)^2 + L^2 \sin^2 \alpha'} \; , \tag{247}$$

which yields the quadratic equation

$$\begin{aligned} &u_l^2\gamma^2 - u_l 2L(2 \sin \alpha'\gamma^2 - \beta\gamma \cos \alpha') \\ &+ 4L^2\gamma^2 \sin^2 \alpha' - L^2(\gamma^2 \cos^2 \alpha' + \sin^2 \alpha') - 4L^2\beta\gamma \cos \alpha' \sin \alpha' = 0 \end{aligned}$$

with the formal solutions

$$u_{1,2} = \frac{L}{\gamma}\Big[(2\gamma \sin\alpha' - \beta\cos\alpha')$$
$$\pm\sqrt{(2\gamma \sin\alpha' - \beta\cos\alpha')^2 - 4\gamma^2\sin^2\alpha' + \gamma^2\cos\alpha' + \sin^2\alpha' + 4\beta\gamma\sin\alpha'\cos\alpha'}\Big]. \tag{248}$$

Since the square root yields the value 1, we again can drop the second solution from physical reasons and get

$$u_l = \frac{L}{\gamma}\left(2\gamma\sin\alpha' - \beta\cos\alpha' + 1\right) . \tag{249}$$

The star is seen in Σ', if $u_r = u_L$. From (246) and (249), this yields the equation for the angle of aberration

$$\gamma\sin\alpha' = \beta\cos\alpha' ,$$

hence

$$\tan\alpha' = \frac{\beta}{\gamma} \quad\longrightarrow\quad \sin\alpha' = \frac{v}{c_L} , \tag{250}$$

which in Σ_o is measured as

$$\tan\alpha = \frac{\gamma^2 b}{a} = \frac{v}{c_L} . \tag{251}$$

Equation (250) defines the exact relativistic angle of aberration, which we have derived by observing it from Σ_o. Therefore, we did not make any use of simultaneity in the system Σ'.

However, if we want to describe this effect with the help of observations made in the system Σ', we need a definition for the synchronisation of clocks there. The most appropriate one is realised by LORENTZ transformation. On the other hand, we also can assume an absolute simultaneity according to REICHENBACH. In this case, the coordinates of an event, observed in Σ_o and Σ' respectively, must be converted with the help of REICHENBACH transformation (143).[2]

[2]Nevertheless, from a theoretical point of view, it is more simple to derive the same result along the well-known procedure using the relativistic invariance of phase on the basis of LORENTZ transformation (177). Even in the classical case, it may be helpful to use the linear approximation of EINSTEIN's definition of simultaneity (156), see p. 138, as was discussed by LIEBSCHER [58] and BROSCHE: Considered in Σ', according to (155), the rays arrive simultaneously at C and B. As a consequence of an isotropic light propagation in Σ' they again meet at O'.

Chapter 20
Tachyons and Causality

For the first part of this chapter, we will keep the EINSTEINian Special Theory of Relativity with its critical velocity, the speed of light c_L. In 'A Supplement to the Oxford English Dictionary', (vol.VI, Oxford 1986) the tachyon is explained as 'A hypothetical particle that travels faster than light and has imaginary mass'. We however know that no particle can ever be accelerated to the speed of light. If we accelerate a particle moving with a velocity $v < c_L$ in the frame Σ_o, then all we actually do is to give the particle a velocity Δv in its actual rest system Σ' during a short time interval Δt. According to the composition of velocities (194) (with c_L in place of c_o for the outside observer), the velocities v and Δv result in the velocity w observed in Σ_o, with which the particle moves after having being accelerated,

$$w = \frac{v + \Delta v}{1 + v \ \ \Delta v / c_L^2} \ ,$$

which always is smaller than c_L, as one can easily verify. This is true in any case, even if we repeat this process as often as we wish asserting as much energy as possible on a particle.

The remaining logical possibility would be that tachyons must move with super light velocity exclusively, as we have already stated in Chap. 17. However, it is not to see how to get hold of such particles. And would not the tachyons perhaps moving with 'infinite velocity' through space give us a chance, hereby completely ignoring the whole Special Theory of Relativity, of carrying out synchronisation of clocks in an arbitrary frame by 'instantaneous action' without making use of any velocity at all? One other important objection that needs to be clarified will be illustrated by means of a short detective story.

H. Günther, *Elementary Approach to Special Relativity*,
https://doi.org/10.1007/978-981-15-3168-2_20

In a reference system Σ_o, an event E_o ($x_o = 0, t_o = 0$) occurs: At time $t_o = 0$, the tachyon physicist DR FAST leaves a prison located at $x_o = 0$, where he was imprisoned, after a highly debatable trail, for several years for committing a murder using a tachyon machine. DR FAST did not lift the secret of this machine during the trail, cf. also Fig. 30.1.

Now, a further event E_1 (x_1, t_1) occurs. His former adversary, the lawyer STUS, drops dead at the location $x_1 = L$ at time $t_1 = \frac{L}{2c_L}$. DR FAST, who is once again in possession of his tachyon machine, is arrested under suspicion of murder.

During this time period, SHERLOCK HOLMES experiments with his new telescopes in the special compartment of the MERCURY Express. The train rushes with 80% light velocity along the rails. This train is our reference system Σ'. HOLMES observes the release of his client DR FAST and registers the coordinates of this event E_o in his reference system Σ' as $x'_o = 0$ and a time $t'_o = 0$, thus

$$E_o : \left.\begin{matrix} x_o = 0 \ , & t_o = 0 \ , \\ x'_o = 0 \ , & t'_o = 0 \ . \end{matrix}\right\} \tag{252}$$

HOLMES also observed the other event E_1, the death of his former adversary STUS, and discovers the coordinates x'_1 and the time t'_1 for this event. We can calculate with the help of the transformation (151), here we have to replace c_o with c_L, the values that HOLMES determined. With $t_1 = \frac{L}{2c_L}$, $x_1 = L$ and $v = \frac{4}{5}c_L$, we find for the time that HOLMES determined as the time of death of MR STUS

$$t'_1 = \frac{t_1 - v x_1/c_L^2}{\sqrt{1 - v^2/c_L^2}} = \frac{\frac{1}{2}\frac{L}{c_L} - \frac{4}{5}c_L\, L/c_L^2}{\sqrt{1 - 16/25}} = \frac{\frac{L}{c_L}\left(\frac{1}{2} - \frac{4}{5}\right)}{\sqrt{9/25}} = \frac{L}{c_L}\frac{5-8}{10}\frac{5}{3} = -\frac{3}{10}\frac{5}{3}\frac{L}{c_L} \ ,$$

$$t'_1 = -\frac{1}{2}\frac{L}{c_L}$$

and thus also for the space coordinate x'_1,

$$x'_1 = \frac{x - v\,t_1}{\sqrt{1 - v^2/c_L^2}} = \frac{L - \frac{4}{5}c_L\frac{L}{2c_L}}{\frac{3}{5}} = \frac{5}{3}\left(1 - \frac{4}{10}\right)L = \frac{5}{3}\frac{6}{10}L = L$$

and thus,

$$E_1 : \left.\begin{matrix} x_1 = L \ , & t_1 = \frac{1}{2}\frac{L}{c_L} \ , \\ x'_1 = L \ , & t'_1 = -\frac{1}{2}\frac{L}{c_L} \ . \end{matrix}\right\} \tag{253}$$

(The identity of the primed and unprimed space coordinates is purely coincidental and is just a result of the chosen numerical example).

According to HOLMES' observations, DR FAST was still in prison at event E_1, because of $t_1' < 0$, and was not in possession of his tachyon machine. DR FAST was released at $t_o' = 0$. HOLMES had thus proven that MR STUS was already dead before DR FAST was released from prison! A tachyon produced by DR FAST at the moment he was released from prison could therefore not be the cause of MR STUS' death. If this were so, then the observations made by HOLMES in the reference system Σ', cause and result, would be completely reversed. This would be in contradiction to our experiences made using our causal constructed world. Do tachyons teach us something else? Do we have to reconsider our proven definitions of causality if we accept the existence of tachyons? Somehow tachyons do not fit into our conceptions. They seem to belong more to the world of metaphysics and are surrounded with a touch of mysticism. We will continue our detective story in Chap. 30.

We now return to the world of our crystal and investigate the question of tachyons from the empirical viewpoint of our internal observer in our infinite crystal, for which a Special Theory of Relativity exists, however with the sine-GORDON equation's critical velocity c_o as the limiting velocity.

Let us remind ourselves of a fact. For internal observers, the solutions of the sine-GORDON equation are objects of experience.

We will firstly calculate that the function $\mathrm{q}^{\mathrm{T}} = \mathrm{q}^{\mathrm{T}}(\mathrm{x}, \mathrm{t})$,

$$\mathrm{q}^{\mathrm{T}}(\mathrm{x}, \mathrm{t}) = 4 \arctan \exp \left[\frac{\mathrm{t} - \mathrm{v}\,\mathrm{x}}{\sqrt{1 - v^2}} \right] + \pi \, , \tag{254}$$

is a solution of the sine-GORDON equation (95),

$$\frac{\partial^2 \mathrm{q}^{\mathrm{T}}}{\partial \mathrm{x}^2} - \frac{\partial^2 \mathrm{q}^{\mathrm{T}}}{\partial \mathrm{t}^2} = \sin \mathrm{q}^{\mathrm{T}} \, , \tag{95}$$

written using the dimensionless variables x and t. We once again use the notation $e^x := \exp[x]$ for the exponential function. Taking (89), (91), (93) and (96) into account, i.e. $q = \frac{a}{2\pi}\,\mathrm{q}$, $\mathrm{x} = x/\lambda_o$, $\mathrm{t} = t/\tau$, $c_o = \lambda_o/\tau_o$ and the resulting dimensionless velocity v,

$$\mathrm{v} = \frac{v}{c_o} \, , \tag{255}$$

then the function $q^T = q^T(x, t)$ is

$$q^T(x, t) = \frac{a}{2\pi}\,\mathrm{q}^{\mathrm{T}}\left(\frac{x}{\lambda_o}, \frac{t}{\tau_o}\right) = \frac{2a}{\pi} \arctan \exp \left[\frac{\dfrac{t}{\tau_o} - \dfrac{v}{c_o}\dfrac{x}{\lambda_o}}{\sqrt{1 - v^2/c_o^2}} \right] + \frac{a}{2} \, ,$$

$$q^T(x,t) = \frac{2a}{\pi}\,\arctan\exp\left[\frac{\frac{c_o}{\lambda_o}\left(t - \frac{v\,x}{c_o^2}\right)}{\sqrt{1 - v^2/c_o^2}}\right] + \frac{a}{2}\,, \tag{256}$$

which is a solution of the sine-GORDON equation (88).

We introduce a velocity u according to

$$u = \frac{c_o^2}{v} \tag{257}$$

and a constant κ,

$$\kappa = \frac{u}{c_o}\sqrt{1 - \frac{v^2}{c_o^2}}\,, \tag{258}$$

thus,

$$\kappa = \operatorname{sign} u \cdot \sqrt{\frac{u^2}{c_o^2} - 1} = \pm\sqrt{\frac{u^2}{c_o^2} - 1}\,, \tag{259}$$

where

$$\operatorname{sign} x = \left.\begin{matrix} +1 & & x > 0\,, \\ & \text{for} & \\ -1 & & x < 0\,, \end{matrix}\right\} \tag{260}$$

so that $\kappa > 0$ for $u > 0$ and $\kappa < 0$ for $u < 0$.

Hence, we can write the function (256) as

$$q^T(x,t) = \frac{2a}{\pi}\,\arctan\exp\left[\frac{-\frac{c_o}{\lambda_o}\frac{v}{c_o^2}\left(x - \frac{c_o^2}{v}\,t\right)}{\sqrt{1 - v^2/c_o^2}}\right] + \frac{a}{2}\,,$$

$$q^T(x,t) = \frac{2a}{\pi}\,\arctan\exp\left[\frac{-\frac{1}{\lambda_o}\left(x - u\,t\right)}{\frac{u}{c_o}\sqrt{1 - v^2/c_o^2}}\right] + \frac{a}{2}\,,$$

or, respectively, according to (99) with our elementary length $L_o = \pi\,\lambda_o\,(= \sqrt{\frac{\pi\,a\,\sigma}{2D}})$ according to (91))

$$q^T(x,t) = \frac{2a}{\pi}\,\arctan\exp\left[\frac{-\pi(x - u\,t)}{L_o\,\kappa}\right] + \frac{a}{2}\,. \tag{261}$$

The function $q^T(x, t)$ is thus a solution of the sine-GORDON equation (88) if the function $\mathrm{q^T(x, t)}$ is a solution of the sine-GORDON equation (95). We know that the function

$$\mathrm{q^I(x, t) = 4 \arctan \exp\left[\frac{x - v\,t}{\sqrt{1 - v^2}}\right]} \tag{262}$$

fulfils Eq. (95), because (262) is nothing other than the Eq. (111) transcribed to the dimensionless variables q, x, v and t. Using (254), it is

$$\mathrm{q^T(x, t) = q^I(t, x) + \pi} \tag{263}$$

and thus

$$\begin{aligned}\frac{\partial^2 \mathrm{q^T(x,t)}}{\partial \mathrm{x}^2} - \frac{\partial^2 \mathrm{q^T(x,t)}}{\partial \mathrm{t}^2} &= \frac{\partial^2 \mathrm{q^I(t,x)}}{\partial \mathrm{x}^2} - \frac{\partial^2 \mathrm{q^I(t,x)}}{\partial \mathrm{t}^2} = -\left(\frac{\partial^2 \mathrm{q^I(t,x)}}{\partial \mathrm{t}^2} - \frac{\partial^2 \mathrm{q^I(t,x)}}{\partial \mathrm{x}^2}\right)\\ &= -\sin \mathrm{q^I(t, x)} \qquad\qquad = \sin\left(\mathrm{q^I(t, x)} + \pi\right)\\ &= \sin \mathrm{q^T(x, t)}\,,\end{aligned}$$

which is exactly what we wanted to show.

What does this new solution (261), $q^T(x, t)$, of the sine-GORDON equation (88) look like? See Fig. 20.1. The solution $q^T(x, t)$ has similarities with our kink solution (111); see also Figs. 9.1 and 10.2. There are however important differences. Firstly, q^T is displaced by $a/2$ in the direction of increasing q-values. This dislocation as an edge line of a lattice plane ending inside of a crystal is located at $q = a/2$ or $q = 3a/2$ when $x \longrightarrow +\infty$ or $x \longrightarrow -\infty$; hence, it is located on the unstable maximum of the lattice potential $D \sin\left(\frac{2\pi}{a} q(x, t)\right)$; see Eq. (86).

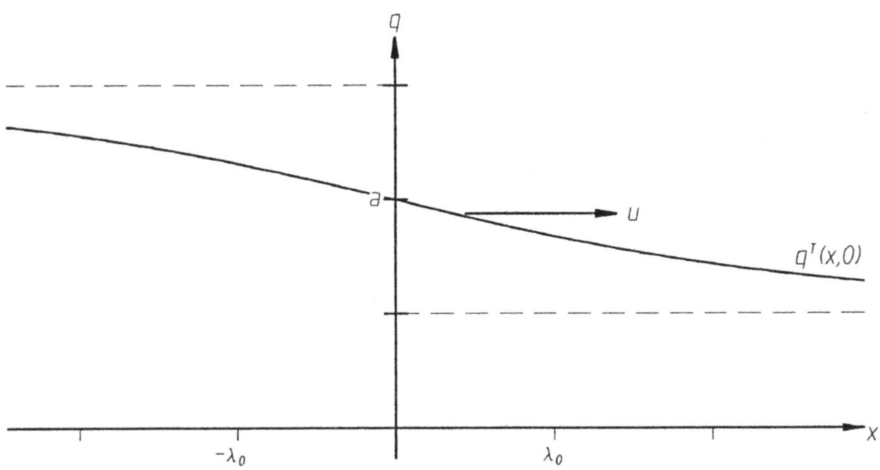

Fig. 20.1 Tachyon solution (261), $q^T(x, t) = \frac{2a}{\pi} \arctan \exp[\frac{-\pi(x-u\,t)}{L_o\,\kappa}] + \frac{a}{2}$ for $t = 0$ and $u = 2\,c_o$

The function q^T has no position of rest, whereas the kink q^I does. The velocity u with which this figure moves in the direction of increasing x-values is, according to (257), principally always larger than the critical velocity c_o,

$$u > c_o \quad \text{for} \quad v < c_o \ . \tag{264}$$

The interval on the x-axis in which the function q^T changes from $q = 3a/2$ to $q = a/2$ is determined by the quantity $L_o\,\kappa$. With exception of a small limit $c_o < u \approx 1,4\,c_o$ this interval is always larger than the analogue interval of an arbitrary kink solution where this changes from one minimum of the lattice potential to the neighbouring; see Figs. 9.1, 10.2 and 18.7. Notice that only very high energy tachyons are attributed velocities near the limit c_o, as we will explicitly see in Chap. 29, (442). Generally, the function q^T changes, in comparison to the kink q^I cf. (111), along an extended region, this region extending infinitely with increasing velocity u. EILENBERGER [11] was the first to point at the tachyon character of this solution (261), $q^T(x,t)$. There are numerous such tachyon states inside of a crystal; see GÜNTHER [37]. Our derivation of the sine-GORDON equation, developed along the same lines as our wave equation, shows that a 'disturbance' running through a medium does in fact represent an energy transfer. In Chaps. 22 and 23, we will discuss the definitions of an energy density e, of stress $\mathbf{t}$, as well as momentum density $\mathbf{p}$ and a density of energy current $\mathbf{s}$ for the solutions of the sine-GORDON equation; see also GÜNTHER [37]. We will also discuss the particle aspect of a sine-GORDON field in Chaps. 22 and 23. In Chaps. 27–29, we will discuss the characteristics of a tachyon, and based upon this we will then analyse the questions of a violation of causality for those problems that we can actually calculate mathematically.

We will show that a tachyon moving with a permanent increasing velocity u results in a decreasing energy density moving through 'space', thus through a crystal, until a 'vanishing energy density with infinite velocity' moves through the crystal so that the only thing remaining is a permanent state of stress.

Does the internal observer with his tachyons come into conflict with his Special Theory of Relativity?

Let us consider the extreme case $u \longrightarrow \infty$. The solution $q^T(x,t)$ transforms into a solution $q^T_\infty(t)$. With (261) and

$$\kappa = \frac{u}{c_o}\sqrt{1 - \frac{c_o^2}{u^2}}$$

we find for this function

$$q^T_\infty(t) = \lim_{u\to\infty} q^T(x,t) = \frac{2a}{\pi}\,\text{arctan}\,\text{exp}\left[\frac{\pi c_o}{L_o}\,t\right] + \frac{a}{2} \ . \tag{265}$$

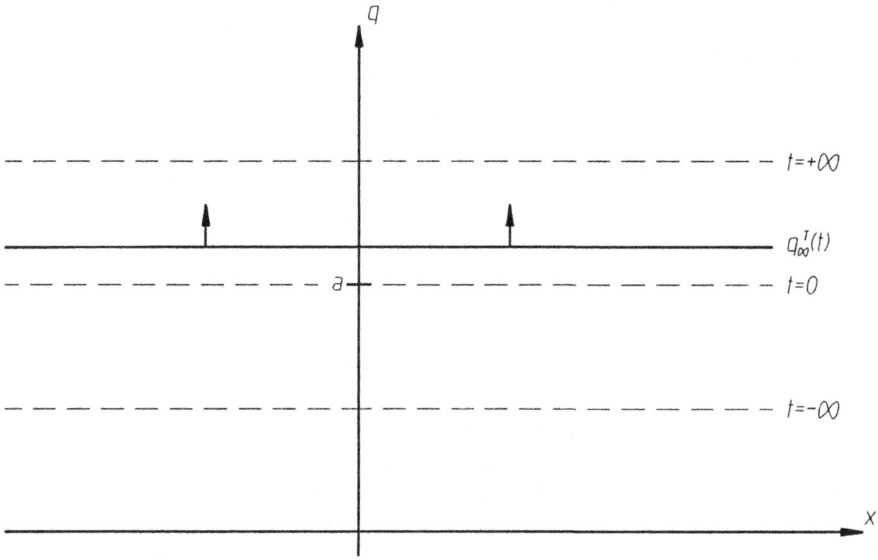

Fig. 20.2 Tachyon solution (265), $q^T_\infty(t) = \frac{2a}{\pi}$ arctan exp$\left[\frac{-\pi\, c_o}{L_o}\, t\right] + \frac{a}{2}$. The limiting positions for $t = -\infty$ and $t = +\infty$, as well as the position of the solutions at $t = 0$ are shown using a dashed line. The arrows show the direction of motion

The internal observer in a reference system Σ_o then registers that a certain state of stress $\mathbf{t}$ simultaneously for all x-values and underlies a change through time according to function (265); see Fig. 20.2.

We will now assume that the observer in Σ_o could use these facts in order to synchronise his clocks that he has distributed throughout his frame with the help of a tachyon. In order to achieve this, he would have to give the following order: All clocks are started, their setting starting at $t = 0$, as soon as the state of stress $\mathbf{t}$ reaches its maximum value. With the help of Eqs. (276) and (290) for function q^T (according to (265)), one can explicitly check that the state of stress defined by function (265) has its maximum value at $t = 0$. Thus, a starting setting can be determined for all clocks in Σ_o by a signal with an infinitely large velocity. This starting position, as seen in Chap. 10, does not change anything in the pace of clocks moving relatively to one another. A reference system Σ', in which all events are attributed space and time coordinates x' and t', possesses the velocity v if observed from the reference system Σ_o. As we have seen in Chap. 13, the comparison of the space and time measurements in both reference systems shows the relation between the time t' of the clock of Σ' at the position x' $\big($an event $E(x', t')\big)$, and the time t that the clock of Σ_o shows at the position x (the same event $E(x, t)$) according to Eq. (151a),

$$t = \frac{t' + v\,x'/c_o^2}{\sqrt{1 - v^2/c_o^2}}\,,$$

and the observer in Σ' registers according to (265) a function $q^T(x', t')$ for the tachyon observed in Σ_o. For this function, we find, analogue to (261),

$$q^T(x', t') = \frac{2a}{\pi} \arctan \exp \left[\frac{\frac{\pi c_o}{L_o}\left(t' + \frac{v\,x'}{c_o^2}\right)}{\sqrt{1 - v^2/c_o^2}} \right] + \frac{a}{2} \,,$$

thus

$$q^T(x', t') = \frac{2a}{\pi} \arctan \exp \left[\frac{\pi(x' + u\,t')}{L_o\,\kappa} \right] + \frac{a}{2} \,. \tag{266}$$

We therefore find the tachyon (265) that is observed in Σ_o. The observer in Σ', who when observed from Σ_o moves in the direction of positive x-values with the velocity v, notices a tachyon moving in the direction of negative x-values with the velocity $-u = -c_o^2/v$. In this respect, the state of the tachyon $q_\infty^T(t)$ is the exact opposite of the static kink solution $q^I(x)$, see (105), so that $q^T(x', t')$ corresponds to the moving kink $q^I(x', t')$. For the observer moving with the velocity v with respect to Σ_o, the kink resting in Σ_o moves with the velocity $-v$ and shows a LORENTZ contraction with the factor γ. For tachyons, we have to replace v with $u = c_o^2/2$ and γ with $\kappa = \gamma\,u/c_o$. We will furthermore assume that the observer in Σ' can also synchronise his clocks using this tachyon. He would give, as if he were synchronising his clocks using a sound signal, the following instructions: When the tachyon arrives at the clock positioned at x', then this clock must be started, according to the running time of the tachyon, from the coordinate origin to x' with the hand setting t' according to

$$t' = \frac{x'}{-u} = \frac{-v\,x'}{c_o^2} \,. \tag{267}$$

Just as with Eq. (113a) $x' = x/\sqrt{1 - v^2/c_o^2}$ applies and therefore

$$t' = \frac{-v\,x/c_o^2}{\sqrt{1 - v^2/c_o^2}} \,. \tag{267a}$$

The internal observers in both reference systems would then have synchronised their clocks using the same tachyon (here we have once again assumed that both clocks had the hand setting 0 at their origin of coordinates). The result of this synchronisation would be: If all clocks in Σ_o have the setting $t = 0$, then all clocks in Σ' have the setting t' according to (267a). This is however nothing other than LORENTZ transformation (151) for $t = 0$, or exactly the hand setting that we found in Chap. 13, Eq. (142) (see also Fig. 12.7), from the different pace of two clocks moving with respect to one another. Tachyons would therefore be ideally suited for synchronising clocks. We therefore cannot talk of a contradiction to the Special Theory of Relativity.

This ideal experiment should not lead us to the hasty conclusion that we have finally, with the help of tachyons, found a signal that allows us to transmit information with infinite velocity. We therefore, until we have discussed this in detail in Chaps. 28–30, formulate the following provisional thesis:

The tachyons of the sine-GORDON equation are not signals.

Causality is the cause of the conflict if we accept the existence of tachyons.

We will firstly convince ourselves that causality is in fact violated if we could transmit signals using tachyons. In order to do this, we examine three reference systems, Σ_o with the coordinates x and t, Σ' with the coordinates x' and t' and $\overline{\Sigma}$ with the coordinates $\bar{x}$ and $\bar{t}$, where Σ' and $\overline{\Sigma}$ possess the velocities $+v$ and $-v$, respectively, measured from Σ_o.

A tachyon (265) may exist in Σ_o,

$$\Sigma_o : \quad q^T_\infty(t) = \frac{2a}{\pi} \arctan \exp\left[\frac{\pi\, c_o}{L_o}\, t\right] + \frac{a}{2}\,, \tag{268}$$

that is registered in Σ' as $q^T(x', t')$ according to (266),

$$\Sigma' : \quad q^T(x', t') = \frac{2a}{\pi} \arctan \exp\left[\frac{\pi\,(x' + u\, t')}{L_o\, \kappa}\right] + \frac{a}{2}\,, \tag{268a}$$

and registered in $\overline{\Sigma}$ as $q^T(\bar{x}, \bar{t})$,

$$\overline{\Sigma} : \quad q^T(\bar{x}, \bar{t}) = \frac{2a}{\pi} \arctan \exp\left[\frac{\pi\,(\bar{x} - u\, \bar{t})}{L_o\, \kappa}\right] + \frac{a}{2}\,, \tag{268b}$$

where we receive the last equation by simply replacing v with $-v$ in (266). We could also replace u with $-u$, but we would then additionally have to replace κ with $-\kappa$ because of $\kappa = \gamma\, u/c_o$.

For all simultaneous events in Σ_o, according to $t = 0$, at arbitrary positions along the x-axis, the observer in Σ' registers, in dependence on his positions x' according to (267), the times $t' = -v\, x'/c_o^2$, whilst the observer in $\overline{\Sigma}$ registers the times $\bar{t} = +v\, \bar{x}/c_o^2$,

$$\left.\begin{aligned} \Sigma_o &: \quad t = 0\,, \\ \Sigma' &: \quad t' = \frac{-v\, x'}{c_o^2}\,, \\ \overline{\Sigma} &: \quad \bar{t} = \frac{+v\, \bar{x}}{c_o^2}\,. \end{aligned}\right\} \tag{269}$$

We observe two simultaneous events E_1 and E_2 in Σ_o,

$$\Sigma_o : \quad \begin{cases} E_1 : & t_1 = 0, \quad 0 < x_1 , \\ E_2 : & t_2 = 0, \quad x_1 < x_2 . \end{cases} \qquad \boxed{t_1 = t_2} \tag{270}$$

These events are observed from Σ' as follows. Firstly, the distance x from the origin of coordinates in Σ_o is the LORENTZ contracted length x', thus $x = x' \sqrt{1 - v^2/c_o^2}$. For the events E_1 and E_2 to following can be derived from (269), in other words applying (151),

$$\Sigma_o : \quad \begin{cases} E_1 : & t_1' = \dfrac{-v\, x_1}{c_o^2 \sqrt{1 - v^2/c_o^2}} , \quad 0 < x_1 , \\ E_2 : & t_2' = \dfrac{-v\, x_2}{c_o^2 \sqrt{1 - v^2/c_o^2}} , \quad x_1 < x_2 . \end{cases} \qquad \boxed{t_2' < t_1'} \tag{270a}$$

The observer in $\overline{\Sigma}$ also finds using $x = \bar{x} \sqrt{1 - v^2/c_o^2}$, and if we only replace v with $-v$,

$$\overline{\Sigma} : \quad \begin{cases} E_1 : & \bar{t}_1 = \dfrac{+v\, x_1}{c_o^2 \sqrt{1 - v^2/c_o^2}} , \quad 0 < x_1 , \\ E_2 : & \bar{t}_2 = \dfrac{+v\, x_2}{c_o^2 \sqrt{1 - v^2/c_o^2}} , \quad x_1 < x_2 . \end{cases} \qquad \boxed{\bar{t}_1 < \bar{t}_2} \tag{270b}$$

According to (270b), a detective in $\overline{\Sigma}$ would observe the event E_2 occurring after event E_1. For a detective in $\overline{\Sigma}$ the event E_2 (maybe a murder, for example the death of MR STUS) could in fact be the result of event E_1 (DR FAST's fatal shot caused by the tachyon). A detective in the reference system Σ' who relies on the principle of causality would contradict this energetically. He states according to (270a) that event E_1 takes place after event E_2 that MR STUS was already dead (E_2) before (E_1), the fatal shot, even occurred. Those relying on the principle of causality will see the observations of a SHERLOCK HOLMES in Σ' as undeniable proof of DR FAST's innocence.

Only the strict formalist (who hopefully does not act as a juror) would object, stating that this time the principle of causality did not hold, according to the motto—because I watered the plants today they started blooming yesterday. We do not want to continue philosophising about this here. However, because of the physics of tachyons, or at least as much as we can mathematically discuss it in the frame of our crystalline solid, the hypothesis of an acausal interaction cannot be supported. The transmission of a signal is always combined with the transference of energy, be it as minute as possible. Now, we have already stated that with tachyons an energy density does in fact move through space (in our case through the crystal). This alone does not suffice. This energy has to be able to be released. In Chap. 30, we will show using an elementary calculation that an 'elastic collision' between a particle and a tachyon results in nothing. This, at least, is one way that a tachyon does not release energy to a particle, and therefore, also does not transmit a signal.

Inelastic collisions between particles and tachyons, where the transfer of energy occurs, are however compatible with the energy–momentum conservation laws.

However, we need an equation allowing tachyon solutions as well as such a particle tachyon collision as a solution. Only in that case, we could speak of the physical reality of such processes. In our case, this is the sine-GORDON equation, but in this equation we can find no such processes. We will delve deeper into this problem in Chaps. 28–30, with the result:

> The tachyons of the sine-GORDON equation do not realise interactions, they realise a correlation.

We will now show that the property $|u| > c_o$ of the tachyons discussed by us is an immediate consequence of the relativity of simultaneity of the internal observers of the crystal, if such a solution (265) q^T_∞ applies in the crystal. Nevertheless, these tachyons exist inside of the crystal just as they do for the outside observer.

In the reference system Σ_o, the tachyon $q^T_\infty(x,t)$ represents a state that depends according to (268) only on time t and not on x, so that everywhere in Σ_o the same original change in state occurs simultaneously. One can show, see Chap. 28, that this is a state of stress $\mathbf{t}$ that changes synchronically with hand settings of the clocks in Σ_o. If all clocks in Σ_o have the hand setting t, then the clocks of an observer in Σ', who moves with the velocity v with respect to Σ_o, show the hand setting t' at the position x' in Σ' according to our Eq. (151a), thus

$$t = \frac{t' + v\,x'/c_o^2}{\sqrt{1 - v^2/c_o^2}} \,.$$

The observer in Σ' registers a state $q^T(x',t')$ due to his clock settings according to

$$q^T(x',t') = \frac{2a}{\pi} \arctan\exp\left[\frac{\dfrac{\pi c_o}{L_o}\left(t' + \dfrac{v\,x'}{c_o^2}\right)}{\sqrt{1 - v^2/c_o^2}} \right] + \frac{a}{2}\,.$$

We can however rewrite this using (257), (258) and (259) with the result

$$\Sigma': \quad q^T(x',t') = \frac{2a}{\pi} \arctan\exp\left[\frac{\pi\,(x' + u\,t')}{L_o\,\kappa} \right] + \frac{a}{2}\,. \tag{268a}$$

If we explicitly choose $u = -2c_o$, then we get the plot in Fig. 20.1. Therefore, whilst the observer in Σ_o registers the constant function according to Fig. 20.2, the observer in Σ' registers the function, based on the effect of the relativity of simultaneity, as plotted in Fig. 20.1. An outside observer could state that the observations made from Σ' are artificial and thus ignore them, donating only the observations made in Σ_o any meaning. Can we conclude that the state registered by the observer in Σ', a state produced only using a peculiar time measuring method, resulting in an artificial operand, that this operand has no physical meaning? Certainly not! The observer in Σ' also does have a solution $q^T_\infty(t')$,

$$\Sigma' : \quad q^T_\infty(t') = \frac{2a}{\pi} \arctan \exp\left[\frac{\pi c_o}{L_o} t'\right] + \frac{a}{2}, \tag{271}$$

in other words, the function shown in Fig. 20.2! Once again, the outside observer may ignore the representations of the observer in Σ', stating that these are artificial, and follows the results of the internal observers in Σ_o. However, for the uniform hand settings t' of the clocks in Σ', the observer in Σ_o positioned at x discovers a hand setting of t for his clocks according to the equation (notice that Σ_o has the velocity $-v$ with respect to Σ')

$$t' = \frac{t + -v\,x/c_o^2}{\sqrt{1 - v^2/c_o^2}} \ .$$

We insert this expression in place of t' into Eq. (271), and after the same procedure as above the observer in Σ_o discovers the function; see also (268b),

$$\Sigma_o : \quad q^T(x,t) = \frac{2a}{\pi} \arctan \exp\left[\frac{-\pi\,(x - u\,t)}{L_o\,\kappa}\right] + \frac{a}{2} \tag{271a}$$

instead of the constant function of the observer in Σ'.

This is however a tachyon moving precisely with the velocity $u > c_o$ (261), whose shape has been illustrated for $t = 0$ and $u = 2c_o$. The function (271a) is a change of state inside the crystal measured by the internal observer and confirmed by the outside observer. We recapitulate as follows:

> The property $|u| > c_o$ of tachyons is founded on an internal relativistic simultaneity effect inside of a crystal.

This once again illustrates the relevance of the internal standards, the measuring-rods and clocks, discussed in detail by us, for the physics of a crystal. We will come back to this in Chaps. 22 and 23 where we will discuss the dynamic properties of the solutions of the sine-Gordon equation and will then be able to base the inertia of energy on this. Tachyons also play an important role for a large number of solutions of the sine-Gordon equation, sf. also Günther [37].

Here, we once again refer to the exact Lorentz symmetry for the structural parts of the eigen stresses of dislocations in case of arbitrary crystal symmetry. We will go into this in the appendix, Chap. 27.

Chapter 21
Violation of Relativity—The Rediscovered Crystal

For an internal observer of a crystal, all of the phenomena observed here, such as plastic and certain elastic deformations of a crystal, represent motions in a relativistic spacetime continuum. We discovered this in Chap. 15 and confirmed this in Chap. 18 where we discussed the DOPPLER effect. For the outside observer, the crystal is just the place where all these phenomena occur, and he can understand these phenomena with the help of the crystal. Furthermore, the crystal makes him clear the internal observer's explanations and descriptions for these phenomena. This crystal has, from the internal observer's point of view, no state of motion and is principally nothing other than a relativistic spacetime continuum. This is valid as long as all assumptions that are needed for this ideal model description are fulfilled. These assumptions are as follows:

1. The primary discrete distribution of the elastically coupled masses is replaced with a continuous distribution according to (84).
2. We calculate using the linearised theory of elasticity according to (85) and also using the equation $c_T = c_o$ if we include transversal sound waves.
3. The principal dependence $m = m_o/\sqrt{1 - v^2/c_o^2}$ of the lattice atom's mass m on their velocity v according to the EINSTEIN's Special Theory of Relativity is ignored. The velocity v of the oscillating atoms does not become larger than the speed of sound c_T, of which we assume that it is many times smaller than the speed of light c_L, so that $v^2/c_L^2 < c_T^2/c_L^2 \ll 1$.

We thus work with non-relativistic mechanics for the lattice atoms. Structures on this lattice, kinks for example, show a relativistic behaviour, however, where the sound velocity is the limit, but not the light velocity.

The law of motion (79) for dislocations together with Eqs. (80) and (81) for interaction forces results in the sine-GORDON equations (88) and (95) with the critical velocity c_o.

We will stay here and consider for a whilst the area, defined by the strict model assumption **1–3**. We will alternatively change our position from that of an internal

H. Günther, *Elementary Approach to Special Relativity*,
https://doi.org/10.1007/978-981-15-3168-2_21

observer to that of an outside observer, who obviously knows more than his internal colleague.

It is fascinating to realise that for the internal observer, a kink (or other configurations of dislocations) is an object that possesses energy, that moves through space, an object that does exactly that what a particle does, a particle with the characteristic property of individual identity. This individuality is guaranteed by the attribution of its position described by the space coordinate x at a certain time t. The particle moves through space according to a well-defined trajectory. We will go into the mathematical details of the physical particle concept for these micro-plastic deformations, especially for kinks, in Chaps. 22 and 23, see also GÜNTHER [37].

The outside observer, who can fixate the crystal and its lattice atoms, sees this completely differently. This supposed particle, the kink, is nothing other than a certain structure of missing lattice positions, a quasi-particle as he would state. If this quasi-particle is displaced through the lattice, then what happens is that the same structure is constructed by other atoms. The outside observer only sees a moving structure. His individual particles, the lattice atoms, from his point of view, remain stationary at their original positions. It becomes clear to us that it makes no sense to attribute a structure an individual identity as the internal observer did.

To illustrate this, we will consider the collision of two kinks which move directly towards each other with the constant velocity v, just as two balls in snooker would. This case can be precisely calculated. We however only want to discuss the result and refer the reader to the calculations made by SEEGER [89]. After a certain amount of time spent by the kinks in rather complicated disorder, a state is achieved in which both kinks move away from each other, as snooker balls would after a collision. It would be completely meaningless to ask which of the two kinks move to the right—as in snooker, the one that came from the right, or the other? Due to the fact that a structure can be built up from completely different individual lattice atoms, we, as outside observers, do not see the need to attribute this structure an individuality, as we do with snooker balls. The internal observer, who has just constructed his concept of particle individuality, sees this completely differently. Only with difficulty could he realise that the presumed individuality of his kink is, strictly seen, not founded on any physical reality, or is at the most founded on a very restricted reality. The two states—reflection or pass each other—are indistinguishable, because the 'particles' are indistinguishable, and they are identical. We remind ourselves that even in NEWTON's mechanics concerning the collision of two particles, such a decision cannot be made without additional information. Both solutions (51) and (52) are equally possible.

Maybe this example of kinks in crystals, the 'particles' of the internal observer, is a consolation for those who still cannot accept that even for us 'outside observers' the idea of an individual particle cannot be sensibly upheld indefinitely that we also move into regions, namely if quantum phenomena become relevant, such as for cold electron and neutron motions for example, where our good old classical idea of individual particles does not properly reproduce our physical reality.

We will now go one step further and state that we have, with our physics of the internal observer inside of an infinite crystal, caught ourselves up in v.WEIZSÄCKER's theory of so-called Uralternatives. In his essay, 'Matter, Energy, Information', C. F. v. WEIZSÄCKER [98] writes 'All forms »consist of« combinations of the »last« simple alternatives. ...Matter is form. Today we understand matter as elementary particles. These are constructed from Uralternatives. Uralternatives are the last elements of real forms. ...According to the simplest model of a massive particle, its restmass is the number of Uralternatives needed to construct a particle at rest'....

It is a fact that a dislocation in a crystal is indeed nothing other than a succession of decisions concerning the occupation of a possible lattice position with atoms. These are the last elements of possible forms, i.e. C. F. v. WEIZSÄCKER's Uralternatives. In line with this, we can calculate, in Chap. 23, an expression using Eq. (299) for the inertial mass m_a of a dislocation along the length a of a lattice parameter. KRÖNER's [50] hypothesis on the inertial mass of a dislocation finds in C. F. v. WEIZSÄCKER's analysis its philosophical confirmation. Our general equation of motion (79) for dislocations can thus be seen from a more general standpoint and is thus better founded.

Let us consider a kink on a dislocation line, the change of a dislocation line from one potential minimum to the neighbouring, our solution q^I of the sine-GORDON equation, cf. (111). Such a kink is nothing other than a sequence of decisions concerning the occupation of possible lattice positions, the last element of possible forms for the dislocation line. Hence, the inertial mass m_o of a kink relative to the crystal lattice is a direct consequence of C. F. v. WEIZSÄCKER's theory of Uralternatives. In Chap. 23, see Eqs. (296) and (300), we will calculate this inertial mass m_o and prove that the kink does in fact possess all properties of a relativistic particle with respect to the crystal lattice.

In Chap. 8, we used NEWTON's axiomatics to explain the motion of dislocations relative to the crystal lattice and thus actually discovered the sine-GORDON equation. Dislocations, the 'linear mispositioning of the ideal lattice', possess NEWTONian inertia with respect to the lattice. If we therefore consider agglomerates of such mispositioning, or parts of these that represent the objects, the 'particles' of the internal observers, then these also possess inertia with respect to the lattice. If one takes the elastic deformations, which originate from such dislocation structures, into consideration, then one can calculate that these structures actually generate forces that energetically influence one another, with the aim of reaching a minimum energy state. These forces are called configurational forces in the field of continuum mechanics and can be calculated according to a general law found by J. D. ESHELBY [19].

One can also construct, in the boundaries of our model assumptions for internal observers, equations of motion for the particles of the internal observers. We do not however intend here on going into the details and problems of such equations of motion. However, no matter how such equations of motion may look like in detail, they will be, in the boundaries of our model assumptions, differential equations of such a type that the motion of a single kink, for example under the influence of a stress field of other surrounding dislocation structures, would satisfy a certain path-time law, just as we could exactly calculate the trajectory of a stone thrown up into the air. We can calculate the position of any particle at any point of time if we know sufficiently exact the initial conditions. Here, we inevitably get caught up in the fangs of the Laplacian demon: The former knows the state of the world at an arbitrary point of time, because he exactly knows his initial state. We however now make an astounding discovery concerning the world of the internal observers. The particles of the internal observers are structures, structures built up from the positions of lattice atoms. These lattice atoms themselves lay, as we know, outside of the internal observers empirical world. They make up their space—their vacuum—the 'ether'. As a consequence, all the statistical fluctuations of the lattice atom positions, which can even vary at different crystal positions depending on temperature, completely fall out of the strict causal laws of motion that the internal observers register. The Laplacian demon has been overpowered. The exact from its law of motion calculated position of a kink cannot be accepted due to the statistical fluctuations of the lattice atom positions. This calculated value only corresponds with the statistical mean of many measurements. The internal observer registers for a single measurement a statistical uncertainty, which was principally not reckoned with. Taken aback by this the internal observer states, God plays dice'. The more we heat up the crystal, the more the internal observer's God licentiously plays dice.

What if the internal observers learn to make far more precise measurements and are able to move to far greater frequencies using ever-increasing energies[1]? The internal observers would quickly arrive at the boundaries of validity of the model assumptions formulated at the beginning of this chapter.

Even if we drop all three assumptions, we would stay to find, for those structures on the lattice that find our interest, an equation, whose solutions are comparable to those of the sine-Gordon equation—with one important difference: A dispersion is created. We will now discuss this for the first of our three assumptions.

Equation (39) for the dispersion free linear chain, is bound to the condition that many atoms are involved in a single oscillation, so that we can replace the deflection of individual atoms with a continuous wave. If however the frequency is very large, resulting in the wavelength becoming very small, so small that it moves into the region of becoming the size of a distance between two atoms in a linear chain, then only a few atoms are used in the construction of a wavelength. The approximation (36), $n \ll N$, is not fulfilled anymore, and we cannot replace the cosine term in the strict Eq. (35) with the first two terms of its Taylor series. The angular velocity ω_n is not proportional to the wave number $k_n = n\,2\pi/L$ anymore, and we must calculate the signal velocity on the chain using Eq. (41), with the result (44). The wave's velocity of propagation $c_K^{(n)}$ now depends on its frequency ω_n. A dispersion is the result. We can no longer speak of a uniform signal velocity, especially if we include the non-linear theory of elasticity, as well as the relativistic (with the speed of light c_L) change of mass of the lattice atoms.

The sine-Gordon equation, the central item of all our considerations, begins to lose cohesion with increasing energies and frequencies. Without a uniform critical velocity c_o for all frequencies in the sine-Gordon equation, we would not receive the

[1]The energy can only be increased by the velocity of motion and thus by the number of passages across the zero position per second, in other words the frequency, because the amplitude of an oscillation is restricted by the crystal's geometry.

universal length contraction of moving measuring rods, nor would we get the time dilatation of moving clocks.

Without length contraction and time dilatation, no principle of relativity for signal propagation!

Special Relativity, the absolute equality of all internal observer's reference systems moving with uniform velocity with respect to each other, is broken. The internal observers are able to, thanks to their high energy physics, track the background of their specific relativistic equations. They examine their space, their vacuum and are finally in the position of discovering the crystalline origin of their vacuum, the crystal lattice, and maybe even the NEWTONian physics of this lattice, which, as we know, is a primary theory for all mechanical processes inside of a crystal.

This does not mean that all of our prior examinations, the results of Chaps. 8–20, are simply false; it means that these are just an approximation—as everything in physics is. In the region of not too large energies, everything remains as it was. The internal observers then exist in a continuous structured space, for which, in the boundaries of their 'everyday experiences', the Special Theory of Relativity is valid: All reference systems moving with a uniform velocity with respect to each other are physically indistinguishable. It is only when the internal observers start to experiment with extreme energies, which does not constitute 'everyday experiences', that they are in the position of coming to the conclusion that the space in which they dwell is actually of a lattice construction. One can make it clear for oneself that a reference frame Σ_o is preferred by a lattice parameter a, the former also being the reference system resting in the crystal: Σ_o is simply the reference system in which a lattice parameter a has the maximal length. Nevertheless, in the regions of minute energies, in other words for relatively large wavelengths λ, which are very large in comparison with this lattice parameter, $\lambda \gg a$, so that the crystal behaves like a continuum, in these regions the reference system Σ_o would not distinguish itself from the others.

Does this in fact therefore mean that the length and time measurements introduced in Chaps. 9 and 10 for internal observers are unnecessary? Of course not! The meaning of the internal crystal's time, a time that is completely different to that of the outside observers, is simply because the speed of light is so much larger than the sound velocity, and this internal crystal's time controls the complete dynamics of its defects, the dynamics of dislocation configurations. We will go into questions connected with this in Chaps. 22 and 23. It is important to bring into mind that the masses of dislocation configurations, of kinks, breathers, etc., are completely different than the masses of the lattice atoms of the crystal. Their dynamics follows different laws. The easiest way to see this is to recognise the different critical velocities, on the one hand sound velocity, on the other the speed of light.

There is a fundamental connection between the relativistic dynamics of particles and relativistic time. The connection is closer than our examinations have shown up to date. In our representation, the relativistic dynamics of particles is a direct consequence of the relativistic spacetime structure, as the path shown in EINSTEIN's [14–16] papers states. One can however show that the reverse path is possible. If we primarily have the relativistic mass' dependency on velocity—this could experimentally be given for example—then one can get the relativistic spacetime structure from this. This approach to the Special Theory of Relativity was first explicitly shown by Liebscher [57] with the help of a so-called dynamic definition of simultaneity. To be able to better classify this method and fit it into the axiomatics of the Special Theory of Relativity, see Chap. 16. At the end of Chap. 23, our relativistic concept of particles with its characteristic dependency on the velocity of inertial masses will be complete, so that this approach to the Special Theory of Relativity will also be open to the internal observers in the infinite crystal lattice.

Chapter 22
Particles and Fields

The typical characteristic for a system of particles is the number of particles contained in the system. Every single particle has a velocity v and an inertial mass m attributed to it. The typical physical parameters of a single particle that play a crucial role in interaction between the particles are momentum $p = m\,v$ and energy E. The latter is determined by the mass in a relativistic theory contained in the EINSTEINian formula $E = m\,c^2$, and it is expressed using the momentum and the mass in the non-relativistic approximation according to $E_{kin} = \frac{p^2}{2m}$. Characteristic for an interaction between two particles is a collision. During this process, all of the particle's total energy and momentum are available for exchange—upholding the laws of energy and momentum conservation for these quantities; see the next chapter.

What do the solutions of the sine-GORDON equation $q = q(x, t)$ have in common with particles? The most important statement in Chap. 5 was that NEWTON's mechanics does not make any statement concerning the number of masses involved in a motion. On the basis of NEWTON's axiomatics, we were able to, during our process of deriving the sine-GORDON equation, immediately move on to the limit of infinitely many particles. We thus got the field $q = q(x, t)$ enabling us to describe motion of the positions of the dislocations line in the proximity of a straight line. If we revert, in respect of the above, back to talking about particles, of single masses and velocities, then these particles can only correspond to those masses that formed the starting point of our considerations. These however would be the masses m_α of the NEWTONian equations (79). These masses do not refer to our (outside) physical space, they are inertias with respect to the crystal lattice, as we discussed in detail in Chap. 8. If we therefore, with respect to the solutions of the sine-GORDON equation, talk of particles, then these are the particles of the internal observers of the crystal, which has our full attention.[1] The question that we have to answer here is: Does the field $q = q(x, t)$ that is composed of many, or even infinitely many particles behave as if it were one

[1] Such particles with respect to the lattice in physics are called quasi-particles. They are called so in order to differentiate them from the particles of our outside physical space.

H. Günther, *Elementary Approach to Special Relativity*,
https://doi.org/10.1007/978-981-15-3168-2_22

particle—or using the expression of mechanics, as a 'solid'—with one mass and one single velocity, or maybe as several such particles or solids?

The function $q = q(x, t)$ is a field that has the value q for every x and every t—just as the solutions of MAXWELL's equations for electric and magnetic field strengths are attributed to specific values in space at every point of time. Physical fields are different with respect to pure mathematical functions in that they are carriers of energy and momentum, which are distributed in a certain density throughout space. Exchange processes involving energy and momentum occur throughout space when two fields interact. Only under certain conditions, describing the interactions between fields, the net result of energy and momentum can be represented as if it were a collision between particles. As is well known, the energy and momentum distribution of an electromagnetic field according to MAXWELL's theory only can be described using a system of particles if the space actually contains no charge carriers. In this way, EINSTEIN was able to introduce his photons, the 'light particles' for the electromagnetic field in vacuum. We will now examine the connection between particles and fields for the sine-GORDON equation. We will first of all agree upon a simplification of notation.

Mathematical quantities, whose properties do not just serve to attribute a number, for example vectors and tensors, have been distinguished using bold characters; we therefore write $\mathbf{F}$ for a force vector in comparison with the length L of a rod, or the temperature T. The boldface type also has the role of generally informing the reader that the corresponding quantity has to be described using more than one numerical date, for example using the components F_x and F_y of a force vector $\mathbf{F} = (F_x, F_y)$ on the x-y-plane. However, we only have to deal with one spatial dimension when discussing the sine-GORDON equation. In this case, spatial vectors and tensors are always described using only one number, $\mathbf{F} = (F_x)$. Even a one-dimensional vector cannot be completely described using only one numerical date. If the direction of the axis is reversed, then the vector component F_x changes its mathematical sign, and the length L or the temperature T stays unchanged. The one-dimensional force vector also has a direction, whereas temperature does not. This special vector property will only play one important role later on in Chap. 29 during our discussion about the tachyon momentum and the velocity of tachyons. We will therefore not to use boldface type for vectors or tensors composed of only one component, so that $(F_x) = F$. We will use boldface type only for quantities made up of two or more components; see for example (276).

A field always creates four quantities that are defined in space:

1. The energy density e describes the energy ΔE present in a (sufficiently small) volume ΔV. In our one-dimensional case of the sine-GORDON equation, $\Delta V = \Delta x$ and thus

$$e = e(x, t) = \frac{\Delta E}{\Delta x} \, . \qquad \text{Energy density of the field } q(x,t) \qquad (272)$$

2. This energy density does not generally remain stationary. It changes according to the energy flux density s, which in our one-dimensional case just simply describes the energy ΔE, which streams out of the 'volume element' Δx during the time interval Δt with a positive mathematical sign if s flows in the direction of increasing x-values,

$$s = s(x,t) = \frac{\Delta E}{\Delta t}\,. \qquad \text{Energy flux density of the field } q(x,t) \qquad (273)$$

3. Our field now produces a momentum density p. This is the momentum ΔP of the field contained in the volume Δx divided by Δx, thus

$$p = p(x,t) = \frac{\Delta P}{\Delta x}\,. \qquad \text{Momentum density of the field } q(x,t) \qquad (274)$$

Here one has to take note that ΔP is only the momentum of our q-field on the length Δx and not the momentum of the dislocation masses m_α according to Eq. (79) in Chap. 8. ΔP only incorporates the increase of the momentum based on the q-field.
4. This momentum density generally changes because of a stress t in the field. The stress t can also be physically understood as the negative field momentum flux density. In our one-dimensional case, this stress is simply reduced to a force F that acts from the particular neighbouring 'volume element' Δx on the particular end point of the 'volume element' Δx. Here, traditionally the stress is calculated as positive if the force at the right-hand side of Δx points in a positive direction,

$$t = t(x,t) = F\,. \qquad \text{Stress of the field } q(x,t) \qquad (275)$$

These four quantities can be combined into the energy–momentum tensor $\mathbf{T}$ of a field $q(x,t)$ according to[2]

$$\mathbf{T} = \begin{pmatrix} -t & p \\ -s & -e \end{pmatrix}. \qquad \text{Energy–momentum tensor of the field } q(x,t) \qquad (276)$$

If the energy and the momentum of the field $q(x,t)$ are only redistributed through time and not transferred on the outside or passed on to other objects, then the following equations must be upheld.

A reduction in energy $-\Delta e\,\Delta x$ during the time lapse Δt only occurs in the volume Δx if the positive amount of energy $s(x+\Delta x,t)\,\Delta t$ flows to the right and the positive amount of energy $s(x,t)\,\Delta t$ flows into the volume from the left, therefore applying Taylor's formula for $s(x+\Delta x,t)$,

$$-\Delta e\,\Delta x = s(x+\Delta x,t)\Delta t - s(x,t)\Delta t = \left(s(x,t) + \frac{\partial s(x,t)}{\partial x}\Delta x\right)\Delta t - s(x,t)\,\Delta t\,,$$

[2]We have written the mixed components of the energy–momentum tensor $\mathbf{T} = T_a{}^b$ in the notation of (here two-dimensional) Minkowski space, with the coordinates $x_1 = x$ and $x_2 = t$ and the metric $(1, -c_o^2)$, so that $\mathrm{div}\mathbf{T} = \frac{\partial}{\partial x} T_a{}^1 + \frac{\partial}{\partial t} T_a{}^2$, $a = 1, 2$, see (278).

and thus

$$\frac{\partial e}{\partial t} = -\frac{\partial s}{\partial x} \,. \tag{277}$$

An increase of momentum $\Delta p \, \Delta x$ during the time lapse Δt in the volume Δx also only occurs if the force pulse $t(x + \Delta x, t) \, \Delta t$ acts at the right end of the volume Δx, and if the force pulse $t(x, t) \, \Delta t$ acts at the left end of the volume Δx , therefore once again applying TAYLOR's formula,

$$\Delta p \, \Delta x = t(x + \Delta x, t)\Delta t - t(x, t)\Delta t = \left(t(x, t) + \frac{\partial t(x, t)}{\partial x} \Delta x \right) \Delta t - t(x, t) \, \Delta t \,,$$

hence

$$\frac{\partial p}{\partial t} = \frac{\partial t}{\partial x} \,. \tag{277a}$$

One can therefore summarise and state, for Eqs. (277) and (277a), that the energy–momentum tensor $\mathbf{T}$ is solenoidal,

$$\operatorname{div} \mathbf{T} = 0 \; : \quad \left. \begin{aligned} -\frac{\partial t}{\partial x} + \frac{\partial p}{\partial t} &= 0 \,, \\ -\frac{\partial s}{\partial x} + \frac{\partial e}{\partial t} &= 0 \,. \end{aligned} \right\} \tag{278}$$

We note that the basic element of NEWTONian mechanics, namely the Second Axiom, is contained in the field theory: According to (275), the gradient of the stress t is a force density, $\partial t/\partial x = \partial F/\partial x = f$, so that $\partial p/\partial t = f$.

A connection between particle and field can be made by considering an extremely simple model of a particle, which is automatically both a particle and a field. We picture a single particle with its mass m extended through space that has the uniform velocity v. This particle should also possess no internal motion, thus approaches the model of a rigid body in such a form that the observer sitting on this particle notices no change of state through time.

The energy E of such a particle is distributed through space with density $e = e(x, t)$, where $E = \int_{-\infty}^{+\infty} e \, dx$ and flows with the particle velocity v through space. If this is so, then (as shown in Fig. 2.1) $e(x, t) = e(x - v\,t)$ must apply. This defines an energy flux density s according to

$$s = s(x, t) = e(x, t) \cdot v = e(x - v\,t)\, v \,,$$

and therefore

$$\frac{\partial s}{\partial x} = \frac{\partial}{\partial x}(e\,v) = v \cdot \frac{\partial e(x - v\,t)}{\partial x} = -\frac{\partial e}{\partial t} \,,$$

in other words Eq. (277) is valid.

The mass m of this particle is distributed with a density ρ in the same fashion, $m = \int_{-\infty}^{+\infty} \rho\, dx$ and flows with the same velocity v through space. Once again, $\rho = \rho(x,t) = \rho(x - v\,t)$ is valid. This mass density is attributed to a momentum density p according to $p = p(x,t) = \rho(x - v\,t)\, v$. Once again, this momentum density p flows with the velocity v through space and generates a field momentum flux density $v\, p = v\, v\, \rho(x - v\,t)$, which is identical to the negative stress t, thus $t = -v\, v\, \rho(x - v\,t)$ so that

$$\frac{\partial t}{\partial x} = -\frac{\partial}{\partial x}\Big(v^2\, \rho(x - v\,t)\Big) = -v\, v\, \frac{\partial \rho(x - v\,t)}{\partial x} = v\, \frac{\partial \rho(x - v\,t)}{\partial t} = \frac{\partial p}{\partial t}\ ,$$

in other words, Eq. (277a) is valid.

We have therefore received the following result. A particle with the density ρ, the energy density e and possessing the velocity v extended through space can be attributed to an energy–momentum tensor $\mathbf{T}$ according to

$$\mathbf{T} = \begin{pmatrix} v^2 \cdot \rho & v \cdot \rho \\ -v \cdot e & -e \end{pmatrix}\ . \qquad (279)$$

Energy–momentum tensor of an extended particle

For the case considered here, namely that of a very simple mass density $\rho = \rho(x - v\,t)$ of the particle, $\operatorname{div} \mathbf{T} = 0$ immediately follows. There are however also more complicated objects, namely those that also have internal motions as well as their motion with velocity v as a whole. The relation between particle and field then demands a larger mathematical investigation, and we must refer to the literature on the subject, see for example B. D. Ivanenko [43] and A. Sokolow. For the special case of a particle at rest, we receive from (279)

$$\mathbf{T} = \begin{pmatrix} 0 & 0 \\ 0 & -e_o \end{pmatrix} \quad \text{with} \quad \frac{\partial e_o}{\partial t} = 0\ . \qquad (280)$$

From an energy–momentum tensor of the form (279), we can immediately read the velocity v of the particle, as well as its mass by integration,

$$m = \int_{+\infty}^{-\infty} \rho\, dx \qquad (281)$$

and furthermore both particle parameters E and P according to

$$\left.\begin{aligned} P &= \int_{-\infty}^{+\infty} p\,dx = v\,m \;, \\ E &= \int_{-\infty}^{+\infty} e\,dx \;. \end{aligned}\right\} \tag{282}$$

Formally, one could in any case integrate the two components e and p of the energy–momentum tensor (276) (assuming the integrals exist) in order to define the two quantities E and P according to (282). The actual question whether these two quantities E and P are really parameters of a particle remains. Up to now we have shown: If the energy–momentum tensor of a field $q = q(x, t)$ has the form (279), then this field is actually equivalent to a particle with the velocity v, energy E and momentum P according to (282). The question which other possibilities exist for the energy–momentum tensor so that this field is equivalent to a particle will not be answered here; see IVANENKO/SOKOLOV l.c.

If in our case we really do find such a field $q = q(x, t)$ that leads to an energy–momentum tensor (279), then our field theory immediately delivers us more. This theory also delivers the relationship between the mass density ρ and the energy density e, as well as the relationship between the mass m and the velocity v of this particle. We are therefore on the verge of discovering the relativistic particle mechanics from our sine-GORDON equation. However, we are not that far yet. We first have to check whether there are any cases where the energy–momentum tensor of a sine-GORDON field has the mathematical form (279). We then have to decide how to calculate the energy–momentum tensor belonging to a certain solution $q = q(x, t)$ of the sine-GORDON equation.

The field equation for $q(x, t)$, the sine-GORDON equation puts us in the position of possessing all the necessary information about our field $q = q(x, t)$. The energy–momentum tensor of this field can be directly won from the field equation. (The energy–momentum tensor of the electromagnetic field can also be derived from MAXWELL's equations for example.)

The mathematically more comfortable path uses the LAGRANGian, more precisely the LAGRANGian density L. This is a function of the fields and their derivatives from which one can, depending on certain rules, attain the field equations, as well as all important field quantities such as the energy–momentum tensor for example. We will use this method for our purposes, because the direct derivation of the energy–momentum tensor from the sine-GORDON equation does not supply anything new. The LAGRANGE formalism is founded in point mechanics and is equivalent to NEWTON's equations of motion. This method was then applied to continuum mechanics and is presently used in every field theory, for example in electrodynamics. We will not explain the LAGRANGE formalism here. For an explanation, see e.g. H. GOLDSTEIN [29], IVANENKO [43] and SSSSOKOLOW. We founded the sine-GORDON equation on Eq. (79) of NEWTON's point mechanics and can therefore apply the equivalent LAGRANGE formalism. In mechanics, the LAGRANGian is defined as the difference between kinetic and potential energy. Correspondingly, the LAGRANGian density L is in the field theory identical to the difference between the potential and kinetic energy densities, $T - W$.

The LAGRANGian density L for the field $q = q(x, t)$ of the sine-GORDON equation is, according to RUBINSTEIN [84]

$$L = L\left(q\,,\frac{\partial q}{\partial x}\,,\frac{\partial q}{\partial t}\right) = -\frac{\alpha}{2}\left(\frac{\partial q}{\partial x}\frac{\partial q}{\partial x} - \frac{1}{c_o^2}\frac{\partial q}{\partial t}\frac{\partial q}{\partial t}\right) + A\left(\cos\left(\frac{2\pi}{a}q\right) - 1\right) . \tag{283}$$

The parameters α and A guarantee that we are dealing with the LAGRANGian density of the sine-GORDON equation for a dislocation inside of a crystal; see below. Where do the individual terms of this LAGRANGian density come from?

In order to find this out, we need to calculate the contributions to the potential and kinetic energy densities of the field q.

Firstly the potential energy density: The relative strain $\varepsilon = \partial q/\partial x$ generates, according to our Eq. (86), the stress τ of the linear chain on the basis of the modulus of elasticity of this chain, which we will designate, in order to prevent mistakes in identity with the particle energy E calculated according to (282), as the line tension σ, which it is identical with (see Chaps. 6 and 8), thus

$$\tau = \sigma\ \varepsilon = \sigma\,\frac{\partial q}{\partial x} \ .$$

This stress τ is, as we know, in the one-dimensional case simply a force acting at the position x. We wish to calculate the work $\Delta W_1\,\Delta x$ needed to increase the displacement $q(x, t)$ by the displacement Δq, which is also dependent on its location, between the position x and $x + \Delta x$ with an existing stress state $\tau = \tau(x, t)$. The quantity ΔW_1 is the increase of energy density on the length Δx caused by Δq. On the piece between x and $x + \Delta x$, we approximately describe the initial state q using the first term of its TAYLOR series,

$$q = q(x, t) + \frac{\partial q}{\partial t}\,\Delta x \ .$$

A displacement $\Delta q(x, t)$ at the left-hand side of Δx changes the localised energy $\Delta W_1\,\Delta x$ on the piece Δx by $-\sigma\,\frac{\partial q}{\partial x}\,\Delta q(x, t)$. The displacement at the right-hand side brings a change of $+\sigma\,\frac{\partial q}{\partial x}\,\Delta q(x + \Delta x, t)$. All together, this would bring,

$$\Delta W_1\,\Delta x = -\sigma\,\frac{\partial q}{\partial x}\Delta q(x, t) + \sigma\,\frac{\partial q}{\partial x}\Delta q(x + \Delta x, t) \ ,$$

$$\Delta W_1\,\Delta x = -\sigma\,\frac{\partial q}{\partial x}\Delta q(x, t) + \sigma\,\frac{\partial q}{\partial x}\Delta q(x, t) + \sigma\,\frac{\partial q}{\partial x}\frac{\partial}{\partial x}\Delta q(x, t)\Delta x \ .$$

We can exchange $\partial/\partial x$ and Δ and receive

$$\Delta W_1 = \sigma \frac{\partial q}{\partial x} \Delta \frac{\partial q}{\partial x}$$

and thus in the limit of arbitrarily small displacements

$$dW_1 = \tau(x,t)\, d\, \frac{\partial q(x,t)}{\partial x} = \tau\, d\varepsilon = \sigma\, \varepsilon\, d\varepsilon \ .$$

Starting from the original state $\varepsilon = 0$, the following results from integration

$$W_1 = \sigma \int\limits_0^{\varepsilon} \tilde{\varepsilon}\, d\tilde{\varepsilon} = \sigma \frac{1}{2} \varepsilon^2 \ ,$$

because in our linear model $\sigma =$ const. is valid.

Writing again the relative strain $\varepsilon = \partial q/\partial x$ the formation of a space-dependent stress τ supplies a contribution W_1 to the potential energy density according to

$$W_1 = \frac{\sigma}{2} \frac{\partial q}{\partial x} \frac{\partial q}{\partial x} \ . \tag{284}$$

On the basis of our original state $\varepsilon = 0$ a further contribution W_2 to the potential energy density originates from out of the position of the dislocation portion Δx in the lattice. According to our assumption (81) with the limit (85), this amount W_2 sums up as

$$W_2 = -\frac{a\, D}{2\pi} \left[\cos\left(\frac{2\pi}{a} q(x,t) \right) - 1 \right] . \tag{284a}$$

The constant -1 ensures the normalisation of this contribution to the potential energy to zero at the equilibrium positions $q = n \frac{a}{2\pi}$ (cf. also (97)). The factor is chosen so that $\frac{\partial W_2}{\partial q} \Delta x$ is equal to the force of the lattice on the dislocation portion of the length Δx according to (85).

According to our linear approximation, we insert $v = \partial q/\partial t$ and according to (88) $\rho_o = \sigma/c_o^2$ into the density for kinetic energy $T = \frac{1}{2}\rho_o v^2$, so that

$$T = \frac{1}{2} \frac{\sigma}{c_o^2} \frac{\partial q}{\partial t} \frac{\partial q}{\partial t} \ . \tag{284b}$$

From (284)–(284b), we actually see for the LAGRANGian density $L = T - (W_1 + W_2)$ the RUBINSTEIN expression (283) confirmed. Furthermore, we determined the parameters α and A in such a manner that we can describe dislocations inside of a crystal, namely

$$\left.\begin{aligned} \alpha &= \sigma\,, \\ A &= \frac{a\,D}{2\pi}\,. \end{aligned}\right\} \tag{285}$$

If we also introduce our measuring-rod L_o according to (103), thus

$$\frac{A}{\sigma} = \frac{a^2}{4\,L_o^2}\,, \tag{286}$$

we discover for the LAGRANGian density of our sine-GORDON field in a crystal

$$L = -\frac{\sigma}{2}\left(\frac{\partial q}{\partial x}\frac{\partial q}{\partial x} - \frac{1}{c_o^2}\frac{\partial q}{\partial t}\frac{\partial q}{\partial t}\right) + \frac{\sigma\,a^2}{a\,L_o^2}\left(\cos\left(\frac{2\pi}{a}q\right) - 1\right)\,. \tag{287}$$

The field equation for $q = q(x,t)$ can be determined from L using the following rule,

$$\frac{\partial L}{\partial q} - \frac{\partial}{\partial x}\left(\frac{\partial L}{\partial \frac{\partial q}{\partial x}}\right) - \frac{\partial}{\partial t}\left(\frac{\partial L}{\partial \frac{\partial q}{\partial x}}\right) = 0\,. \tag{288}$$

We check that (288) together with L does in fact give us our sine-GORDON equation, doing this using the parameters α and A out of reasons of simplicity,

$$\begin{aligned}
\frac{\partial L}{\partial q} &= \frac{2\pi\,A}{a}\sin\left(\frac{2\pi}{a}q\right)\,. \\
\frac{\partial L}{\partial \frac{\partial q}{\partial x}} &= -\alpha\,\frac{\partial q}{\partial x} \quad\Longrightarrow\quad -\frac{\partial}{\partial x}\left(\frac{\partial L}{\partial \frac{\partial q}{\partial x}}\right) = \alpha\,\frac{\partial^2 q}{\partial x^2}\,, \\
\frac{\partial L}{\partial \frac{\partial q}{\partial x}} &= \frac{\alpha}{c_o^2}\frac{\partial q}{\partial x} \quad\Longrightarrow\quad -\frac{\partial}{\partial t}\left(\frac{\partial L}{\partial \frac{\partial q}{\partial t}}\right) = -\frac{\alpha}{c_o^2}\frac{\partial^2 q}{\partial t^2}\,,
\end{aligned}$$

hence finally

$$-\frac{2\pi A}{a}\sin\left(\frac{2\pi}{a}q\right) + \alpha\,\frac{\partial^2 q}{\partial x^2} - \frac{\alpha}{c_o^2}\frac{\partial^2 q}{\partial t^2} = 0\,. \tag{88$'$}$$

This is actually our sine-GORDON equation (88) if we insert the parameters from Eq. (285) in place of α and A.

The individual components of the energy–momentum tensor (276) can be determined directly from the LAGRANGian density according to the following rule:

$$\left.\begin{aligned} -t &= L + \sigma\,\frac{\partial q}{\partial x}\frac{\partial q}{\partial x}\,, \quad p = -\frac{\sigma}{c_o^2}\frac{\partial q}{\partial x}\frac{\partial q}{\partial t}\,, \\ -s &= \sigma\,\frac{\partial q}{\partial x}\frac{\partial q}{\partial t}\,, \qquad e = L - \frac{\sigma}{c_o^2}\frac{\partial q}{\partial t}\frac{\partial q}{\partial t}\,. \end{aligned}\right\} \tag{289}$$

We therefore receive the energy–momentum tensor **T** belonging to the solution $q = q(x,t)$ of the sine-GORDON equation according to

$$\mathbf{T} = \begin{pmatrix} \frac{\sigma}{2}\left(\frac{\partial q}{\partial x}\frac{\partial q}{\partial x} + \frac{1}{c_o^2}\frac{\partial q}{\partial t}\frac{\partial q}{\partial t}\right) + A\cos\left(\frac{2\pi}{a}q\right) - A & -\frac{\sigma}{c_o^2}\frac{\partial q}{\partial x}\frac{\partial q}{\partial t} \\ \sigma\frac{\partial q}{\partial x}\frac{\partial q}{\partial t} & -\frac{\sigma}{2}\left(\frac{\partial q}{\partial x}\frac{\partial q}{\partial x} + \frac{1}{c_o^2}\frac{\partial q}{\partial t}\frac{\partial q}{\partial t}\right) + A\cos\left(\frac{2\pi}{a}q\right) - A \end{pmatrix} . \tag{290}$$

We check that the energy–momentum tensor, which is built from the field equation belonging to the LAGRANGian density L, in other words the sine-GORDON equation (88) fulfils the energy–momentum conservation (278),

$$\operatorname{div}\mathbf{T} = 0 \ . \tag{291}$$

Indeed,

$$\frac{\partial}{\partial x}\left[\frac{\sigma}{2}\left(\frac{\partial q}{\partial x}\frac{\partial q}{\partial x} + \frac{1}{c_o^2}\frac{\partial q}{\partial t}\frac{\partial q}{\partial t}\right) + A\cos\left(\frac{2\pi}{a}q\right) - A\right] + \frac{\partial}{\partial t}\left[-\frac{\sigma}{c_o^2}\frac{\partial q}{\partial x}\frac{\partial q}{\partial t}\right] =$$

$$\sigma\left(\frac{\partial q}{\partial x}\frac{\partial^2 q}{\partial x^2} + \frac{1}{c_o^2}\frac{\partial q}{\partial t}\frac{\partial^2 q}{\partial x\partial t}\right) - \frac{2\pi A}{a}\frac{\partial q}{\partial x}\sin\left(\frac{2\pi}{a}q\right) - \frac{\sigma}{c_o^2}\left(\frac{\partial q}{\partial x}\frac{\partial^2 q}{\partial t^2} + \frac{\partial q}{\partial t}\frac{\partial^2 q}{\partial x\partial t}\right) =$$

$$\frac{\partial q}{\partial x}\left[\sigma\left(\frac{\partial^2 q}{\partial x^2} - \frac{1}{c_o^2}\frac{\partial^2 q}{\partial t^2}\right) - \frac{2\pi A}{a}\sin\left(\frac{2\pi}{a}q\right)\right] = 0 \ ,$$

Here, we have used the sine-GORDON equation (88) and taking (285) into consideration.

The following is just as valid,

$$\frac{\partial}{\partial x}\left[\sigma\frac{\partial q}{\partial x}\frac{\partial q}{\partial t}\right] + \frac{\partial}{\partial t}\left[-\frac{\sigma}{2}\left(\frac{\partial q}{\partial x}\frac{\partial q}{\partial x} + \frac{1}{c_o^2}\frac{\partial q}{\partial t}\frac{\partial q}{\partial t}\right) + A\cos\left(\frac{2\pi}{a}q\right) - A\right] =$$

$$\sigma\left(\frac{\partial q}{\partial x}\frac{\partial^2 q}{\partial x\partial t} + \frac{\partial q}{\partial t}\frac{\partial^2 q}{\partial x^2}\right) - \sigma\left(\frac{\partial q}{\partial x}\frac{\partial^2 q}{\partial x\partial t} + \frac{1}{c_o^2}\frac{\partial q}{\partial t}\frac{\partial^2 q}{\partial t^2}\right) - \frac{2\pi A}{a}\frac{\partial q}{\partial t}\sin\left(\frac{2\pi}{a}q\right) =$$

$$\frac{\partial q}{\partial t}\left[\sigma\left(\frac{\partial^2 q}{\partial x^2} - \frac{1}{c_o^2}\frac{\partial^2 q}{\partial t^2}\right) - \frac{2\pi A}{a}\sin\left(\frac{2\pi}{a}q\right)\right] = 0 \ .$$

We now only need to pick out the individual solutions of (88), insert them into (290) and check if a tensor of the form (279) is generated.

Chapter 23
A Particle Solution—The Inertia of Energy

We will now turn to the solution $q = q^I(x, t)$, our kink solution (111), that according to (98a), (91) and (99) gave us our natural, stationary measuring-rods L_o and also our moving measuring rods L' and the including, all so important LORENTZ contraction (112). Here the following questions will be considered: Do these internal measuring-rods, on which the internal observers base their experiences, mechanically behave as we outside observers expect our outside measuring-rods to behave? Are we in the position of attributing these measuring-rods as a whole a single mass and a single velocity? Is an internal observer's single measuring-rod L' a single object in the sense of NEWTONian mechanics? In order to answer these questions we need to calculate the tensor (290) for function (111) and to control whether we receive a mathematical expression that fulfils (279). We will now calculate this:

$$q^I(x, t) = \frac{2a}{\pi} \arctan \exp\left[\frac{\pi(x - vt)}{L_o\gamma}\right] \quad , \gamma = \sqrt{1 - \frac{v^2}{c_o^2}} \ .$$

Using

$$\cos\alpha = \frac{1}{\sqrt{1 + \tan^2\alpha}} \quad , \quad \cos(4\alpha) - 1 = -8\frac{\tan^2\alpha}{(1 + \tan^2\alpha)^2} \ ,$$

the following is derived,

$$\cos\left(\frac{2\pi}{a} q^I\right) - 1 = \cos\left(4 \arctan \exp\left[\frac{\pi(x - vt)}{L_o\gamma}\right]\right) - 1 =$$

H. Günther, *Elementary Approach to Special Relativity*,
https://doi.org/10.1007/978-981-15-3168-2_23

$$= -8 \frac{\tan^2\left(\arctan\exp\left[\frac{\pi(x-vt)}{L_o\gamma}\right]\right)}{\left(1+\tan^2\left(\arctan\exp\left[\frac{\pi(x-vt)}{L_o\gamma}\right]\right)\right)^2} ,$$

$$\cos\left(\frac{2\pi}{a} q^I\right) - 1 = -8 \frac{\exp\left[2\frac{\pi(x-vt)}{L_o\gamma}\right]}{\left(1+\exp\left[2\frac{\pi(x-vt)}{L_o\gamma}\right]\right)^2} ,$$

and from

$$\frac{\partial q^I}{\partial x} = \frac{2a}{L_o\gamma} \frac{\exp\left[\frac{\pi(x-vt)}{L_o\gamma}\right]}{1+\exp\left[2\frac{\pi(x-vt)}{L_o\gamma}\right]} \quad \text{as well as} \quad \frac{\partial q^I}{\partial t} = -v\,\frac{\partial q^I}{\partial x}$$

we find, taking (285) and (286) into consideration,

$$\frac{\sigma}{2}\left(\frac{\partial q^I}{\partial x}\frac{\partial q^I}{\partial x} + \frac{1}{c_o^2}\frac{\partial q^I}{\partial t}\frac{\partial q^I}{\partial t}\right) + \frac{\sigma a^2}{4L_o^2}\left(\cos\left(\frac{2\pi}{a} q^I\right) - 1\right) =$$

$$= \frac{\sigma}{2}\frac{4a^2}{L_o^2\gamma^2}\left(1+\frac{v^2}{c_o^2}\right)\frac{\exp\left[2\frac{\pi(x-vt)}{L_o\gamma}\right]}{\left(1+\exp\left[2\frac{\pi(x-vt)}{L_o\gamma}\right]\right)^2} - \frac{\sigma a^2}{4L_o^2}\, 8 \frac{\exp\left[2\frac{\pi(x-vt)}{L_o\gamma}\right]}{\left(1+\exp\left[2\frac{\pi(x-vt)}{L_o\gamma}\right]\right)^2}$$

$$= \frac{\exp\left[2\frac{\pi(x-vt)}{L_o\gamma}\right]}{\left(1+\exp\left[2\frac{\pi(x-vt)}{L_o\gamma}\right]\right)^2}\left(\frac{2\sigma a^2}{L_o^2\gamma^2}\left(1-\frac{v^2}{c_o^2}+2\frac{v^2}{c_o^2}\right) - \frac{2\sigma a^2}{L_o^2}\right)$$

$$= \frac{v^2}{c_o^2}\frac{4\sigma a^2}{L_o^2\gamma^2}\frac{\exp\left[2\frac{\pi(x-vt)}{L_o\gamma}\right]}{\left(1+\exp\left[2\frac{\pi(x-vt)}{L_o\gamma}\right]\right)^2} .$$

If we introduce another function $\rho = \rho(x - v\,t)$ according to

$$\rho(x-vt)=\frac{1}{c_o^2}\frac{1}{\gamma^2}\frac{4\,\sigma a^2}{L_o^2\gamma^2}\frac{\exp\left[2\frac{\pi(x-vt)}{L_o\gamma}\right]}{\left(1+\exp\left[2\frac{\pi(x-vt)}{L_o\gamma}\right]\right)^2}\ , \tag{292}$$

we receive for the first component $-t$ of the energy–momentum tensor

$$-t=v^2\rho(x-vt)\ . \tag{293}$$

We also find

$$-\frac{\sigma}{c_o^2}\frac{\partial q^I}{\partial x}\frac{\partial q^I}{\partial t}=-\frac{1}{c_o^2}(-v)\frac{4\,\sigma a^2}{L_o^2\gamma^2}\frac{\exp\left[\frac{\pi(x-vt)}{L_o\gamma}\right]}{\left(1+\exp\left[2\frac{\pi(x-vt)}{L_o\gamma}\right]\right)^2}$$

and thus

$$p=v\,p \tag{293a}$$

and according to (290) also

$$-s=-v\,c_o^2\,\rho\ . \tag{293b}$$

Finally, the following is derived

$$-\frac{\sigma}{2}\left(\frac{\partial q^I}{\partial x}\frac{\partial q^I}{\partial x}+\frac{1}{c_o^2}\frac{\partial q^I}{\partial t}\frac{\partial q^I}{\partial t}\right)+\frac{\sigma a^2}{4L_o^2}\left(\cos\left(\frac{2\pi}{a}q^I\right)-1\right)$$

$$=-\frac{\sigma}{2}\frac{4a^2}{L_o^2\gamma^2}\left(1+\frac{v^2}{c_o^2}\right)\frac{\exp\left[2\frac{\pi(x-vt)}{L_o\gamma}\right]}{\left(1+\exp\left[2\frac{\pi(x-vt)}{L_o\gamma}\right]\right)^2}-\frac{\sigma a^2}{4L_o^2}\,8\frac{\exp\left[2\frac{\pi(x-vt)}{L_o\gamma}\right]}{\left(1+\exp\left[2\frac{\pi(x-vt)}{L_o\gamma}\right]\right)^2}$$

$$=\frac{\exp\left[2\frac{\pi(x-vt)}{L_o\gamma}\right]}{\left(1+\exp\left[2\frac{\pi(x-vt)}{L_o\gamma}\right]\right)^2}\left(\frac{2\sigma a^2}{L_o^2\gamma^2}\left(1-\frac{v^2}{c_o^2}-2\right)-\frac{2\sigma a^2}{L_o^2}\right)$$

$$=-\frac{4\,\sigma a^2}{L_o^2\gamma^2}\frac{\exp\left[2\frac{\pi(x-vt)}{L_o\gamma}\right]}{\left(1+\exp\left[2\frac{\pi(x-vt)}{L_o\gamma}\right]\right)^2}$$

and thus once again with the function $\rho = \rho(x - v\,t)$ according to (292)

$$- e = -c_o^2\,\rho\ . \tag{293c}$$

From (293)–(293c), we can determine the energy–momentum tensor $\mathbf{T}^I$ of the sine-GORDON field q^I with $\rho = \rho(x - v\,t)$ according to (254), in other words of the field defined by our natural measuring-rods L_o and L', respectively,

$$T^I = \begin{pmatrix} v^2\rho & v\rho \\ -v\,c_o^2\,\rho & -c_o^2\,\rho \end{pmatrix}\ . \qquad \text{Energy-momentum tensor of the field } q^I(x,t) \tag{294}$$

We register: The tensor $\mathbf{T}^I$ is the energy–momentum tensor of a particle according to (279). In order to prepare the discussion of this remarkable result, we firstly calculate the inertial mass m_a that a dislocation portion possesses along the length a of a lattice parameter. m_a is therefore the inertial mass Δm_α on the length a in Eq. (79).

According to the second Eq. (88) where we use σ for the modulus of elasticity, $c_o = \sqrt{\sigma/\rho_o}$ and (corresponding to our expression (284b) for the kinetic energy density $T = \frac{\rho_o}{2}(\frac{\partial q}{\partial t})^2$) the formula $\rho_o = \sigma/c_o^2$ applies for the mass density ρ_o of the dislocation and therefore after multiplication with the lattice parameter a,

$$m_a = \frac{a\,\sigma}{c_o^2} \tag{295}$$

is valid for the mass m_a on the length a.

According to (281), we can calculate using (292) the inertial mass m of the object registered by the internal observer belonging to (294),

$$m = \frac{1}{c_o^2}\frac{1}{\gamma^2}\frac{4\,\sigma a^2}{L_o^2}\int\limits_{-\infty}^{+\infty}\frac{\exp\left[2\dfrac{\pi(x-vt)}{L_o\gamma}\right]dx}{\left(1+\exp\left[2\dfrac{\pi(x-vt)}{L_o\gamma}\right]\right)^2} = \frac{1}{c_o^2}\frac{1}{\gamma^2}\frac{4\,\sigma a^2}{L_o^2}\int\limits_{-\infty}^{+\infty}\frac{\exp\left[\dfrac{2\pi x}{L_o\gamma}\right]dx}{\left(1+\exp\left[\dfrac{2\pi x}{L_o\gamma}\right]\right)^2}$$

$$= \frac{1}{c_o^2}\frac{1}{\gamma^2}\frac{4\,\sigma a^2}{L_o^2}\frac{L_o\gamma}{2\pi}\int\limits_{-\infty}^{+\infty}\frac{e^x}{\left(1+e^x\right)^2}dx$$

$$= \frac{1}{\gamma}\frac{1}{c_o^2}\frac{2\,\sigma a^2}{\pi L_o}\left[\frac{-1}{1+e^x}\right]_{-\infty}^{+\infty} = \frac{1}{\gamma}\frac{1}{c_o^2}\frac{2\,\sigma a^2}{\pi L_o}\ .$$

Taking (295) into consideration, we can write for this

$$\left.\begin{aligned} m &= \frac{m_o}{\sqrt{1-v^2/c_o^2}} , \\ m_o &= f m_a , \quad f = \frac{2}{\pi}\frac{a}{L_o} , \quad m_a = \frac{a\sigma}{c_o^2} . \end{aligned}\right\} \tag{296}$$

For the energy E and momentum P of the particle, the following immediately results,

$$P = \frac{m_o}{\sqrt{1-v^2/c_o^2}} , \qquad E = \frac{m_o}{\sqrt{1-v^2/c_o^2}} c_o^2 . \tag{297}$$

In Eq. (296), the dependence of an inertial mass m on its velocity v demanded by EINSTEIN's Special Theory of Relativity can be seen, and in Eq. (297), EINSTEIN's famous equivalence between the energy E and inertial mass m of a particle can be seen. All of this is simply a consequence of our sine-GORDON equation. The solution q^I of this equation does not just simply illustrate any arbitrary particle, it also illustrates a strictly relativistic one. In other words, the measuring-rods L_o and L' of the internal observers constructed from these are single objects in the sense of the relativistically extended NEWTONian axiomatics, i.e. with a velocity-dependent mass according to (296). The velocity v of this object always stays smaller than the signal velocity c_o. An increase of velocity to close vicinity of the critical velocity would lead, according to (296) to an unlimited increase of the mass of the object. This critical velocity therefore cannot ever be achieved by an object with $m \neq 0$.

We also note supplementarily: We observe the particle defined by Eq. (297) from two reference systems. An observer in the system Σ' with the coordinates x', t' determines the energy $E' = m_o c_o^2/\sqrt{1-v'^2/c_o^2}$ and the momentum $P' = m_o v'/\sqrt{1-v'^2/c_o^2}$. An observer in the reference system Σ_o using the coordinates x, t determines the velocity of the system Σ' as v. He determines for the particle the values $E = m_o c_o^2/\sqrt{1-v_o^2/c_o^2}$ and $P = m_o v_o/\sqrt{1-v_o^2/c_o^2}$. How large is v_o? We know: Due to the fact that a particle moves with the velocity v' with respect to Σ' and the whole system Σ' moves with the velocity v with respect to Σ_o, the velocity v_o of the particle measured from Σ_o can be calculated according to the composition of velocities (194),

$$v_o = \frac{v+v'}{1+v\,v'/c_o^2} , \tag{194'}$$

which we received as an immediate consequence from the LORENTZ transformation (151). From this, we can reach the conclusion that for the values P' and E' in Σ' and P and E in Σ_o, the equations of the LORENTZ transformation (151) are also valid if one replaces x with P and $c_o t$ with E/c_o, as well as x' with P' and $c_o t'$ with E'/c_o, respectively. Therefore, the following is valid,

$$\left.\begin{aligned} P' &= \frac{P - v\,E/c_o^2}{\sqrt{1 - v^2/c_o^2}}\,, & & & P &= \frac{P' - v\,E'/c_o^2}{\sqrt{1 - v^2/c_o^2}}\,, \\ & & \longleftrightarrow & & & \\ E' &= \frac{E - vP}{\sqrt{1 - v^2/c_o^2}}\,, & & & E &= \frac{E' - vP'}{\sqrt{1 - v^2/c_o^2}}\,. \end{aligned}\right\} \tag{298}$$

We find it satisfactory just to verify the last Eq. (298). Using (194)′, we get

$$1 - \frac{v_o^2}{c_o^2} = 1 - \frac{\dfrac{(v+v')^2}{c_o^2}}{\left(1 + \dfrac{vv'}{c_o^2}\right)^2} = \frac{1 + 2\dfrac{vv'}{c_o^2} + \dfrac{v^2v'^2}{c_o^4} - \dfrac{v^2}{c_o^2} - 2\dfrac{vv'}{c_o^2} - \dfrac{v'^2}{c_o^2}}{\left(1 + \dfrac{vv'}{c_o^2}\right)^2}$$

$$= \frac{1 - \dfrac{v^2}{c_o^2} - \dfrac{v'^2}{c_o^2}\left(1 - \dfrac{v^2}{c_o^2}\right)}{\left(1 + \dfrac{vv'}{c_o^2}\right)^2} = \frac{\left(1 - \dfrac{v^2}{c_o^2}\right)\left(1 - \dfrac{v'^2}{c_o^2}\right)}{\left(1 + \dfrac{vv'}{c_o^2}\right)^2}\,,$$

therefore

$$\sqrt{1 - \frac{v_o^2}{c_o^2}} = \frac{\sqrt{1 - \dfrac{v^2}{c_o^2}}\sqrt{1 - \dfrac{v'^2}{c_o^2}}}{1 + \dfrac{vv'}{c_o^2}}\,,$$

thus for E from (284a),

$$E = \frac{\dfrac{m_o c_o^2}{\sqrt{1 - v'^2/c_o^2}} + v\dfrac{m_o v'}{\sqrt{1 - v'^2/c_o^2}}}{\sqrt{1 - v^2/c_o^2}} = \frac{m_o c_o^2\left(1 + \dfrac{vv'}{c_o^2}\right)}{\sqrt{1 - v'^2/c_o^2}\sqrt{1 - v^2/c_o^2}}\,,$$

so that

$$E = \frac{m_o c_o^2}{\sqrt{1 - v_o^2/c_o^2}}\,,$$

with v_o according to (194)′, what is exactly what we wanted to prove. One verifies the expression for P in the same manner. Equations (298) and (298a), respectively, are *that* physical characteristic for a relativistic particle that we have exactly proven with this for the object of the internal observer with his signal velocity c_o described by the solution $q^I(x, t)$.

In order to estimate the magnitude of the mass m_o, we use a value for the line tension that we get from the calculation of line energy for the crystal niob according to F. ACKERMANN, H. MUGRABI and A. SEEGER [90], here $\sigma \approx 3 \cdot 10^{-10}$ Nm. Together with a lattice parameter of $a \approx 2,86 \cdot 10^{-10}$ m, as well as an approximative value for c_o to sound velocity according to $c_o \approx 4,5 \cdot 10^3$ ms^{-1}, we receive an estimate rate for the dislocation mass m_a of

$$m_a \approx 5 \cdot 10^{-27}\,\text{kg} \ . \tag{299}$$

The mass of the niob's lattice atoms lies in the region of $m_{Nb} \approx 154 \cdot 10^{-27}$ kg. Hence, the mass m_a of the dislocation chain on one lattice parameter a has around 3% of the mass m_{Nb} of the niob atoms.

Lets us assume an estimate of four lattice parameters for the length L_o, thus $L_o \approx 4\,a$, then the factor in (296) would result in $f \approx 0,16$. We would receive an estimate rate for the kink mass m_o,

$$m_o \approx 0,8 \cdot 10^{-27}\,\text{kg} \ , \tag{300}$$

i.e. for the restmass m_o of the object defined by our kink solution q^I. This would make up around 0, 5% of the mass of surrounding lattice atoms.

Such a comparison between inertial masses of dislocations and the inertial masses of the lattice atoms can easily lead to false conclusions. In fact, in order to be precise, this should be: The inertia m_a of a dislocation portion of the length a with respect to the lattice totals around 3% of the inertia of niob atoms with respect to our 'outside' space, just as the inertia m_o of the 'body' which we can, according to our previously made calculations, call a kink takes up around 0, 5% of the inertia of niob atoms with respect to our 'outside' space.

The difference in the corresponding energies is however, compared to this, far larger. Our object with the mass m_o possesses according to (297), if we once again assume a value of $c_o \approx 4,5 \cdot 10^3$ ms^{-1}, the stored-up energy of $E_o = m_o c_o^2$, thus $E_o = 0,8 \cdot 10^{-27} \cdot 4,5^2 \cdot 10^6$ kgm^{2}s^{-2} and therefore $E_o \approx 1,6 \cdot 10^{-20}$ Nm. However, for the stored-up energy of the niob atoms, the speed of light $c_L = 3 \cdot 10^8$ ms^{-1} is valid. We therefore receive for the stored-up energy E_{Nb} of a stationary niob atom $E_{Nb} = m_{Nb} \cdot c_L^2 \approx 154 \cdot 10^{-27} \cdot 9 \cdot 10^{16}$ kgm^{2}s^{-2}, thus $E_{Nb} \approx 1,4 \cdot 10^{-9}$ Nm. The stored-up energy of our object, defined with respect to our crystal, with the mass m_o contains around 10^{-9}% of the above value—this is an enormous difference![1] It can also be proven that other solutions of the sine-GORDON equation can be seen as particles or objects in a mechanical sense respectively using the above-described fields. For example, this would apply to our field (117), the oscillating breather

[1] There is also another important difference between the inertial mass of a kink with respect to the lattice and the inertial mass of a niob atom to our physical space. For a niob atom, this inertial mass is identical, according to the principle of equivalence of the General Theory of Relativity, to its gravitational mass, which the niob atom underlies according to universal gravity. There is no analogy for the internal observer's particles of our infinite crystal for the property of 'gravity' of a particle. We just use the everyday term of mass for this property.

$q = q^{III}(x, t)$, with whose help we were able to define the clocks of the internal observers in Chaps. 9 and 10. The breather not only possesses velocity as an object, but it also possesses internal motion, namely its own oscillations. The calculations for the verification of the particle property of a breather solution would therefore be somewhat more complicated. We therefore refer the reader to the original paper, cf. GÜNTHER [37].

Without actually delving deeper into the problem, we note that non-linear equations generally contain the laws of motion of their particle solutions. The sum of two solutions of the sine-GORDON equation (88) in contrast to electrodynamics because of the non-linear term $\sin(\frac{2\pi}{a} q)$ cannot simply just be another solution of this equation. Two solutions can only be successfully combined to another solution if one determines their 'correct' motions (and the mutually caused deformations) with respect to one another. Keeping this in mind, the sine-GORDON equation contains the dynamics of its particles. The MAXWELLian equations are linear, and one only has to supplementarily postulate the relativistically corrected NEWTONian mechanics in order to be able to calculate the electromagnetic field of two charged particles through time. The relativistic, non-linear sine-GORDON equation with its critical velocity c_o inside of a crystal on the one hand corresponds to the system composed of MAXWELLian equations and the relativistic, mechanical equations of motion with the speed of light c_L on the other hand. The equivalence, in the boundaries of our one-dimensional model between EINSTEIN's Special Theory of Relativity of our physical spacetime and the Special Theory of Relativity on a solid is complete.

Chapter 24
The Michelson Experiment

We here describe the simple idea of the Michelson experiment as illustrated below. A source S emits a wave train which arrives at a semi silver-coated glass plate P. There the wave train is split into two coherent wave trains moving along the arms l_1 and l_2 of Michelson's interferometer. The waves were reflected at the ends l_1 and l_2 by the mirrors M_1 and M_2, respectively, and finally give rise to an interference figure observed at P.

Let us assume that the speed of light c_L has the same amount in an arbitrary direction if it is measured in the preferred frame Σ_o. There is also another reference system Σ' that has the velocity v with respect to Σ_o.

We firstly consider the case where Michelson's interferometer is at rest in Σ_o, cf. Fig. 24.1. The propagation of the wave trains is isotropic in Σ_o. Hence, the time needed for the wave trains to traverse the distance along the arms l_1 and l_2 and back are $t_1^o = 2l_1/c_L$ and $t_2^o = 2l_2/c_L$, respectively. The interference figure at P is determined by the difference Δt^o in the running times,

$$\Delta t^o = t_2^o - t_1^o = \frac{2l_1}{c_L} - \frac{2l_2}{c_L} . \tag{301}$$

The idea of the experiment is to rotate the interferometer by $\pi/2$, cf. Fig. 24.1b. Taking the isotropy of light propagation in Σ_o into account, the difference $\Delta t^o_{\frac{\pi}{2}}$ in the running times $t^o_{1,\frac{\pi}{2}}$ and $t^o_{2,\frac{\pi}{2}}$ does not change during this rotation,

$$\Delta t^o_{\frac{\pi}{2}} = t^o_{2,\frac{\pi}{2}} - t^o_{1,\frac{\pi}{2}} = \frac{2l_2}{c_L} - \frac{2l_1}{c_L} = \Delta t^o . \tag{302}$$

Let us define a quantity δ as the difference between the running time differences of the coherent wave trains before and after rotation of the interferometer,

$$\delta := \Delta t_{\frac{\pi}{2}} - \Delta t \ . \tag{303}$$

H. Günther, *Elementary Approach to Special Relativity*,
https://doi.org/10.1007/978-981-15-3168-2_24

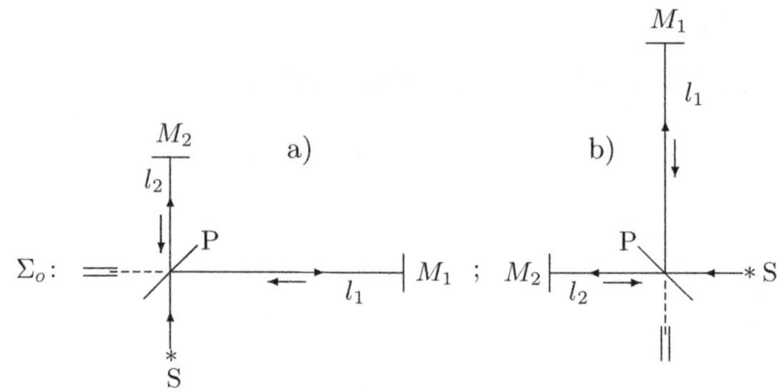

Fig. 24.1 **a** Diagrammatic representation of Michelson's interferometer. **b** The interferometer is rotated by $\pi/2$

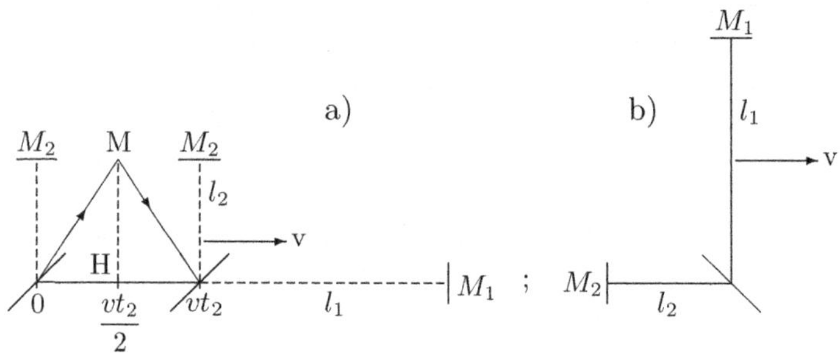

Fig. 24.2 **a** The interferometer has the velocity v with respect to the preferred frame Σ_o. **b** The moving interferometer is rotated by $\pi/2$

This quantity δ is a measure for a possible change of the observed interference figure during the rotation.

Hence, if the interferometer is situated in the preferred frame Σ_o, we get

$$\delta^o = \Delta t^o_{\frac{\pi}{2}} - \Delta t^o = 0 \ . \qquad \text{Resting interferometer} \tag{304}$$

The interference figure does not change during the rotation of the interferometer in Σ_o.

Next, we assume that MICHELSON's interferometer is at rest in the reference system Σ'. This means the interferometer has the velocity v with respect to the preferred frame Σ_o. This experiment, as the other before, is observed and described by an observer resting in Σ_o, see Fig. 24.2.

On the way from P to M_1, the relative velocity of the wave train is $c_L - v$ with respect to the mirror M_1, on the way back the relative velocity is $c_L + v$ with respect to P. Hence, the observer in Σ_o measures a total running time t_1 along the arm l_1

according to

$$t_1 = \frac{l_1}{c_L - v} + \frac{l_1}{c_L + v} = \frac{l_1}{c_L}\left(\frac{1}{1 - v/c_L} + \frac{1}{1 + v/c_L}\right) ,$$

$$t_1 = \frac{2l_1}{c_L} \frac{1}{1 - v^2/c_L^2} . \tag{305}$$

The running time along the arm l_2 has the same value $t_2/2$ for both journeys, see Fig. 24.2a. The wave train has the velocity c_L; the interferometer has the velocity v. Hence, we conclude from the triangle OMH

$$\left(\frac{c_L t_2}{2}\right)^2 = \left(\frac{v t_2}{2}\right)^2 + l_2^2 ,$$

$$\frac{t_2^2}{4} \left(c_L^2 - v^2\right) = l_2^2 ,$$

$$t_2^2 = \frac{l_2^2 4}{c_L^2 - v^2} = \frac{4 l_2^2}{c_L^2} \frac{1}{1 - v^2/c_L^2} ,$$

$$t_2 = \frac{2 l_2}{c_L} \frac{1}{\sqrt{1 - v^2/c_L^2}} . \tag{306}$$

Hence, the difference Δt in the running times t_1 and t_2 is

$$\Delta t = t_2 - t_1 = \frac{2 l_2}{c_L} \frac{1}{\sqrt{1 - v^2/c_L^2}} - \frac{2l_1}{c_L} \frac{1}{1 - v^2/c_L^2} . \tag{307}$$

Once again, we rotate the interferometer by $\pi/2$, cf. Fig. 24.2b. Let $t_{1,\frac{\pi}{2}}$ and $t_{2,\frac{\pi}{2}}$ be the running times to traverse the distances along the arms l_1 and l_2, respectively. In order to calculate the difference $\Delta t_{\frac{\pi}{2}}$ in these running times as measured of an observer resting in Σ_o, we only have to replace l_1 with l_2 in equation (305) and l_2 with l_1 in equation (306) in order to get $t_{2,\frac{\pi}{2}}$ and $t_{1,\frac{\pi}{2}}$, respectively. Hence

$$\Delta t_{\frac{\pi}{2}} = t_{2,\frac{\pi}{2}} - t_{1,\frac{\pi}{2}} = \frac{2 l_2}{c_L} \frac{1}{1 - v^2/c_L^2} - \frac{2l_1}{c_L} \frac{1}{\sqrt{1 - v^2/c_L^2}} . \tag{308}$$

The difference $\delta = \Delta t_{\frac{\pi}{2}} - \Delta t$ is a measure for a possible change of the observed interference figure during rotation. We receive

$$\delta = \Delta t_{\frac{\pi}{2}} - \Delta t =$$

$$= \frac{2\,l_2}{c_L}\,\frac{1}{1 - v^2/c_L^2} - \frac{2l_1}{c_L}\,\frac{1}{\sqrt{1 - v^2/c_L^2}} - \left(\frac{2\,l_2}{c_L}\,\frac{1}{\sqrt{1 - v^2/c_L^2}} - \frac{2l_1}{c_L}\,\frac{1}{1 - v^2/c_L^2}\right)$$

$$= \frac{2\,l_2}{c_L}\left(\frac{1}{1 - v^2/c_L^2} - \frac{1}{\sqrt{1 - v^2/c_L^2}}\right) + \frac{2\,l_1}{c_L}\left(\frac{1}{1 - v^2/c_L^2} - \frac{1}{\sqrt{1 - v^2/c_L^2}}\right),$$

hence

$$\delta = \left(\frac{2\,l_1}{c_L} + \frac{2\,l_2}{c_L}\right)\left(\frac{1}{1 - v^2/c_L^2} - \frac{1}{\sqrt{1 - v^2/c_L^2}}\right). \quad \text{Moving interferometer} \tag{309}$$

With the TAYLOR series expansions $1/(1-x^2) \approx 1 + x^2$ and $1/\sqrt{1-x^2} \approx 1 + (1/2)x^2$ we get with $x = v/c_L$

$$\delta = \frac{2}{c_L}\,(l_1 + l_2)\left(1 + \frac{v^2}{c_L^2} - 1 - \frac{1}{2}\,\frac{v^2}{c_L^2}\right) = \frac{2}{c_L}\,(l_1 + l_2)\,\frac{1}{2}\,\frac{v^2}{c_L^2}\,,$$

hence

$$\delta = \frac{l_1 + l_2}{c_L}\,\frac{v^2}{c_L^2}\,. \tag{310}$$

This quantity δ determines the shift of the interference band resulting from the $\pi/2$-rotation of the interferometer.

We assume that our laboratory on earth, with its rigidly installed interferometer, determines the reference system Σ'. The trajectory velocity of earth is around $v = 30\,000\,\text{m/s}$. This is the velocity of Σ' with respect to the preferred frame Σ_o. We assume, roughly speaking, that the rest of the solar system determines Σ_o. It was arranged using repeated reflections of wave trains in MICHELSON's historical experiment, cf. BLEYER et al. [5], that $l_1 + l_2 = 30\,\text{m}$. Applying a sodium lamp determines a wavelength of $\lambda = 6 \cdot 10^{-7}\,\text{m}$ for the interfering light. Hence, with $c_L = 3 \cdot 10^8\,\text{m/s}$, the period of oscillation is

$$\tau = \frac{\lambda}{c_L} = \frac{6 \cdot 10^{-7}}{3 \cdot 10^8}\,\text{s} = 2 \cdot 10^{-15}\,\text{s}\,.$$

On the other hand, we receive

$$\delta = \frac{30}{3 \cdot 10^8}\left(\frac{3 \cdot 10^4}{3 \cdot 10^8}\right)^2 \text{s} = 10^{-15}\,\text{s}$$

for the change in running times differences of the interfering wave trains before and after rotation, according to (310). I.e.

$$\delta = \Delta t_{\frac{\pi}{2}} - \Delta t = \frac{1}{2}\tau \; . \qquad \text{Moving interferometer} \tag{311}$$

Due to this result, $\delta = \tau/2$, a total displacement of the bands should be observed.

Nevertheless, nothing occurs: Neither during the first experiment in 1881, nor during any other experiment anytime, anyplace.

LORENTZ' [62] simple explanation is: 'We are therefore led to suppose that the influence of a translation on the dimension (of the separate electrons and of a ponderable body as a whole) is confined to those that have the direction of the motion, these becoming k times smaller than they are in the state of rest', where k is the factor $k = \sqrt{1 - v^2/c_L^2}$.

Indeed, replacing the length of the arm l_1 with $l_1 \sqrt{1 - v^2/c_L^2}$ in (307) and the length of the arm l_2 with $l_2 \sqrt{1 - v^2/c_L^2}$ in (308), we receive $\Delta t = \Delta t_{\frac{\pi}{2}}$, so that according to $\delta = 0$ no change of the interference figure can be expected.

Chapter 25
Elastic Displacements and Waves

We now sketch out the case of the three-dimensional ideal lattice. Our starting point is once again NEWTON's equations, (53) from which we already derived the wave equation (61), and which we will now formulate for small masses Δm_i. We will once again ignore the external forces $\mathbf{f}_a$ for the sake of simplicity. We will also, once again, begin with the supposition that there is a simultaneous position of equilibrium for all masses of the solid, of such a type that every solid's strain state different from this position can in fact be achieved by a simultaneous displacement of all the masses of this solid. We will also discover further on that exactly this supposition considerably restricts the number of possible physical stress states. We therefore move towards the so-called classical theory of elasticity. For a detailed analysis of this theory, we refer the reader to the textbook of L. D. LANDAU [54] and E. M. LIFSCHITZ. The spatial displacements out of this position of equilibrium are called $\mathbf{s}_i$, therefore

$$\left.\begin{aligned} \frac{d}{dt}\left(\Delta m_i \frac{d}{dt}\mathbf{s}_i\right) &= \sum_{k\neq i} \mathbf{f}_{ki} \ , \\ \mathbf{f}_{ik} &= -\mathbf{f}_{ki} \ . \end{aligned}\right\} \tag{312}$$

The time-dependent displacements $\mathbf{s}_i(t)$ we replace with the function $\mathbf{s}(x, t)$ in the limit to the continuum. We have to note for this limit that the term of inertia on the left-hand side of the NEWTONian equations describes the acceleration of a certain mass Δm, which according to definition changes its position. In other words, the volume ΔV_m occupied by this mass changes its position as well as its form. The possibly large number of masses in the volume ΔV_m we replace with a spatial density ρ (not to be mistaken with the density ρ_o which we constructed from our linear chain according to (84)), thus

$$\Delta m_i \longrightarrow \rho \Delta V_m \ , \qquad \mathbf{s}_i(t) \longrightarrow \mathbf{s}(\mathbf{x}, t) \ , \tag{313}$$

where $\mathbf{x} = (x, y, z) = (x_1, x_2, x_3)$.

H. Günther, *Elementary Approach to Special Relativity*,
https://doi.org/10.1007/978-981-15-3168-2_25

For the transition from the NEWTONian equations for single masses to a field theoretic description of a continuum, we need to fix the volume ΔV_m at a time t at position x. This gives us a fixed volume that we call ΔV. Particles continuously flow into and back out of this volume ΔV. We now calculate how the NEWTONian term of inertia, which is primarily defined for the masses in the volume ΔV_m, can be expressed with the help of the localised volume ΔV. In analysis it is shown how an integral over a moving volume ΔV_m can be differentiated with respect to a parameter (here time t), if the boundaries of this volume also depend on this parameter. In our case, the position and form of the volume ΔV_m are defined by the particles that move according to the NEWTONian equations. Here, one speaks of the so-called material volume. The following then applies for the calculation of the NEWTONian inertia term of the particles that occupy this volume,

$$\frac{d}{dt}(\Delta m\, \mathbf{v}) = \frac{d}{dt}\iiint\limits_{\Delta V_m} \rho\, \mathbf{v}\, dV = \iiint\limits_{\Delta V} \frac{\partial}{\partial t}(\rho\, \mathbf{v})\, dV + \iint\limits_{(\Delta V)} \rho\, \mathbf{v}\, \mathbf{v}\cdot d\mathbf{A} \ .$$

Here, (ΔV) stands for the surface of the volume ΔV. The surface integral is calculated with the help of the vectorial surface element $d\mathbf{A}$.[1] This integral takes into consideration that particles with the velocity $\mathbf{v}$ move into and out of the volume ΔV. We can, as shown in analysis by repeated partial integration, transform the surface integral into a volume integral with the help of the GAUSSian theorem by noting the components according to

$$\frac{d}{dt}(\Delta m\, v_k) = \iiint\limits_{\Delta V} \frac{\partial}{\partial t}(\rho\, v_k)\, d^3x + \iiint\limits_{\Delta V} \sum_{r=1}^{3} \frac{\partial}{\partial x_r}(\rho\, v_k v_r)\, d^3x \ . \qquad (314)$$

Due to the fact that we are only interested in the masses constituting the volume ΔV_m, $\frac{d}{dt}\Delta V_m = 0$ applies by definition. Analogous to (314) the following also applies,

$$\frac{d}{dt}(\Delta m) = \iiint\limits_{\Delta V} \frac{\partial}{\partial t}(\rho)\, d^3x + \iiint\limits_{\Delta V} \sum_{r=1}^{3} \frac{\partial}{\partial x_r}(\rho\, v_r)\, d^3x = 0 \ . \qquad (314a)$$

For a sufficiently small volume ΔV, we can write for (314)

$$\frac{d}{dt}(\Delta m\, v_k) = \Delta V \left(\frac{\partial}{\partial t}(\rho\, v_k) + \sum_{r=1}^{3} \frac{\partial}{\partial x_r}(\rho\, v_k v_r) \right) \ , \qquad (315)$$

and for (314a) it follows, if we cancel ΔV,

---

[1] $d\mathbf{A}$ is the vector whose amount is equal to the surface area of the observed surface element, and whose direction is perpendicular to this surface element, directed into the outside space of the volume ΔV. Therefore, the quantity $\rho\, \mathbf{v}\cdot d\mathbf{A}$ measures the mass per unit of time of particles moving out of the surface element $d\mathbf{A}$.

$$\frac{\partial}{\partial t}\rho + \sum_{r=1}^{3} \frac{\partial}{\partial x_r}(\rho\, v_r) = 0 \ . \tag{315a}$$

The last equation is called continuity equation and simply means that no particles disappear during their motion.

Taking into consideration (315a), we carry out the differentiation in (315) and finally get

$$\frac{d}{dt}(\Delta m\, v_k) = \Delta V\, \rho \left(\frac{\partial}{\partial t}\, v_k + \sum_{r=1}^{3} \frac{\partial v_k}{\partial x_r}\, v_r \right) \ . \tag{316}$$

The brackets on the right-hand side contain the so-called material or total time derivative d/dt. This is the change through time that a comoving observer determines, if he were to move with the velocity $\mathbf{v}$,

$$\frac{d}{dt} = \frac{\partial}{\partial t} + \sum_{r=1}^{3} \frac{dx_r}{dt}\frac{\partial}{\partial x_r} = \frac{\partial}{\partial t} + \sum_{r=1}^{3} v_r \frac{\partial}{\partial x_r} \ .$$

Hence, for (316) we can simply write

$$\frac{d}{dt}(\Delta m \mathbf{v}) = \Delta V\, \rho \frac{d\mathbf{v}}{dt} \ . \tag{316a}$$

From the change in time of the momentum $\mathbf{P} = \Delta m\, \mathbf{v}$ according to the NEWTONian equation of motion, only the change in time of the matter velocity $\mathbf{v}$ remains for the momentum density $\mathbf{p} = \rho\, \mathbf{v}$. Here, we explicitly note that: Because the momentum density $\mathbf{p}$, as well as the mass density ρ are measurable, the material velocity $\mathbf{v}$ is also an elementary measurable quantity, whether a displacement field $\mathbf{s} = \mathbf{s}(\mathbf{x}, t)$ exists or not. We also especially note that: In comparison to the one-dimensional continuum, an elastic displacement field $\mathbf{s} = \mathbf{s}(\mathbf{x}, t)$ is generally not defined and therefore not measurable for a three-dimensional (and also a two-dimensional) continuum. Nonetheless, the matter velocity $\mathbf{v}$ is measurable. These more complicated facts will be discussed and explained in the next chapter.

We now turn to the linearised theory of elasticity. This means that we ignore the non-linear terms on the right-side of (316). Hence, the NEWTONian term of inertia is simplified for the limit to the continuum according to

$$\frac{d}{dt}(\Delta m\, \mathbf{v}) \ \longrightarrow \ \Delta V\, \rho \frac{\partial \mathbf{v}}{\partial t} \ . \tag{317}$$

We now consider the forces. Inside the volume ΔV, all interaction forces $\mathbf{f}_{ik}$ cancel each other due to the reaction axiom. Due to the presupposed hypothesis of continuous action (cf. Chap. 6), only the forces $\Delta\mathbf{f}_{ik}$ acting through the surface of ΔV are left over from the whole sum over the interaction forces (312). We write for the limit of a continuous mass and force distribution,

$$\sum_{k \neq i} \mathbf{f}_{ki} = \sum \Delta \mathbf{f}_{ki} \longrightarrow \iint_{(\Delta V)} d\mathbf{f} \ . \tag{318}$$

(ΔV) once again stands for the surface of the volume ΔV, $d\mathbf{f}$ is the force acting on the vectorial surface element $d\mathbf{A}$. This defines the *stress tensor* $\boldsymbol{\sigma}$ according to

$$d\mathbf{f} = \boldsymbol{\sigma} \cdot d\mathbf{A} \ , \quad \text{i. e.} \quad \begin{pmatrix} df_x \\ df_y \\ df_z \end{pmatrix} = \begin{pmatrix} \sigma_{xx} & \sigma_{xy} & \sigma_{xz} \\ \sigma_{yx} & \sigma_{yy} & \sigma_{yz} \\ \sigma_{zx} & \sigma_{zy} & \sigma_{zz} \end{pmatrix} \begin{pmatrix} dA_x \\ dA_y \\ dA_z \end{pmatrix} \ . \tag{319}$$

In place of the stress τ of the rod, the far more complicated quantity $\boldsymbol{\sigma}$ arises when we deal with three-dimensional problems. According to definition (319), $\boldsymbol{\sigma}$ is a force per unit area. Written down in detail, Eq. (319) looks like

$$\left. \begin{aligned} df_x &= \sigma_{xx}\, dA_x + \sigma_{xy}\, dA_y + \sigma_{xz}\, dA_z \ , \\ df_y &= \sigma_{yx}\, dA_x + \sigma_{yy}\, dA_y + \sigma_{yz}\, dA_z \ , \\ df_z &= \sigma_{zx}\, dA_x + \sigma_{zy}\, dA_y + \sigma_{zz}\, dA_z \ . \end{aligned} \right\} \tag{320}$$

We have indicated all quantities with x, y, z in order to enable easier recognition. We use the equivalent indicating with numbers, especially when dealing with simplifications of sums over individual components according to $1 \leftrightarrow x, 2 \leftrightarrow y, 3 \leftrightarrow z$, thus df_1 in place of df_x, σ_{23} in place of σ_{yz} etc.

The individual components of $\boldsymbol{\sigma}$ have the following meaning: If one dissects the continuum along the area $d\mathbf{A}$, then one has to use the force $d\mathbf{f}$ so that the intersection planes do not move. It is accepted for the sign that a traction is positive and a pressure is negative. If one chooses for example that $d\mathbf{A}$ is orthogonal to the x-axis, then σ_{xx}, σ_{xy} and σ_{xz} are the CARTESian components of $d\mathbf{f}$ divided by the amount of the area $|d\mathbf{A}|$.

An important property of the stress tensor $\boldsymbol{\sigma}$ still has to be discussed. In order to do this we consider a small, cubic volume element ΔV positioned parallel to the axes. We assume at the areas of the cubes surface $\sigma_{xy} > 0$ as well as $\sigma_{yx} > 0$. Then σ_{xy} causes a clockwise rotation of the cube along the z-axis and σ_{yx} a rotation in the opposite direction, in other words, for $\sigma_{xy} - \sigma_{yx} \neq 0$ the cube begins to rotate. However, we received $\boldsymbol{\sigma} = \boldsymbol{\sigma}\,(\mathbf{x}, t)$ as the limit of the interaction forces on an arbitrarily chosen mass point at $\mathbf{x}$, in other words $\boldsymbol{\sigma} = \boldsymbol{\sigma}(\mathbf{x}, t)$ describes the motion of the mass point if we let ΔV go to zero. A point cannot however rotate. The following important law follows[2]:

---

[2]For the case of elastic coupled mass points which we consider exclusively here, this property for the stress tensor $\boldsymbol{\sigma}$ of the equivalent continuum is necessary. There are however other models where the symmetry of the stress tensor is lost. This is the case for example, if one calculates using small rigid solids instead of mass points (for example, as a model of approximation for molecules). For the resulting continua new limits have to be reformulated.

The stress tensor $\boldsymbol{\sigma}$ is symmetrical,

$$\left.\begin{aligned} \sigma_{xy} &= \sigma_{yx} \ , \\ \sigma_{xz} &= \sigma_{zx} \ , \\ \sigma_{yz} &= \sigma_{zy} \ . \end{aligned}\right\} \tag{321}$$

Applying again Gauss' theorem on the surface integral in (318) we can write

$$\iint\limits_{(\Delta V)} d\mathbf{f} = \iint\limits_{(\Delta V)} \boldsymbol{\sigma} \cdot d\mathbf{A} = \iiint\limits_{\Delta V} \operatorname{div} \boldsymbol{\sigma} \, dV \ ,$$

therefore, if we make the volume ΔV small enough,

$$\iiint\limits_{\Delta V} \operatorname{div} \boldsymbol{\sigma} \, dV \ \longrightarrow \ \operatorname{div} \boldsymbol{\sigma} \, \Delta V \ .$$

Here div $\boldsymbol{\sigma}$ is a vector with the components

$$\left.\begin{aligned} \operatorname{div} \sigma_x &= \frac{\partial \sigma_{xx}}{\partial x} + \frac{\partial \sigma_{xy}}{\partial y} + \frac{\partial \sigma_{xz}}{\partial z} \ , \\ \operatorname{div} \sigma_y &= \frac{\partial \sigma_{yx}}{\partial x} + \frac{\partial \sigma_{yy}}{\partial y} + \frac{\partial \sigma_{yz}}{\partial z} \ , \\ \operatorname{div} \sigma_z &= \frac{\partial \sigma_{zx}}{\partial x} + \frac{\partial \sigma_{zy}}{\partial y} + \frac{\partial \sigma_{zz}}{\partial z} \ . \end{aligned}\right\} \tag{322}$$

For the limit of the Newtonian equations for single masses to the continuum, we replace, taking into consideration the aimed linearisation of the velocities $\mathbf{v}$,

$$\mathbf{v} = \frac{d\mathbf{s}}{dt} = \frac{\partial \mathbf{s}}{\partial t} + \sum_{r=1}^{3} \frac{\partial \mathbf{s}}{\partial x_r} v_r$$

in (317), the material time derivative of the displacement vector $\mathbf{s} = \mathbf{s}(\mathbf{x}, t)$ with the partial time derivative,

$$\mathbf{v} = \frac{d\mathbf{s}}{dt} \ \longrightarrow \ \frac{\partial \mathbf{s}}{\partial t} \ . \tag{323}$$

We therefore summarise and write,

$$\left.\begin{aligned} &\frac{d}{dt}\left(\Delta m_i \frac{d}{dt} s_i\right) \longrightarrow \rho \, \Delta V \, \frac{\partial}{\partial t} \frac{\partial s(\mathbf{x}, t)}{\partial t} \ , \\ &\sum_{i \neq k} f_{ki} \qquad\qquad \longrightarrow \operatorname{div} \boldsymbol{\sigma} \, \Delta V \ . \end{aligned}\right\} \quad \text{Continuous distributions} \tag{324}$$

Equation (66), of the one-dimensional case dealt with in Chap. 6 can be received from (324) using $\Delta V \rightarrow \Delta x$, if we notice that we have in this case only one space coordinate x.

Using (324) we discuss the NEWTONian equations (312) and get by cancelling out ΔV,

$$\rho \frac{\partial}{\partial t} \frac{\partial s(\mathbf{x}, t)}{\partial t} = \operatorname{div} \boldsymbol{\sigma} \ . \tag{325}$$

If we again introduce the external forces $\mathbf{f}_a$ from Eq. (53) into (312), then these appear in (325) as an external volume force density, thus generally $\mathbf{f} = \mathbf{f}(\mathbf{x}, t)$ and (325a) results from (325),

$$\rho \frac{\partial}{\partial t} \frac{\partial s(\mathbf{x}, t)}{\partial t} = \operatorname{div} \boldsymbol{\sigma} + \mathbf{f} \ . \tag{325a}$$

For the general case, where no displacement field exists, we must replace (325a) with (see also (316a) and (317)),

$$\rho \frac{\partial}{\partial t} \mathbf{v} = \operatorname{div} \boldsymbol{\sigma} + \mathbf{f} \ . \tag{326}$$

Equation (325) generalises Eq. (67), of the one-dimensional lattice. Just as we did in Chap. 6, we now want to formulate the basic assumptions of the linear theory of elasticity. In order to do this, we need to consider what type of mathematical quantity the relative elastic strain $\boldsymbol{\varepsilon}$ is in the three-dimensional case. As a comparison, we will once again describe the elastic strain ε of the rod: The absolute elastic displacement $s = s(x, t)$ at x and the corresponding displacement $s_o = s(x_o, t)$ at x_o allows for small $\Delta x = x - x_o$ the TAYLOR series expansion

$$s = s_o + \frac{\partial s}{\partial x} \Delta x \ . \tag{327}$$

Together with $s - s_o = \Delta s$, the relative elastic strain ε follows according to

$$\Delta s = \frac{\partial s}{\partial x} \Delta x \longrightarrow \varepsilon = \frac{\partial s}{\partial x} \ . \tag{328}$$

We will now reconstruct this process for the three-dimensional case. In order to do this, we need that quantity $\boldsymbol{\varepsilon}$ that describes the relative elastic deformation of a small volume ΔV. Starting from a uniform state of equilibrium we assumed that the complete elastic deformation, inside of the continuum, can be described by an elastic displacement vector $\mathbf{s} = \mathbf{s}(\mathbf{x}, t)$. We will see further on that this assumption does not in fact have to be fulfilled. If this assumption is however valid, then we are dealing with a space-dependent vector $\mathbf{s} = \mathbf{s}(\mathbf{x}, t)$ that describes the displacements of the points $\mathbf{x_0}$ and $\mathbf{x}$ of ΔV. The absolute elastic displacement $\mathbf{s} = \mathbf{s}(\mathbf{x}, t)$ at $\mathbf{x}$ and the corresponding displacement $\mathbf{s}_o = \mathbf{s}(\mathbf{x}_o, t)$ at $\mathbf{x}_o$ allows a TAYLOR expansion or every component of $\mathbf{s}$ if we are dealing with small $\Delta\mathbf{x} = \mathbf{x} - \mathbf{x}_o$ (small ΔV),

$$\left.\begin{aligned} s_x &= s_{ox} + \frac{\partial s_x}{\partial x}\,\Delta x + \frac{\partial s_x}{\partial y}\,\Delta y + \frac{\partial s_x}{\partial z}\,\Delta z\ , \\ s_y &= s_{oy} + \frac{\partial s_y}{\partial x}\,\Delta x + \frac{\partial s_y}{\partial y}\,\Delta y + \frac{\partial s_y}{\partial z}\,\Delta z\ , \\ s_z &= s_{oz} + \frac{\partial s_z}{\partial x}\,\Delta x + \frac{\partial s_z}{\partial y}\,\Delta y + \frac{\partial s_z}{\partial z}\,\Delta z\ . \end{aligned}\right\} \tag{329}$$

With $\mathbf{s} - \mathbf{s}_o = \Delta\mathbf{s}$ we write

$$\begin{pmatrix} \Delta s_x \\ \Delta s_y \\ \Delta s_z \end{pmatrix} = \begin{pmatrix} \frac{\partial s_x}{\partial x} & \frac{\partial s_x}{\partial y} & \frac{\partial s_x}{\partial z} \\ \frac{\partial s_y}{\partial x} & \frac{\partial s_y}{\partial y} & \frac{\partial s_y}{\partial z} \\ \frac{\partial s_z}{\partial x} & \frac{\partial s_z}{\partial y} & \frac{\partial s_z}{\partial z} \end{pmatrix} \begin{pmatrix} \Delta x \\ \Delta y \\ \Delta z \end{pmatrix} \tag{330}$$

and summarising,

$$\Delta\mathbf{s} = \boldsymbol{\beta}^T \cdot \Delta\mathbf{x} \qquad \text{with} \qquad \boldsymbol{\beta}^T = \begin{pmatrix} \frac{\partial s_x}{\partial x} & \frac{\partial s_x}{\partial y} & \frac{\partial s_x}{\partial z} \\ \frac{\partial s_y}{\partial x} & \frac{\partial s_y}{\partial y} & \frac{\partial s_y}{\partial z} \\ \frac{\partial s_z}{\partial x} & \frac{\partial s_z}{\partial y} & \frac{\partial s_z}{\partial z} \end{pmatrix}\ . \tag{330a}$$

(Due to the fact that in dislocation theory the matrix $\boldsymbol{\beta}$ is considered, where the rows and columns are exchanged, we have indicated the matrix using a 'T' for 'transposed'). $\boldsymbol{\beta}^T$ is however still not the looked for elastic strain. The reason for this is simply that all deformations and shifts are contained in (330), including those that represent pure rotations of the volume element ΔV. These rotations do not represent elastic deformations of ΔV. None of the internal spring forces are required for that (see however, Chap. 26, where we refer to the relevance of relative rotations). In order to split off the rotations from $\boldsymbol{\beta}^T$, so that only the pure elastic strain $\boldsymbol{\varepsilon}$ remains, we write

$$\left.\begin{array}{l}\beta^T = \omega^T + \varepsilon^T = \omega^T + \varepsilon \ , \\ \text{with} \\ \omega^T = \begin{pmatrix} 0 & \frac{1}{2}\left(\frac{\partial s_x}{\partial y} - \frac{\partial s_y}{\partial x}\right) & \frac{1}{2}\left(\frac{\partial s_x}{\partial z} - \frac{\partial s_z}{\partial x}\right) \\ \frac{1}{2}\left(\frac{\partial s_y}{\partial x} - \frac{\partial s_x}{\partial y}\right) & 0 & \frac{1}{2}\left(\frac{\partial s_y}{\partial z} - \frac{\partial s_z}{\partial y}\right) \\ \frac{1}{2}\left(\frac{\partial s_z}{\partial x} - \frac{\partial s_x}{\partial z}\right) & \frac{1}{2}\left(\frac{\partial s_z}{\partial y} - \frac{\partial s_y}{\partial z}\right) & 0 \end{pmatrix} , \\ \varepsilon = \begin{pmatrix} \frac{\partial s_x}{\partial x} & \frac{1}{2}\left(\frac{\partial s_x}{\partial y} + \frac{\partial s_y}{\partial x}\right) & \frac{1}{2}\left(\frac{\partial s_x}{\partial z} + \frac{\partial s_z}{\partial x}\right) \\ \frac{1}{2}\left(\frac{\partial s_y}{\partial x} + \frac{\partial s_x}{\partial y}\right) & \frac{\partial s_y}{\partial y} & \frac{1}{2}\left(\frac{\partial s_y}{\partial z} + \frac{\partial s_z}{\partial y}\right) \\ \frac{1}{2}\left(\frac{\partial s_z}{\partial x} + \frac{\partial s_x}{\partial z}\right) & \frac{1}{2}\left(\frac{\partial s_z}{\partial y} + \frac{\partial s_y}{\partial z}\right) & \frac{\partial s_z}{\partial z} \end{pmatrix} . \end{array}\right\} \tag{331}$$

We now write

$$\Delta \mathbf{s} = \omega^T \cdot \Delta \mathbf{x} + \varepsilon \cdot \Delta \mathbf{x} \ . \tag{332}$$

If one constructs a vector $\vec{\omega}$ from the three independent components of ω according to

$$\begin{aligned}\vec{\omega} &= \tfrac{1}{2}\left(\tfrac{\partial s_z}{\partial y} - \tfrac{\partial s_y}{\partial z} \, , \, \tfrac{\partial s_x}{\partial z} - \tfrac{\partial s_z}{\partial x} \, , \, \tfrac{\partial s_y}{\partial x} - \tfrac{\partial s_x}{\partial y}\right) \\ &= (\omega_{yz} \, , \, \omega_{zx} \, , \, \omega_{xy}) := (\omega_x \, , \, \omega_y \, , \, \omega_z)\end{aligned} \tag{333}$$

then one can calculate for the first term on the right-hand side of (332) that

$$\omega^T \cdot \Delta \mathbf{x} = \vec{\omega} \times \Delta \mathbf{x} = \left(\omega_y \, \Delta z - \omega_z \, \Delta y \, , \ \omega_z \, \Delta x - \omega_x \, \Delta z \, , \ \omega_x \, \Delta y - \omega_y \, \Delta x\right) . \tag{334}$$

The vector product on the right-hand side of (334) describes nothing else than a rotation of the whole volume element ΔV in the direction defined by $\vec{\omega}$. This part of the shift of ΔV must be ignored and this leads us to the tensor ε of elastic strain according to (331),

$$\varepsilon = \begin{pmatrix} \frac{\partial s_x}{\partial x} & \frac{1}{2}\left(\frac{\partial s_x}{\partial y} + \frac{\partial s_y}{\partial x}\right) & \frac{1}{2}\left(\frac{\partial s_x}{\partial z} + \frac{\partial s_z}{\partial x}\right) \\ \frac{1}{2}\left(\frac{\partial s_y}{\partial x} + \frac{\partial s_x}{\partial y}\right) & \frac{\partial s_y}{\partial y} & \frac{1}{2}\left(\frac{\partial s_y}{\partial z} + \frac{\partial s_z}{\partial y}\right) \\ \frac{1}{2}\left(\frac{\partial s_z}{\partial x} + \frac{\partial s_x}{\partial z}\right) & \frac{1}{2}\left(\frac{\partial s_z}{\partial y} + \frac{\partial s_y}{\partial z}\right) & \frac{\partial s_z}{\partial z} \end{pmatrix} . \tag{335}$$

We recognise from (335):

The strain tensor ε is symmetrical,

$$\begin{aligned} \varepsilon_{xy} &= \varepsilon_{yx} \ , \\ \varepsilon_{xz} &= \varepsilon_{zx} \ , \\ \varepsilon_{yz} &= \varepsilon_{zy} \ . \end{aligned} \tag{336}$$

We are now in a position where we can precisely formulate the three basic assumptions of the linear theory of elasticity for the three-dimensional case.

1. In the NEWTONian term for inertia, the total time derivative for the momentum **P** of the moving masses is replaced with the partial derivative of the material velocity **v**.
2. The spatial distributed mass density ρ is considered as a constant parameter.
3. HOOKE's law is valid, in other words the stress tensor $\boldsymbol{\sigma}$ is directly proportional to the tensor of relative strain ε at position **x** and point of time t. This results in the equations

$$\begin{aligned} \frac{d}{dt}\mathbf{P} &= \rho \frac{\partial}{\partial t}\mathbf{v} \ , \\ \rho &= \text{const.} \ , \\ \boldsymbol{\sigma}(\mathbf{x}, t) &= \mathbf{C} \cdot \cdot \, \boldsymbol{\varepsilon}(\mathbf{x}, t) \ . \end{aligned} \tag{337}$$

Equations (337) formally look exactly like the corresponding Eqs. (67), for the one-dimensional case. This however should not distract the reader from the fact that HOOKE's law is now far more complicated. The quantities $\boldsymbol{\sigma}$ and ε are tensors of the second order that we have to describe as matrices with three rows and columns. HOOKE's tensor **C**, the tensor of the coefficients of elasticity, then has to be a fourth order tensor. Both points in $\mathbf{C} \cdot \cdot \, \varepsilon$ should point towards the fact that a double sum must be taken. This relationship can be written as

$$\sigma_{ik} = \sum_{r=1}^{3} \sum_{s=1}^{3} C_{ikrs}\, \varepsilon_{rs} \ . \tag{338}$$

For the following, we will adopt the so-called EINSTEINian summation convention, with which many equations can be easier formulated. Above all doubly occurring subscripts, 1–3 will be summed up without expressively writing the summation symbol Σ. For Eq. (338), we therefore simply write

$$\sigma_{ik} = C_{ikrs}\, \varepsilon_{rs} \ . \tag{338a}$$

For a comparison with other representations of the theory of elasticity, we draw the reader's attention to a notation introduced by W. VOIGT. In VOIGT's notation, two subscripts are joined into one, which then goes from 1 to 6. The the stress–strain relation (338) according to VOIGT reads

$$\sigma_I = \sum_{K=1}^{6} C_{IK}\, \varepsilon_K \qquad \text{VOIGT's notation} \tag{339}$$

with $\sigma_1 = \sigma_{11}$, $\sigma_2 = \sigma_{22}$, $\sigma_3 = \sigma_{33}$, $\sigma_4 = \sigma_{23}$, $\sigma_5 = \sigma_{13}$, $\sigma_6 = \sigma_{12}$, likewise for ε_I, hence, e.g., $\sigma_1 = C_{11}\,\varepsilon_1 + C_{12}\,\varepsilon_2 + C_{13}\,\varepsilon_3 + C_{14}\,\varepsilon_4 + C_{15}\,\varepsilon_5 + C_{16}\,\varepsilon_6$. We will here not use this notation.

HOOKE's tensor **C**, the tensor of the coefficients of elasticity, possesses some symmetry properties that reduce the maximum number of its independent components, see also L. D. LANDAU [54] and E. M. LIFSCHITZ. The following is always valid,

$$C_{ikrs} = C_{kirs} = C_{rsik} \ . \tag{340}$$

With the help of (340), one can check that the tensor **C** can still have a maximum of 21 independent components. The higher the symmetry in space of the observed crystal, the lower the number of independent moduli of elasticity of HOOKE's tensor. For the cubic crystal, only three independent moduli of elasticity of HOOKE's tensor remain. If one goes one step further and assumes that we are dealing with a homogeneous and isotropic continuum, with a continuum that looks the same at every location **x** and in every direction, then only two independent elastic moduli remain. These two moduli, as we will soon see, make sure that in comparison with the one-dimensional case, at least two different wave equations exist for the relative elastic strain ε of a three-dimensional continuum.

We insert (337) into (326) and find (instead (61) or (67) for the rod),

$$\rho\,\frac{\partial}{\partial t}\,\mathbf{v}(\mathbf{x},t) = \operatorname{div}\big(\mathbf{C}\cdot\cdot\,\varepsilon(\mathbf{x},t)\big) + \mathbf{f} \ . \tag{341}$$

We explicitly refer the reader to the following concerning the evaluation of Eq. (341).

(1) On the left-hand side of this equation, we find the time rate of a material velocity **v** that we, according to our derivation, discovered to be an elementary measurable quantity. The momentum density $\mathbf{p} = \rho\,\mathbf{v}$ is in our case according to $\mathbf{p} = \rho\,\partial\mathbf{s}/\partial t$ founded on an elastic displacement field $\mathbf{s} = \mathbf{s}(\mathbf{x},t)$. This refers especially to the stress free initial configuration, which we assumed for the derivation of this equation. If an elastic displacement field exists, $\mathbf{s} = \mathbf{s}(\mathbf{x},t)$, as we assumed above, then $\mathbf{v} = \partial\mathbf{s}/\partial t$ is valid in the linear approximation. However, the existence of such an elastic displacement field cannot be derived from this relationship. We will see that such an elastic displacement field generally does not exist.

(2) The situation concerning the tensor of elastic strain ε is similar. We determined this tensor in (335) under the presumption that an elastic displacement field $\mathbf{s} = \mathbf{s}(\mathbf{x},t)$ exists. In this special case, the symmetrical tensor ε, which generally contains six independent components, was already completely determined by the three functions of the elastic displacement field. Immediately measurable are the components of the symmetrical stress tensor $\boldsymbol{\sigma}$. We get six independent components for the general stress state of a solid. By inverting the third equation of (337), we can calculate the six components of the strain tensor ε, which then cannot generally be reduced to the three functions of a displacement field.

In order to comprehend these two statements in an easier manner, we must differentiate between the two different causes for the creation of elastic deformation, namely internal and external stress sources.

All volume forces **f** (see (325a)) and those forces that act through the boundary of a body, surface forces, are typical external stress causes. We can let them influence an object from the outside, or we can remove their influence from the object whenever we choose. An object that is *exclusively* influenced by such external stress sources completely moves into a stress free condition when these stress sources are removed. Such a condition was chosen by us as an initial condition.

This is the characteristic of external stress sources: If only these stress causes are active, then the stress $\boldsymbol{\sigma}$ caused by them results from a state where all mass elements of the object are simultaneously in a position of stress free equilibrium. The change in this state, caused alone by external stress sources, can be described using a displacement vector $\mathbf{s} = \mathbf{s}(\mathbf{x}, t)$ as we have done here. The calculation of this displacement vector **s** is the job of the classical theory of elasticity.

If we therefore express ε in (341) according to (331) using the displacement vector **s** and also insert $\mathbf{v} = \partial\mathbf{s}/\partial t$, then we receive the basic equation of the classical theory of elasticity. These are three equations for the elastic displacement vector **s** with presupposed forces **f** and with known moduli of elasticity of HOOKE's tensor **C**. We will explicitly write down these equations even though we do not intend on using this general form for any calculations,

$$\rho\,\frac{\partial^2 s_i}{\partial t^2} = \frac{1}{2}\sum_{k=1}^{3}\frac{\partial}{\partial x_k}\sum_{r=1}^{3}\sum_{s=1}^{3} C_{ikrs}\left(\frac{\partial s_r}{\partial x_s} + \frac{\partial s_s}{\partial x_r}\right) + f_i \;. \tag{342}$$

Due to the fact that **C** is constant, the derivatives with respect to x_k only apply to the components of **s**.

It is far easier to write down these equations if we use the summation convention introduced earlier on. We introduce a further notation enabling simplification. The partial derivatives with respect the coordinate x_k will only be shown as an index k after a comma, thus for a function $f = f(x_k)$,

$$\frac{\partial f}{\partial x_k} := f_{,k} \;, \qquad \frac{\partial^2 f}{\partial x_k\,\partial x_l} := f_{,kl} \;, \ldots .$$

We can then simply write

$$\rho\,\frac{\partial^2 s_i}{\partial t^2} = \frac{1}{2}\,C_{ikrs}\,(s_{r,sk} + s_{s,rk}) + f_i = C_{ikrs}\,\varepsilon_{rs,k} + f_i \tag{342a}$$

for the mathematically highly complicated Eq. (342). Using (340) we then receive

$$\rho \frac{\partial^2 s_i}{\partial t^2} = C_{ikrs}\, s_{r},_{sk} + f_i \ . \tag{342b}$$

If we have a case where a displacement vector **s** does not exist, we must go back to Eq. (341) and receive,

$$\rho \frac{\partial}{\partial t} v_i = C_{ikrs}\, s_{r},_{sk} + f_i \ . \tag{343}$$

We formulate the following equation by differentiating the above Eq. (343) for later application,

$$\rho \frac{\partial^2}{\partial t^2}\left(\frac{1}{2}(v_{i},_{j} + v_{j},_{i})\right) = \frac{1}{2}(C_{ikrs}\, \varepsilon_{rs},_{kj} + C_{jkrs}\, \varepsilon_{rs},_{ki}) + \frac{1}{2}(f_{i},_{j} + f_{j},_{i}) \ . \tag{343a}$$

The analysis of Eq. (342b) for all possible variants of Hooke's tensor **C** is the matter of classical theory of elasticity. We intend on restricting ourselves to the simplest case that, however, contains all those questions we are interested here, i.e. to the above case of complete isotropy. In this approximation, we assume that the solid has the same mechanical properties in all directions from every observed point. For a description of HOOKE's tensor, one uses the so-called KRONECKER symbol δ_{ik} according to

$$\delta_{ik} = \begin{matrix} 1 \\ 0 \end{matrix} \text{ for } \begin{matrix} i = k \\ i \neq k \end{matrix} \ ,$$

thus $\delta_{11} = \delta_{22} = \delta_{33} = 1$ and $\delta_{12} = \delta_{21} = \delta_{13} = \delta_{31} = \delta_{23} = \delta_{32} = 0$.

One can show, and here we refer the reader to the well-known representations, cf. LANDAU [54] and LIFSCHITZ, that the mechanical isotropy must be described by HOOKE's tensor of the following type,

$$C_{ikrs} = \mu(\delta_{ir}\,\delta_{ks} + \delta_{kr}\,\delta_{is}) + \lambda\,\delta_{ik}\,\delta_{rs} \ . \qquad \text{Isotropy} \tag{344}$$

The isotropic solid is characterised by two moduli of elasticity, here we have chosen the so-called LAMÉ parameters μ and λ. For HOOKE's law (338a), we receive for the isotropic case after a simple calculation

$$\sigma_{ik} = 2\,\mu\,\varepsilon_{ik} + \lambda\,\delta_{ik}\,\varepsilon \ . \tag{345}$$

The *scalar* dilatation ε (not in boldface type) introduced in (345) is, in the case of the existence of a displacement vector **s**, equal to its divergence according to

$$\varepsilon = \sum_{r=1}^{3} \varepsilon_{rr} = \sum_{r=1}^{3} \frac{\partial s_r}{\partial x_r} = s_{r},_{r} = \operatorname{div} \mathbf{s} \ . \tag{346}$$

With (344) Eqs. (342b) for the isotropic case can be greatly simplified. Insertion gives

$$\rho \frac{\partial^2 s_i}{\partial t^2} = [\mu(\delta_{ir}\,\delta_{ks} + \delta_{kr}\,\delta_{is}) + \lambda\,\delta_{ik}\,\delta_{rs}]\, s_{r\,,sk} + f_i \ .$$

If we calculate the sums we receive

$$\rho \frac{\partial^2 s_i}{\partial t^2} = (\mu + \lambda)\, s_{r\,,ri} + \mu\, s_{i\,,rr} + f_i \ . \qquad (347)$$

Notice that

$$s_{i\,,kk} = \sum_{k=1}^{3} \frac{\partial^2}{\partial x_k^2} = \Delta s_i \ , \qquad (348)$$

where Δ represents the LAPLACian. Equations (347) are now in fact equivalent to two D'ALEMBERTian wave equations that we get by appropriate differentiations from (347).

If we differentiate (347) with respect to x_i and sum up over the index i, then taking $s_{i\,,i} = \operatorname{div} \mathbf{s}$ into consideration, a wave equation for div **s** follows, namely

$$\rho \frac{\partial^2}{\partial t^2} \operatorname{div} \mathbf{s} = (2\mu + \lambda)\,\Delta \operatorname{div} \mathbf{s} + f_{i\,,i}$$

or

$$\Delta\,(\operatorname{div} \mathbf{s}) - \frac{1}{c_l^2} \frac{\partial^2}{\partial t^2} (\operatorname{div} \mathbf{s}) = -\, f_{i\,,i} \ . \qquad (349)$$

According to (349) the elastic dilatation $\varepsilon = \operatorname{div} \mathbf{s}$ propagates with the longitudinal sound velocity c_l,

$$c_l = \sqrt{\frac{2\mu + \lambda}{\rho}} \ . \qquad (350)$$

In order to arrive at the second wave equation we need to do the following. We write down Eq. (347) for an index j,

$$\rho \frac{\partial^2 s_i}{\partial t^2} = (\mu + \lambda)\, s_{r\,,rj} + \mu\, s_{j\,,rr} + f_j \ . \qquad (347')$$

We differentiate equation (347) with respect to x_j, Eq. (347)$'$ with respect to x_i and subtract both results from each other. Due to the fact that the second derivatives are interchangeable, the first terms on the right-hand side drop out and the following equation remains, once again using the LAPLACian Δ,

$$\rho \frac{\partial^2}{\partial t^2} (s_{i\,,j} - s_{j\,,i}) = \mu\,\Delta\,(s_{i\,,j} - s_{j\,,i}) + f_{i\,,j} - f_{j\,,i} \ . \qquad (351)$$

If we introduce the vector curl **s** with the three components

$$\text{curl}\,\mathbf{s}_1 = \frac{\partial s_3}{\partial x_2} - \frac{\partial s_2}{\partial x_3}\ , \qquad \text{curl}\,\mathbf{s}_2 = \frac{\partial s_1}{\partial x_3} - \frac{\partial s_3}{\partial x_1}\ , \qquad \text{curl}\,\mathbf{s}_3 = \frac{\partial s_2}{\partial x_1} - \frac{\partial s_1}{\partial x_2}\ , \tag{352}$$

we then see that every one of its components fulfils Eq. (351). After simple readjustment, we receive the wave equation for curl **s**,

$$\triangle\,(\text{curl}\,\mathbf{s}) - \frac{1}{c_T^2}\frac{\partial^2}{\partial t^2}\,(\text{curl}\,\mathbf{s}) = -\text{curl}\,\mathbf{f}\ . \tag{353}$$

This time every component of the vector curl **s** propagates with the transversal sound velocity c_T, where

$$c_T = \sqrt{\frac{\mu}{\rho}}\ . \tag{354}$$

Due to the above-assumed constancy of mass density ρ, the longitudinal sound velocity as well as the transversal sound velocity can be seen as being constant.

In the theory of elasticity, one can show that both sound velocities c_l and c_T only could coincide if YOUNG's modulus E would be zero. This would mean that the ratio of stress to strain for a pulled rod would disappear, cf. LANDAU [54] and LIFSCHITZ.[3] This would however mean that we could infinitely elongate a rod without creating any stress at all. Such atomic lattices do not exist. We here also assume that the sound velocities are not infinitely large. Therefore, the following applies to arbitrary isotropic solids,

$$c_T \neq c_l\ . \tag{355}$$

We find the following law confirmed:

There are at least two different sound velocities in every solid.

---

[3]For this, using the equations $\mu = \frac{E}{2(1+\nu)}$, $\lambda = \frac{E\nu}{(1-2\nu)(1+\nu)}$, where ν is POISSON's ratio, one writes $c_l = \sqrt{\frac{E(1-\nu)}{\rho(1+\nu)(1-2\nu)}}$, $c_T = \sqrt{\frac{E}{2\rho(1+\nu)}}$. Now a simple calculation shows that $c_l = c_T$ is impossible for $E \neq 0$.

Chapter 26
Eigen Stresses and Dislocations

One only needs to sufficiently heat an object disproportionately in order to create a state of stress that cannot principally be created or reduced by volume and surface forces. We are dealing here with eigen stresses. According to SOMMERFELD [50], these arise essentially in three ways: by unevenly heating, by magnetostrictive and elektrostrictive influences or by developmental stresses (so-called Werdegangs-spannungen). The were later explaned by dislocations. In fact, every deviation from the homogeneity of the base structure of a continuum can be seen as a source of eigen stresses.

We will explicitly examine the following case: An ideal crystal lattice (constant temperature and a uniform lattice parameter) may take up the ideal, stress free state of a mechanical continuum, the so-called stress free 'initial configuration'. The real, the actual existing lattice now deviates from its ideal base structure because of the dislocations present everywhere throughout the crystal. The crystal, or the continuum approximating the crystal is therefore present in a state where it contains eigen stresses that cannot principally be removed using external forces. There is therefore principally no displacement vector $\mathbf{s} = \mathbf{s}(\mathbf{x}, t)$, through which the former stress state can be created from a uniform stress free initial state according to a displacement of its mass elements. One can indeed create a stress free, in comparison to the ideal initial configuration strain free state for a sufficiently small environment of every point, but not, however, for the whole object. For the continuum containing dislocations, there indeed exists a well-defined strain ε, that once again according to HOOKE's law (315a) is connected to the stress $\boldsymbol{\sigma}$,

$$\frac{\partial}{\partial t}\rho + \sum_{r=1}^{3} \frac{\partial}{\partial x_r}(\rho\, v_r) = 0 \ . \tag{315a}$$

H. Günther, *Elementary Approach to Special Relativity*,
https://doi.org/10.1007/978-981-15-3168-2_26

The difference to the state of stress described in the last chapter is that this symmetrical strain tensor ε cannot be constructed from a displacement vector $\mathbf{s}$ anymore. Equation (335), in other words $\varepsilon_{ij} = \frac{1}{2}(s_{i},_{j} + s_{j},_{i})$ is no longer fulfilled. One says that the strain tensor ε is no longer compatible to a displacement field $\mathbf{s}$. An equation exists with which one can calculate whether a strain tensor is compatible to a displacement field or not. This would be the so-called conditions of compatibility of DE SAINT VENANT that we lay down in the following form,[1]

$$R_{likj} := \varepsilon_{ik},_{lj} - \varepsilon_{ij},_{lk} - \varepsilon_{lk},_{ij} + \varepsilon_{lj},_{ik} = 0 \quad \longleftrightarrow \quad \varepsilon_{ij} = \frac{1}{2}(s_{i},_{j} + s_{j},_{i}) \ . \qquad (356)$$

We will explain the following. An equivalent, compact formulation of above is

$$\operatorname{ink} \varepsilon = 0 \quad \longleftrightarrow \quad \varepsilon = \operatorname{def} \mathbf{s} \ . \qquad (356a)$$

Let us formulate this in words: Exactly then when $\operatorname{ink} \varepsilon = 0$, an *elastic* displacement vector $\mathbf{s}$ occurs, from which the elastic deformation $\varepsilon = \operatorname{def} \mathbf{s}$ is created.

The operator (in other words, the rule of construction) ink stands for 'incompatibility' and def stands for 'deformation'. Here we are dealing with certain differentiation rules that we intend on explaining below. From out of every vector $\mathbf{v}$ using $\operatorname{def} \mathbf{v}$ a tensor, a quantity with two indices is constructed according to

$$(\operatorname{def} \mathbf{v})_{ik} = \frac{1}{2}(v_{i},_{k} + v_{k},_{i}) \ . \qquad (357)$$

The operator ink is acting on a tensor, on a quantity with two indices. In order to define ink we will introduce the so-called LEVI-CIVITA symbol ϵ. This is a thrice indicized quantity that can only take the values $+1$, -1 or 0 according to the following rules,

$$\left.\begin{array}{l} \epsilon_{123} = \epsilon_{231} = \epsilon_{312} = +1 \ , \\ \epsilon_{213} = \epsilon_{132} = \epsilon_{321} = -1 \ , \\ \epsilon_{ijk} = 0 \quad \text{otherwise} \ . \end{array}\right\} \qquad (358)$$

Using this, we can express the curl of a vector $\mathbf{a}$, i.e. the components of the vector $\operatorname{curl} \mathbf{a}$, which is also written as $\nabla \times \mathbf{a}$ as follows:

$$(\operatorname{curl} \mathbf{a})_i \equiv (\nabla \times \mathbf{a})_i = \epsilon_{irs} a_{s},_{r} = \sum_{r=1}^{3} \sum_{s=1}^{3} \epsilon_{irs} a_{s},_{r} \ . \qquad (359)$$

One can also apply the operator curl on a tensor. One then only has to state which index is to underlie differentiation. One would write for the tensor β_{ik} e.g.,

[1] The quantity $\mathbf{R}$ is the so-called curvature tensor with the metric $g_{ik} = \delta_{ik} - 2\epsilon_{ik}$, which is a highly non-linear quantity. Here and further on, we only use its linear approximation. For a detailed relation between non-linear kinematics of dislocations and RICCI-calculus, cf. GÜNTHER [31].

$$(\text{curl}\,\boldsymbol{\beta})_{ik} \equiv (\nabla \times \boldsymbol{\beta})_{ik} = \epsilon_{irs}\,\beta_{sk},_{r} \quad . \tag{359a}$$

This one must differentiate from

$$(\boldsymbol{\beta} \times \nabla)_{ik} = \epsilon_{kpq}\,\beta_{ip},_{q} \quad . \tag{359b}$$

If one does not keep the correct order of indices, errors in the mathematical sign may occur.

The operator ink is now defined as the double application of the operator curl according to (359a) and (359b),

$$\begin{aligned} &\text{ink}\,\boldsymbol{\varepsilon} = \nabla \times \boldsymbol{\varepsilon} \times \nabla \ , \\ &(\text{ink}\,\boldsymbol{\varepsilon})_{ik} \equiv (\nabla \times \boldsymbol{\varepsilon} \times \nabla)_{ik} := \epsilon_{irs}\,\epsilon_{kpq}\,\varepsilon_{sp},_{rq} \quad . \end{aligned} \tag{360}$$

The quantity ink $\boldsymbol{\varepsilon}$ created here is a tensor with two indices. Here, one should note that we are dealing with a quadrupled sum, where however because of (358) the greater part of the terms disappear and others correspond with exception of the mathematical sign. The relationship of ink to the quantity R_{likj} in (356) is concurred in the following manner. We create the expressions below by summation,

$$R_{ik} = R_{rikr} = \varepsilon_{ik},_{rr} - \varepsilon_{rr},_{ik} - \varepsilon_{ri},_{rk} + \varepsilon_{rk},_{ri}$$

and

$$R := R_{ss} = 2\varepsilon_{ss},_{rr} - 2\varepsilon_{rs},_{rs} \quad .$$

Under observation of (358), one checks that

$$(\text{ink}\,\boldsymbol{\varepsilon})_{ik} = R_{ik} - \frac{1}{2}\,\delta_{ik}\,R \ . \tag{361}$$

Now $R_{ik} = 0$ follows, due to $\delta_{rr} = 3$ through summation over i and k from $R_{ik} - \frac{1}{2}\,\delta_{ik}\,R = 0$. Through simple introduction of all possible index combinations $(i, j, k, l = 1, 2, 3)$ one can directly calculate that also $R_{likj} = 0 \longleftrightarrow R_{ik} = 0$ and thus

$$R_{likj} = 0 \quad \longleftrightarrow \quad (\text{ink}\,\boldsymbol{\varepsilon})_{ik} = 0 \ . \tag{362}$$

One can easily check that for the special structure $\boldsymbol{\varepsilon} =$ def $\mathbf{s}$ the construction ink $\boldsymbol{\varepsilon}$ always vanishes. The reverse conclusion is: From the validity of the equation ink $\boldsymbol{\varepsilon} = 0$, $\boldsymbol{\varepsilon} =$ def $\mathbf{s}$ inevitably follows. This is somewhat more complicated, and we do not intend to prove it here. It is basically connected to EUCLIDean and non-EUCLIDean geometry, and we refer the reader to KRÖNER's [48] representation, see also GÜNTHER [31]. For the continuum approximating our crystal we can state:

Dislocations are found in our continuum at exactly those positions where ink $\boldsymbol{\varepsilon} = 0$.

We now have to comprehend this statement quantitatively. How can one quantitatively calculate and express the deviation of ink $\varepsilon = 0$ for a given distribution of dislocations? This question was first answered by KRÖNER [47, 48]. We will give a short summary of this answer.

We principally have to differentiate between elastic and plastic displacements $\delta\mathbf{s}$ and $\delta\mathbf{s}^p$ when dealing with the displacement of atoms of a crystal lattice. Both types of displacement generally change with the space coordinates $\mathbf{x}$ and thus cause so-called elastic or plastic distorsions $\boldsymbol{\beta}$ and $\boldsymbol{\beta}^p$ of the lattice, respectively, according to

$$\beta_{ik} = \frac{\delta s_k}{\delta x_i} \,, \qquad \beta_{ik}^p = \frac{\delta s_k^p}{\delta x_i} \,. \tag{363}$$

The total displacement $\mathbf{s}^g$ of the lattice atoms, the sum of the elastic and plastic displacements, is the subject of the NEWTONian equations and therefore always a certain, although usually extremely complicated function of space and time, $\mathbf{s}^g = \mathbf{s}^g(\mathbf{x}, t)$, that we have already referred the reader to at the end of Chap. 8. A single lattice element can always only participate in a total displacement $d\mathbf{s}^g$. This displacement is nothing other than the change of position in space during the time dt, which is well defined in the NEWTONian equations (53), Chap. 5. Only because of its imbedding in the crystalline formation do we distinguish, for a lattice element, between elastic and plastic displacements. This splitting up is defined as a collective phenomenon and thus loses its meaning for an isolated point of mass. The total displacement $d\mathbf{s}^g$ can now be put together in infinite ways using a non-integrable elastic displacement $\delta\mathbf{s}$ and a non-integrable plastic displacement $\delta\mathbf{s}^g$. We must therefore write,

$$d\mathbf{s}^g = \delta\mathbf{s} + \delta\mathbf{s}^g \,, \tag{364}$$

where for the total displacement only the completely integratable differential $d\mathbf{s}^g$ may occur belonging to a finite function $\mathbf{s}^g = \mathbf{s}^g(\mathbf{x}, t)$.

Only in the specific case in the preceding chapter, where we started with an ideal initial state and only observed an elastic displacement, but not a plastic displacement of the mass elements, the elastic displacement completely coincides with the total displacement. There is therefore a well-defined function for the elastic displacement $\mathbf{s} = \mathbf{s}(\mathbf{x}, t)$ of lattice elements, from which we, according to (330) and (330a) constructed the tensor of elastic distorsion $\boldsymbol{\beta}$ (resp. $\boldsymbol{\beta}^T$). If plastic deformations play a role, then there is generally neither a function $\mathbf{s}^p = \mathbf{s}^p(\mathbf{x}, t)$, nor a function $\mathbf{s} = \mathbf{s}(\mathbf{x}, t)$. There are only the infinitesimal quantities $\delta\mathbf{s}^p$ and $\delta\mathbf{s}$ defined in the vicinity of every point $\mathbf{x}$. If we construct a total distorsion according to $\boldsymbol{\beta}^g = \partial\mathbf{s}^g/\partial\mathbf{x}$, then the following immediately occurs,

$$\beta_{ik}^g \equiv \frac{\partial s_k^g}{\partial x_i} = \frac{\delta s_k}{\delta x_i} + \frac{\delta s_k^p}{\delta x_i} = \beta_{ik} + \beta_{ik}^p \,. \tag{365}$$

These are all well-defined functions of space and time. One should not however come to the conclusion that we are already dealing with functions of state of the crystal lattice that could be measured from a present crystal. This is a common error! We cannot generally measure such functions. We will explane this using an illustrative example: Every atom of an ideal gas moves along a certain path through space. However, we cannot exactly determine the position of this atom by measuring the variables of state, namely pressure and volume. The same applies to a crystal lattice underlying plastic deformation. E. KRÖNER succeeded in finding a statement concerning the variables of state from Eq. (365), in other words a statement about the actual measurable quantities. We construct the curl to the left of (365). Due to (359a), the following is valid with (358),

$$(\operatorname{curl}\boldsymbol{\beta}^g)_{ik} \equiv (\nabla \times \boldsymbol{\beta}^g)_{ik} = \epsilon_{irs}\,\beta^g_{sk},_r = \epsilon_{irs}\,s^g_k,_{sr} = 0 \tag{366}$$

and thus

$$\operatorname{curl}\boldsymbol{\beta}_{ik} = -\operatorname{curl}\boldsymbol{\beta}^p_{ik} \equiv -\operatorname{curl}\frac{\delta s^p_k}{\delta x_i}\ .$$

We integrate this equation over a small element of area $\Delta\mathbf{A}$ and find, by applying STOKES' theorem,

$$\iint\limits_{\Delta\mathbf{A}} \operatorname{curl}\boldsymbol{\beta}_{ik}\,dA_i = -\operatorname{curl}\boldsymbol{\beta}_{ik}\,\Delta A_i = -\iint\limits_{\Delta\mathbf{A}} \operatorname{curl}\frac{\delta s^p_k}{\delta x_i}\,dA_i = -\oint\limits_{(\Delta\mathbf{A})} \frac{\delta s^p_k}{\delta x_i}\,dx_i\ ,$$

hence,

$$\operatorname{curl}\boldsymbol{\beta}_{ik}\,\Delta A_i = -\oint\limits_{(\Delta\mathbf{A})} \delta s^p_k\ . \tag{367}$$

Here, KRÖNER's [47, 48] and NEY's [68] physical analysis brings into play. If the path of the line integral (367) encloses a dislocation, then its total BURGERS vector $\Delta\mathbf{b}$ is determined according to

$$-\oint\limits_{(\Delta\mathbf{A})} \delta s^p_k = \Delta b_k\ . \tag{368}$$

In more detail: Even though a local, plastic displacement $\delta\mathbf{s}^p$ that occurs at any arbitrary lattice position cannot be seen afterwards, because the crystal has exactly the same appearance before and after, the sum of all such plastic displacements along a closed path in the crystal can be experimentally determined by the measurement of its BURGERS vector. This is a very distinguished example for the difference between local and global questioning. The simple conclusion that the local plastic displacement $\delta\mathbf{s}^p$ cannot be seen and therefore also that the sum of these along an arbitrary path cannot be seen is a nasty error of which one should take heed. The crystal teaches us that this conclusion is false: The sum of all plastic displacements $\delta\mathbf{s}^p$, multiplied with

-1, along a closed path in the crystal is equal to the total BURGERS vector $\Delta\mathbf{b}$ of the dislocation enclosed by this path. This is the result of KRÖNER's analysis.

For the illustration of these facts, we turn to Figs. 7.7 and 7.8, in Chap. 7. We consider the case that a straight dislocation is created by the removal of the half of a lattice plane. In the ideal experiment, one can depict this as follows. The crystal is cut open along the line of dislocation that is to be created. One removes one half of the lattice plane from this cross section and puts the crystal back together so that the cross section can no longer be seen, see Fig. 26.1 and compare with KRÖNER [47, 48].

The experimental proof of dislocations is possible in many ways. Today, dislocations can be made visible using the electron microscope, see KITTEL [46], Chap. 19 foll.

The rest is just mathematics. We do not just approximate the crystal using a continuum, but also the large number of contain dislocations using a continuous dislocation density $\boldsymbol{\alpha}$. The second-order tensor of dislocation density $\boldsymbol{\alpha}$ is defined according to

$$\Delta A_i\,\alpha_{ik} = \Delta b_k \qquad \text{that is} \qquad \Delta\mathbf{A}\cdot\boldsymbol{\alpha} = \Delta\mathbf{b} \tag{369}$$

due to the sum Δb_k of the BURGERS vectors of all lines of dislocation that perpendicularly penetrate the area ΔA_i.

Equations (367) and (369) together make up the equation of KRÖNER and NEY that describes the complicated relationship between dislocations and elastic distorsions of the lattice in a most simplistic manner, KRÖNER [47] 1958, in English KRÖNER [48] 1980,

$$\epsilon_{irs}\,\beta_{sk\,,r} = \alpha_{ik} \qquad \text{that is} \qquad \operatorname{curl}\boldsymbol{\beta} = \boldsymbol{\alpha}\ . \tag{370}$$

Figures 26.1 and 26.2: Dislocation and plastic deformation. Only the position of the dislocation line pointing perpendicularily out of the drawing plane marked with the sign $\perp$ for an edge dislocation, can be experimentally verified. The dislocation in Fig. 26.2 is created by removing the shaded half lattice plane of Fig. 26.1 from the crystal and then putting the two halves back together again. The dotted lines in Fig. 26.2 show which lattice atoms received new next neighbours after this procedure. Here, the atoms were plastically displaced by $\delta\mathbf{s}_o^p$. The final position of the atoms is achieved by an additional elastic displacement. The partially filled atom has been labelled to allow better orientation in the lattice. On the thick lined orbit, only this atom has experienced a plastic displacement and thus alone adds $\delta\mathbf{s}_o^p$ to the sum $\sum\delta\mathbf{s}^p$. All other atoms in this orbit were only displaced elastically. This results in $\Delta\mathbf{b} = -\sum\delta\mathbf{s}^p = -\delta\mathbf{s}_o^p$ for the BURGERS vector $\Delta\mathbf{b}$ of the dislocation enclosed in the orbit.

From (370), we come to an important conclusion. With the help of (358), one can easily calculate that always

$$\frac{\partial}{\partial x_i}\,\epsilon_{irs}\,\beta_{sk\,,s} = \epsilon_{irs}\,\beta_{sk\,,si} = 0\ .$$

Hence, for an arbitrary distribution of dislocations, the following is always valid,

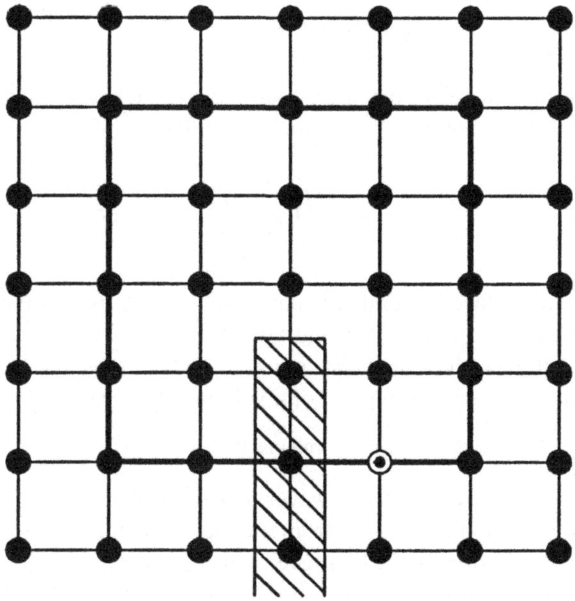

Fig. 26.1 Dislocation and plastic distorsion

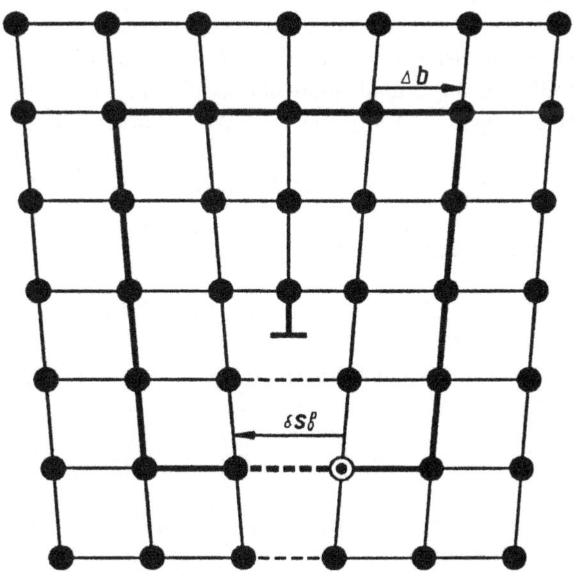

Fig. 26.2 Dislocation and plastic distorsion

$$\frac{\partial}{\partial x_i} \alpha_{ik} \equiv \alpha_{ik},_i = 0 \qquad \text{that is} \qquad \operatorname{div} \boldsymbol{\alpha} \equiv \nabla \cdot \boldsymbol{\alpha} = 0 \ . \tag{371}$$

Equation (371) has an astounding physical meaning. We integrate (371) over an arbitrary volume V and find by applying GAUSS' theorem, as well as our Eq. (369),

$$0 = \iiint\limits_{V} \frac{\partial}{\partial x_i} \alpha_{ik} \, dV = \iint\limits_{(V)} \alpha_{ik} \, dA_i = \iint\limits_{(V)} db_k \ . \tag{372}$$

The surface integral calculated here is the sum of the BURGERS vectors of all dislocations coming out of the volume V. This sum is always zero: In every arbitrary volumes, there are just as many dislocations coming into a volume as there are dislocations coming out of a volume. We discover:

A dislocation can never end inside of a volume.

We have not yet however arrived at our aimed for destination. Calculating the dislocation density $\boldsymbol{\alpha}$ from the plastic distorsions of the lattice, we have found out the one quantity which even can be measured. The dislocation density is a state variable formed from the plastic deformation. The elastic distorsion $\boldsymbol{\beta}$ is not such a measurable quantity as we have already discussed in the last chapter. We need a relationship between the dislocation density $\boldsymbol{\alpha}$ and the elastic strain $\boldsymbol{\varepsilon}$, which we could determine from the stress measurements made according to HOOKE's law (338a). Using a mathematical operation that was also stated by KRÖNER [47, 48], this functions in the following way: We construct at Eq. (370) the curl to the right, thus

$$\nabla \times \boldsymbol{\beta} \times \nabla = \boldsymbol{\alpha} \times \nabla \ .$$

Formulated properly, this gives

$$\epsilon_{irs} \epsilon_{kpq} \, \beta_{sp},_{rq} = \epsilon_{kpq} \, \alpha_{ip},_q \ .$$

We now notice that the elastic strain $\boldsymbol{\varepsilon}$ is the symmetrical part of the elastic distorsion $\boldsymbol{\beta}$,

$$\varepsilon_{ik} = \frac{1}{2} \left(\beta_{ik} + \beta_{ki} \right) \ , \tag{373}$$

and take the special properties (358) of the LEVI-CIVITA symbol ϵ_{ijk} into consideration. One only then needs to calculate that

$$\frac{1}{2} \left(\epsilon_{irs} \epsilon_{kpq} \, \beta_{sp},_{rq} + \epsilon_{krs} \epsilon_{ipq} \, \beta_{sp},_{rq} \right) = \epsilon_{irs} \epsilon_{kpq} \, \varepsilon_{sp},_{rq} \ .$$

Therefore,

$$\epsilon_{irs} \epsilon_{kpq} \, \varepsilon_{sp},_{rq} = \frac{1}{2} \left(\epsilon_{ipq} \, \alpha_{kp},_q + \epsilon_{kpq} \, \alpha_{ip},_q \right) \ . \tag{374}$$

Here, the right-hand side is nothing other than ink ε, and for the left-hand side, KRÖNER introduced the term $\boldsymbol{\eta}$, so that we arrive from Eq. (374) at the famous well-known KRÖNER equation,

$$\text{ink}\,\varepsilon = \boldsymbol{\eta} \tag{374a}$$

with

$$\eta_{ik} = \frac{1}{2}\left(\epsilon_{ipq}\,\alpha_{kp},_q + \epsilon_{kpq}\,\alpha_{ip},_q\right) \ . \tag{375}$$

We have achieved our first goal. Using Eq. (374), it is possible to calculate the eigen stresses that are created by a certain distribution of dislocations. As long as these dislocations are present as a static distribution, where they do not move through the crystal, Eqs. (374) and (375) are the only equations that we need, together with the equations of motion (343) and the material equations, HOOKE's law (338a), in order to calculate the stresses, including the eigen stresses. It is astounding, and these facts will occupy our line of thought that the formulation of equation (374), having only the eigen strain as contents, is completely free of any material parameters. For the elastic strain caused by external forces, a corresponding formulation cannot principally be made.

Before we include the motion of dislocations into our considerations, we will refer the reader to another omission which we will include in our depiction. These are the so-called momentum stresses. It is not completely correct that the real crystal that we approximate using a continuum does not react mechanically to a rotation of its volume elements ΔV, as we assumed in Chap. 25. In fact, because of the presence of dislocations a relative rotation of the volume elements occurs that also reacts energetically. We will not take these effects into consideration and refer the reader to KRÖNER's [53] paper.

We now consider a single straight dislocation with the BURGERS vector **b**, whose line direction is described by us using the unit vector **t**. If $\Delta\mathbf{A}$ is an area element of the size ΔA, which is chosen perpendicular to the line direction **t**, then according to Eq. (369), the following applies: $\Delta A_i\,\alpha_{ik} = \Delta A\,\alpha_{ik} = t_i\,b_k$. For a single straight dislocation on the area ΔA in the **t**-direction and with the BURGERS vector **b** we can write

$$\alpha_{ik} = \frac{t_i\,b_k}{\Delta A} \ . \tag{376}$$

This dislocation may now move with the velocity **V**. It is evident that a motion of a dislocation in its own line direction makes no physical sense. In fact, the line direction must traverse across the area $\Delta\mathbf{F}$ during a motion. The vector $\mathbf{n} = \mathbf{V} \times \mathbf{t}$ stands perpendicular on this area, and is thus a vector **n** with the components $n_i = \epsilon_{irs}\,V_r t_s$. The unit vector $\mathbf{n}_o$ constructed from this is

$$n_{oi} = \frac{\epsilon_{irs}\,V_r\,t_s}{|V|} \ . \tag{377}$$

The single dislocation that moves past the position **x** produces a plastic displacement of the vector **b** in this area $\Delta\mathbf{F}$. For each area element $\Delta\mathbf{A}$, a dislocation may be distributed equally. If equidistant dislocations move with the velocity $\mathbf{V}$, then $|\mathbf{V}|\,dt$ dislocations move past the position **x** during the time dt, so that the plastic displacement produced during this procedure can be calculated from $\delta\mathbf{g} = -\mathbf{b}\,|\mathbf{V}|$ dt, in other words,

$$\delta g_k = -b_k|\mathbf{V}|\,dt \ . \tag{378}$$

The plastic displacement $\delta\mathbf{g}$ is opposite to the dislocation motion, thus the minus sign, see Fig. 26.3.

If we multiply this displacement with the unit vector $\mathbf{n}_o$, then the plastic distorsion $d\beta^p$ is described by $\mathbf{n}_o\,\delta\mathbf{g} = \Delta A\,d\beta^p$ that belongs to the single dislocation on the area element $\Delta\mathbf{A}$, which has the velocity $\mathbf{V}$. It therefore follows using (377) and (378)

$$\Delta A\,d\beta^p_{ik} = n_{oi}\,\delta g_k = \frac{\epsilon_{irs}\,V_r\,t_s}{|V|}(-b_k\,|V|\,dt) \ ,$$
$$d\beta^p_{ik} = -\frac{\epsilon_{irs}\,V_r\,t_s\,b_k\,dt}{\Delta A}$$

and therefore with (376)

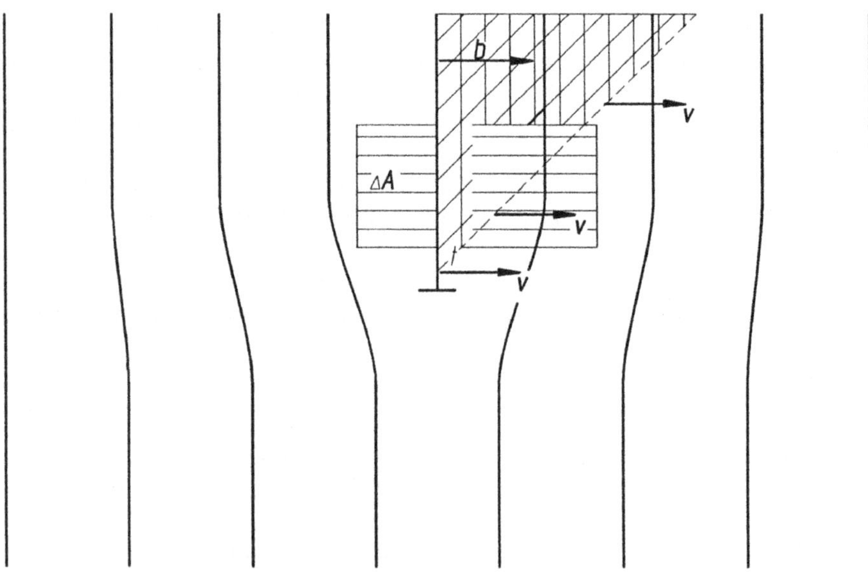

Fig. 26.3 Motion of dislocations and plastic deformation. Illustrated is a straight edge dislocation. The line direction **t** of the dislocation that penetrates the area element $\Delta\mathbf{A}$ perpendicularly, runs perpendicular to the BURGERS vector **b**. During motion with the velocity **V** perpendicular to the line of direction **t**, the dislocation creates, during the time dt, a plastic displacement $\delta\mathbf{g} = -\mathbf{b}\,|\mathbf{V}|dt$

$$\frac{d}{dt}\,\beta^p_{ik} = -\epsilon_{irs}\,V_r\,\alpha_{sk} \ . \tag{379}$$

Due to the fact that not only the dislocation density $\boldsymbol{\alpha}$, but also the dislocation velocity **V** are measurable quantities, the time derivative of the plastic distorsion is a measurable quantity of the solid. In comparison to plastic distorsion $\boldsymbol{\beta}^p$ the time derivative is once again a state variable of our mechanical continuum.

Equation (379) describes the physical relationship between the motion of dislocations and plastic deformation. This equation contains the complete linear dislocation kinematics and was discovered in 1956 by E. KRÖNER [49] and G. RIEDER. Just as the non-linear theory of static dislocations can be evolved from the geometry of the lattice, see KRÖNER [48] and GÜNTHER [31], it also follows the non-linear theory of moving dislocations from the geometrical properties of the crystal lattice, see GÜNTHER [31]. We cannot however delve deeper into this presently.

If the BURGERS vector **b** lies in the plane defined by **V** and **t**, then one talks of a gliding of dislocations. The vector product $\mathbf{V} \times \mathbf{t}$ defines the so-called glide plane. One also denotes this as a conservative motion of dislocations. Here, the material volume and the total mass, the total number of the particles be contained in the volume surrounding the dislocation line remains constant. Moreover, there are also non-conservative motions of dislocations. In a material volume surrounding the dislocation line mass escapes, the number of contained particles changes. One talks of a climbing of the dislocation. The BURGERS vector **b** lies hereby perpendicular to the $\mathbf{V} \times \mathbf{t}$-plane. Because for screw dislocations where **b** lies parallel to **t**, only edge dislocations can climb. We will explicitly consider and deal with the gliding of dislocations, thus only the conservative dislocation motion.

For an explicit formulation of dislocation kinematics, with which one can work, one defines the negative time derivative of the plastic distorsion $\frac{d}{dt}\,\boldsymbol{\beta}^p$ as the second order tensor of the density of dislocation current **J**. Using (379), we get

$$J_{ik} := \frac{d}{dt}\,\beta^p_{ik} = -\epsilon_{irs}\,V_r\,\alpha_{sk} \ . \tag{380}$$

We have now found the second state variable of the plastic deformation of our continuum. We can state:

The dislocation density $\boldsymbol{\alpha}$ and the density of dislocation current **J** are the two state variables of plastic deformation.

From (365), we construct the time derivative and then the curl from left, thus using (366),

$$\epsilon_{irs}\frac{d}{dt}\,\beta^p_{sk},_r + \epsilon_{irs}\frac{d}{dt}\,\beta_{sk},_r = \frac{d}{dt}\,\epsilon_{irs}\beta_{sk},_r + \epsilon_{irs}\frac{d}{dt}\,\beta^p_{sk},_r = 0 \ .$$

If we insert (370) and (380), then

$$\frac{d}{dt}\,\alpha_{ik} - \epsilon_{irs}\,J_{sk},_r = 0 \ . \tag{381}$$

We now observe that the time derivative of the total displacement vector $\mathbf{s}^g$ just about gives us the velocity $\mathbf{v}$ of the mass elements of our continuum,

$$\mathbf{v} = \frac{d}{dt}\,\mathbf{s}^g \qquad \text{that is} \qquad v_i = \frac{d}{dt}\,s_i \ . \tag{382}$$

Here, we wish to note that it is this velocity $\mathbf{v}$ that stands in Eqs. (337), (341), (343) and (382), respectively, cannot be constructed from out of an elastic displacement $\mathbf{s}$. From (365), we construct the sum during the exchange of the indices, in other words

$$\frac{1}{2}\left(\frac{\partial s_k^g}{\partial x_i} + \frac{\partial s_k^g}{\partial x_i}\right) = \frac{1}{2}\left(\beta_{ik} + \beta_{ki}\right) + \frac{1}{2}\left(\beta_{ik}^p + \beta_{ki}^p\right) \ .$$

Due to (337), (382) and (373), this equation's time derivative gives us

$$\frac{d}{dt}\,\varepsilon_{ik} - \frac{1}{2}\left(v_{i,k} + v_{k,i}\right) = \frac{1}{2}\left(J_{ik} + J_{ki}\right) \ . \tag{383}$$

We now have all equations necessary in order to completely determine, from an arbitrary distribution of dislocations $\boldsymbol{\alpha}$ and its arbitrary predetermined motion through the dislocation current $\mathbf{J}$ in an arbitrary continuum with HOOKE's tensor $\mathbf{C}$, as well as arbitrary volume forces $\mathbf{f}$ and taking certain boundary conditions in the case of a finite continuum into consideration, the elastic deformation ε and the velocity $\mathbf{v}$ of the masses of our continuum. Together these would be Eqs. (343), (375), (383), (371), (381), as well as (338a). We will formulate these equations here once again in the above order, whereby we, as a consequence, replace the total time derivative in (383) and (381) with the partial time derivative corresponding to our approximation presumption of a linearised theory of elasticity,

$$\left.\begin{aligned}
\rho\,\frac{\partial}{\partial t}\,v_i - C_{ikrs} \quad &= f_i \ , && (a)\\
\epsilon_{irs}\,\epsilon_{kpq}\,\varepsilon_{sp,rq} \quad &= \frac{1}{2}\left(\epsilon_{ipq}\,\alpha_{kp,q} + \epsilon_{kpq}\,\alpha_{ip,q}\right) \ , && (b)\\
\frac{\partial}{\partial t}\,\varepsilon_{ik} - \frac{1}{2}\left(v_{i,k} + v_{k,i}\right) &= \frac{1}{2}\left(J_{ik} + J_{ki}\right) \ , && (c)\\
0 \quad &= \alpha_{ik,i} \ , && (d)\\
0 \quad &= \frac{\partial}{\partial t}\,\alpha_{ik} - \epsilon_{ipq}\,J_{kq,p} \ , && (e)\\
\sigma_{ik} - C_{ikrs}\,\varepsilon_{rs} \quad &= 0 \ . && (f)
\end{aligned}\right\} \tag{384}$$

With the help of the given right-hand sides of equation (384), the fields on the left-hand sides need to be calculated. We are even in a position of being able to calculate, using Eq. (384), the elastic deformations that are caused by such dislocation configurations, which are solutions of the sine-GORDON equation. The dislocation density $\boldsymbol{\alpha}$ and the density of dislocation current $\mathbf{J}$ must then contain the kinks, breathers etc., as well as their motions.

In order to achieve a better understanding in the mathematical structure of these equations, we consider an isotropic continuum. HOOKE's tensor **C** is then already defined by the two LAMÉ parameters μ and λ according to (344), and for the relationship between stress and strain (345) is valid,

$$\sigma_{ik} = 2\,\mu\,\varepsilon_{ik} + \lambda\,\delta_{ik}\,\varepsilon \ . \tag{318}$$

We intend on completely ignore external forces,

$$f_i = 0 \ .$$

For (384a), we thus receive

$$\rho\,\frac{\partial}{\partial t}\,v_i - \mu\,\varepsilon_{ri},_r - \lambda\,\varepsilon_{rr},_i = 0 \ . \tag{385}$$

By differentiation, we receive (386) instead of equations (343a)

$$\frac{\partial}{\partial t}\,\frac{1}{2}\,(v_i,_k + v_k,_i) - \mu\,(\varepsilon_{ir},_{kr} + \varepsilon_{kr},_{ir}) - \lambda\,\varepsilon_{rr},_{ik} = 0 \ . \tag{386}$$

Equation (384b) we transform to

$$\begin{aligned}\varepsilon_{ik},_{rr} + \varepsilon_{rr},_{ik} - (\varepsilon_{ir},_{rk} + \varepsilon_{kr},_{ri}) = {} & \frac{1}{2}\,(\epsilon_{mni}\,\alpha_{nk},_m + \epsilon_{mnk}\,\alpha_{ni},_m) \\ & + \frac{1}{2}\,(\epsilon_{mni}\,\alpha_{mn},_k + \epsilon_{mnk}\,\alpha_{mn},_i) \ .\end{aligned} \tag{387}$$

The actual equivalence of (387) to (384b) can be verified by inserting all values for $i, k = 1, 2, 3$, whereby one should still take $\alpha_{ri},_r = 0$ into consideration. Equation (384c) can be differentiated with respect to time, so that

$$\frac{\partial^2}{\partial t^2}\,\varepsilon_{ik} - \frac{\partial}{\partial t}\,\frac{1}{2}\,(v_i,_k + v_k,_i) = \frac{\partial}{\partial t}\,\frac{1}{2}\,(J_{ik} + J_{ki}) \ . \tag{388}$$

We insert Eqs. (387) and (388) into (386) and find after some simple calculations,

$$\begin{aligned}\Delta\,\varepsilon_{ik} + \frac{\mu + \lambda}{\mu}\,\varepsilon_{rr},_{ik} - \frac{\rho}{\mu}\,\frac{\partial^2}{\partial t^2}\,\varepsilon_{ik} = {} & -\frac{1}{2}\,(\epsilon_{mni}\,\alpha_{nk},_m + \epsilon_{mnk}\,\alpha_{ni},_m) - \\ & -\frac{1}{2}\,(\epsilon_{mni}\,\alpha_{mn},_k + \epsilon_{mnk}\,\alpha_{mn},_i) - \\ & -\frac{\rho}{\mu}\,\frac{\partial}{\partial t}\,\frac{1}{2}\,(J_{ik} + J_{ki}) \ .\end{aligned} \tag{389}$$

We also find by differentiating (386) with respect to time and including (384c) the field equation for material velocities

$$\Delta v_i + \frac{\mu+\lambda}{\mu} v_{r,ri} - \frac{\rho}{\mu}\frac{\partial^2}{\partial t^2} v_i = -J_{ir,r} - J_{ri,r} - \frac{\lambda}{\mu} J_{rr,i} \ . \tag{390}$$

Equations (389) and (390) show an important property of eigen stresses created by dislocations. Although we observe the most simplistic, the isotropic continuum, the field equations do not reduce themselves to the simplest wave equation with a uniform signal velocity. Apart from the transversal speed of sound $c_T = \sqrt{\mu/\rho}$, a second signal velocity occurs, the longitudinal speed of sound $c_l = \sqrt{(2\mu+\lambda)/\rho}$, which becomes effective especially for the dilatation $\varepsilon = \varepsilon_{rr}$, as well as for the divergence of the velocity vector div $\mathbf{v} = v_{r,r}$. If we sum up Eq. (389) above $i = k$, then following occurs,

$$\Delta \varepsilon_{rr} - \frac{\rho}{2\mu+\lambda} \varepsilon_{rr} = -\frac{2\mu+\lambda}{\mu} (\epsilon_{mnr}\, \alpha_{nr,m} + \epsilon_{mnr}\, \alpha_{mn,r}) - \frac{\rho}{2\mu+\lambda}\frac{\partial}{\partial t} J_{rr} \ . \tag{391}$$

Taking the divergence of (390), we find

$$\Delta v_{r,r} - \frac{\rho}{2\mu+\lambda} v_{r,r} = -\frac{2\mu+\lambda}{\mu} (2J_{rs,rs} + \frac{\lambda}{\mu} J_{rr,ss}) \ . \tag{391a}$$

Two signal velocities superimpose when dealing with eigen stresses.

There is a special situation where only one signal velocity, the transversal speed of sound c_T is effective. For this case, we will construct a solution;in other words,we will calculate the elastic deformation that is created by a certain dislocation motion, which itself causes a plastic deformation.

We will assume that a single straight screw dislocation, lying parallel to the z-axis, with a BURGERS vector of the amount b moves in the x-direction with a constant velocity V.

We also wish to calculate using an unbounded continuum in every direction. The dislocation density $\boldsymbol{\alpha}$ is then defined by the following tensor,

$$\alpha_{ik} = b\,\delta_{i3}\,\delta_{k3}\,\delta(x - V\,t)\delta(y) \ . \tag{392}$$

Here, $\delta(x)$ is the so-called DIRAC's function, an 'improper' function that possesses the following characteristics,

$$\int_{-a}^{+a} \delta(x)\,dx = 1 \ , \qquad \int_{-a}^{+a} \delta(x)\,f(x)\,dx = f(0) \ , \qquad \text{if } a > 0 \,. \tag{393}$$

One also can differentiate the δ-function and work with it as if it were a 'normal' function. Its strict mathematical definition occurs in the theory of distributions. A, for the physicist sufficient, mathematical explanation of the δ-function can be found in IVANENKO [43] and SOLOKOW course book.

The δ-function is used to describe localised densities, for example for point charges. In (350),we used this to describe a dislocation density that consists of a single straight dislocation line lying parallel to the z-axis, which penetrated the x-y-plane at the position $x = V\,t\,,\ y = 0$. Because the dislocation moves with the velocity $V_r = V\,\delta_{r1}$ in the x-direction, a tensor of the density of dislocation current $\mathbf{J}$ belongs to, according to (380), the dislocation (392) according to $J_{ik} = \epsilon_{irs}\,V\,\delta_{r1}\,b\,\delta_{s3}\,\delta_{k3}\,\delta(x - V\,t)\,\delta(y)$, thus taking (358) into consideration,

$$J_{ik} = V\,b\,\delta_{i2}\,\delta_{k3}\,\delta(x - V\,t)\,\delta(y)\ . \tag{394}$$

We once again state that the dislocation velocity V should not be mistaken for the material velocity $\mathbf{v} = \mathbf{v}(\mathbf{x}, t)$ defined throughout our continuum.

The expressions (392) and (394) for the dislocation density and its current are inserted into Eqs. (391) and (391a),and after simple calculations,we discover that the right-and sides exactly disappear. The straight screw dislocations fulfils the equation

$$\Delta\varepsilon_{rr} - \frac{1}{c_l^2}\,\frac{\partial^2}{\partial t^2}\,\varepsilon_{rr} = 0\ , \qquad \Delta v_{r,r} - \frac{1}{c_l^2}\,\frac{\partial^2}{\partial t^2}\,v_{r,r} = 0\ . \tag{395}$$

For our unbounded medium,we demand so-called natural boundary conditions that have as an effect that all fields disappear for $\mathbf{x}$ tending to infinity. (Through this all hidden sources in infinity are excluded just as much as the fields that are not bound to sources, waves e.g.). Equation (395) then only have the trivial solution

$$\varepsilon_{rr} = 0\ , \qquad v_{r,r} = 0\ . \tag{396}$$

Our screw dislocation does not therefore create a dilatation field. This therefore applies to all straight screw dislocations, because the direction of the z-axis is of course arbitrary. We can state:

Straight screw dislocations create no dilatation fields in an isotropic, unbounded medium.

This in turn leads to an important simplification of the field Eqs. (389) and (390). For arbitrary straight screw dislocations $\boldsymbol{\alpha}^S$ and the corresponding currents $\mathbf{J}^S$ we receive the field equations

$$\begin{aligned}\Delta\,\varepsilon_{ik} - \frac{1}{c_T^2}\,\frac{\partial^2}{\partial t^2}\,\varepsilon_{ik} = &-\frac{1}{2}\,(\epsilon_{mni}\,\alpha^S_{nk},_m + \epsilon_{mnk}\,\alpha^S_{ni},_m) - \\ &-\frac{1}{2}\,(\epsilon_{mni}\,\alpha^S_{mn},_k + \epsilon_{mnk}\,\alpha^S_{mn},_i) - \\ &-\frac{1}{c_T^2}\,\frac{\partial}{\partial t}\,\frac{1}{2}\,(J^S_{ik} + J^S_{ki})\end{aligned} \tag{397}$$

and

$$\Delta v_i - \frac{1}{c_T^2}\,\frac{\partial^2}{\partial t^2}\,v_i = -J^S_{ir},_r - J^S_{ri},_r\ . \tag{398}$$

This supplies us with equations familiar to us from the relativistic field theories with one single signal velocity, here the transversal speed of sound c_T. This has far-reaching consequences that we will now illustrate using our above, uniform moving straight screw dislocation, by explicitly calculating the field of this dislocation. In order to do this,we insert (392) and (394) into (397) and (398). After elementary calculations,we find the following field equation, the $'$ always signifies the derivative of the immediately following variables in the brackets,

$$\left(\frac{\partial^2}{\partial x^2}+\frac{\partial^2}{\partial y^2}-\frac{1}{c_T^2}\frac{\partial^2}{\partial t^2}\right)\varepsilon_{ik}=$$
$$=-\frac{b}{2}\begin{pmatrix} 0 & 0 & \delta(x-V\,t)\,\delta(y) \\ 0 & 0 & -(1-\frac{V^2}{c_T^2})\delta'(x-V\,t)\,\delta(y) \\ \delta(x-V\,t)\,\delta'(y)-(1-\frac{V^2}{c_T^2})\delta'(x-V\,t)\,\delta(y) & 0 & 0 \end{pmatrix} \tag{399}$$

and

$$\left(\frac{\partial^2}{\partial x^2}+\frac{\partial^2}{\partial y^2}-\frac{1}{c_T^2}\frac{\partial^2}{\partial t^2}\right)v_i = V\,b\begin{pmatrix} 0 \\ 0 \\ \delta(x-V\,t)\,\delta'(y) \end{pmatrix}. \tag{400}$$

This is mathematically seen, due to the fact that all derivatives with respect to z disappear, a plane problem. For the mathematical discussion of these strict relativistic field equations,we refer the reader to GÜNTHER [34].

The only solution of equations (399) and (400) using natural boundary conditions is,

$$\left.\begin{aligned} \varepsilon_{13}=\varepsilon_{31} &= -\frac{b}{4\pi}\frac{1}{\sqrt{1-V^2/c_T^2}}\frac{\partial}{\partial y}\ln\sqrt{(x-V\,t)^2+(1-\frac{V^2}{c_T^2})y^2}\ , \\ \varepsilon_{23}=\varepsilon_{32} &= -\frac{b}{4\pi}\frac{1}{\sqrt{1-V^2/c_T^2}}\frac{\partial}{\partial x}\ln\sqrt{(x-V\,t)^2+(1-\frac{V^2}{c_T^2})y^2}\ , \\ v_3 &= -\frac{V\,b}{2\pi}\frac{1}{1-V^2/c_T^2}\frac{\partial}{\partial y}\ln\sqrt{(x-V\,t)^2+(1-\frac{V^2}{c_T^2})y^2}\ . \end{aligned}\right\} \tag{401}$$

All other components of ε and $\mathbf{v}$ are zero. Equation (401) is therefore the elastic deformation field, which accompanies the plastic deformation, which in turn creates a straight screw dislocation moving with a constant velocity through the crystal.

One can calculate the elastic energy E that such a dislocation moving through the medium with the velocity V causes.[2] If E_o is the energy of the stationary dislocation, then one discovers the relationship $E=E_o/\sqrt{1-v^2/c_T^2}$. Such a field possesses,

[2]The question concerning the 'cutoff radius' will not answered here.

with respect to the lattice, an inertial mass m, that is connected to the energy E via the EINSTEINian equation $E = m\, c_T^2$. Even for the inertial mass m of the field created by a screw dislocation the strict relativistic equation $m = m_o/\sqrt{1 - v^2/c_T^2}$ applies, if we designate the inertial mass of the field of the stationary dislocation as m_o. These relationships are analogue to those of the electron with its electromagnetic field. One can for example relatively easily calculate that the strain velocity field (401) of a dislocation can be summarised as a LORENTZ tensor, just as the electromagnetic field of an electron can, cf. GÜNTHER [34]. We now have to add the field mass to the bare mass of a dislocation that we numerically approximated and introduced in Chap. 8, as well as in Eq. (295). If we once again accept a correspondence between the characteristic velocity c_o of the sine-GORDON equation and the transversal speed of sound c_T, $c_o = c_T$, then both mass portions cannot be differentiated because of their identical velocity dependencies; these are facts that we exactly find confirmed when dealing with electrons. The field Eqs. (397) and (398) thus describe a physical situation that is completely analogue to the conditions concerning electrons according to the MAXWELL-LORENTZ theory with all of its mathematical consequences. Here,we especially think of those consequences that arise, if quantum theory is taken into account with the consequence of the secondary relativistic effects such as pair creation and vacuum polarisation. The creation of a electron-positron pairs in quantum electrodynamics corresponds to the creation of elementary dislocation loops in the theory of dislocations, cf. GÜNTHER [33, 35].

In Chaps. 9–21,we developed a complete Special Theory of Relativity for the internal observers of our crystal on the basis of the sine-GORDON equation. We did this by restricting the number of classes of competitive mechanical phenomena. We restricted ourselves to the solutions of the sine-GORDON equation and ignored the creation of elastic deformations by dislocations. We now saw that these deformations are determined by those equations according to (397) and (398) that fulfil all demands made by a relativistic theory. Can we therefore include elastic deformations caused by dislocations in our relativistic considerations inside of the crystal? Can the internal observers of our crystal support their measurements and thoughts on elastic deformations, as well as plastic deformations? Such a conclusion would be too far fetched here. Equations (397) and (398) only apply to a strongly restricted class of dislocations, the straight screw dislocations in an isotropic, unbounded continuum. Not even the straight screw dislocation with a kink belong in this class!

But we can show that the part of elastic deformations bound structurally to dislocations, bound to *any* dislocations, is in fact determined by only relativistic wave equations with one single signal velocity. These facts will be examined by us in the next chapter. In the view of further illustrations to this problem,we must refer the reader to GÜNTHER [32, 33].

It has been know for quite a whilst, here we refer the reader to a paper of ESHELBY [19] from 1949 that the eigen stresses of dislocations show certain relicts of relativistic behaviour. These relicts gradually dissipate with the increasing complexity of HOOKE's tensor. We only discover unrestricted relativistic behaviour for straight screw dislocations in an isotropic, unbounded medium, as seen above. It does not good putting edge dislocations into an isotropic, unbounded medium. Various LORENTZ factors are combined in the fields for the elastic deformation ε of a straight edge dislocation moving with a constant velocity V, those based on transversal and those that are based on the longitudinal speed of sound. These are combined in complicated manners resulting in the transference of velocity dependencies of elastic energy, so that one can no longer talk of behaviour in the sense of the Special Theory of Relativity. There is a simple reason for such general complicated picture of dislocations:

In Eq. (384), two elastic fields with completely different symmetry properties, namely a field ε^I structurally coupled to dislocations and an elastic displacement field $\mathbf{s}$, according to $\varepsilon_{ik} = \varepsilon_{ik}^I + \frac{1}{2}(s_{i,k} + s_{k,i})$, simply combined to an elastic field ε, for which Eq. (384) then apply. We will now sketch out that Eq. (384) do in fact arise from the superposition of two, different independent systems of equations. From a system of equation for ε^I and a completely different one for $\mathbf{s}$.

Chapter 27
The Separation of Eigen Stresses

The transformations with which we replaced Eq. (384) with (389) and (390) for the isotropic continuum in the last chapter was made possible by the assumed isotropy, which allowed us to insert Eq. (387) into (386). This cannot be done using the general case of HOOKE's tensor that we will assume now. By differentiating Eq. (384a), we find

$$\rho \frac{\partial}{\partial t} \frac{1}{2} (v_{i,k} + v_{k,i}) - \frac{1}{2} (\sigma_{ri,rk} + \sigma_{rk,ri}) = \frac{1}{2} (f_{i,k} + f_{k,i}) \tag{402}$$

as well as

$$\rho \frac{\partial^2}{\partial t^2} v_i - \frac{\partial}{\partial t} \sigma_{ri,r} = \frac{\partial}{\partial t} f_i \ . \tag{403}$$

We further require Eq. (384c) and its derivative with respect to time (388),

$$\frac{\partial}{\partial t} \varepsilon_{ik} - \frac{1}{2} (v_{i,k} + v_{k,i}) = \frac{1}{2} (J_{ik} + J_{ki}) \ , \tag{384c}$$

$$\frac{\partial^2}{\partial t^2} \varepsilon_{ik} - \frac{\partial}{\partial t} \frac{1}{2} (v_{i,k} + v_{k,i}) = \frac{\partial}{\partial t} \frac{1}{2} (J_{ik} + J_{ki}) \ . \tag{388}$$

We now need the general case of HOOKE's law, as well as its inverse formulation; this means

$$\left. \begin{aligned} \sigma_{ik} &= C_{ikrs} \, \varepsilon_{rs} \ , \\ \varepsilon_{ik} &= S_{ikrs} \, \sigma_{rs} \end{aligned} \right\} \tag{404}$$

with the tensor **S** reciprocal to **C**, thus

$$\mathbf{S} \cdot\cdot \mathbf{C} = \mathbf{1} \qquad \text{that is} \qquad S_{ikrs} \, C_{rspq} = \delta_{ip} \, \delta_{kq} \ . \tag{405}$$

H. Günther, *Elementary Approach to Special Relativity*,
https://doi.org/10.1007/978-981-15-3168-2_27

We then introduce the following decomposition for HOOKE's tensor,

$$\left.\begin{aligned} C_{ikrs} &= C^o_{ikrs} + \widetilde{C}_{ikrs} \quad \text{with} \quad C^o_{ikrs} := p\,(\delta_{ir}\,\delta_{ks} + \delta_{is}\,\delta_{kr} - \delta_{ik}\,\delta_{rs})\ , \\ S_{ikrs} &= S^o_{ikrs} + \widetilde{S}_{ikrs} \quad \text{with} \quad S^o_{ikrs} := \tfrac{1}{4p}\,(\delta_{ir}\,\delta_{ks} + \delta_{is}\,\delta_{kr} - \delta_{ik}\,\delta_{rs})\ , \end{aligned}\right\} \tag{406}$$

where the decomposition for $\mathbf{S}$ is necessitated by those from $\mathbf{C}$. Here, the quantity p is an formally introduced, decomposition parameter, whose value we will later discuss appropriately.

With (404), (406) and (406a), we use the following agglomeration of HOOKE's law,

$$\varepsilon_{ik} = \frac{1}{2p}\,(\sigma_{ik} - \delta_{ik}\,\sigma_{rr}) + \widetilde{S}_{ikpq}\,C_{pqrs}\,\varepsilon_{rs}\ . \tag{407}$$

We take (407) into the three terms $\varepsilon_{rr,ik} - (\varepsilon_{ri,kk} + \varepsilon_{rk,ri})$ of Eq. (387), which are equivalent to (384b), namely $\text{ink}\,\varepsilon = \eta$ and insert the thus transformed Eq. (387), as well as (388) into (402). The result is

$$\begin{aligned} \Delta\,\varepsilon_{ik} &- \frac{\rho}{p}\frac{\partial^2}{\partial t^2}\varepsilon_{ik} + \frac{1}{2p}\left[p\,\frac{\partial}{\partial x_k}\left(-2\frac{\partial}{\partial x_r}\,\widetilde{S}_{irpq} + \frac{\partial}{\partial x_i}\,\widetilde{S}_{rrpq}\right)C_{pqmn}\,\varepsilon_{mn} + \right. \\ &\left. + p\,\frac{\partial}{\partial x_i}\left(-2\frac{\partial}{\partial x_r}\,\widetilde{S}_{krpq} + \frac{\partial}{\partial x_k}\,\widetilde{S}_{rrpq}\right)C_{pqmn}\,\varepsilon_{mn}\right] \\ = &\frac{1}{2}\,(\epsilon_{mni}\,\alpha_{nk,m} + \epsilon_{mnk}\,\alpha_{ni,m}) - \frac{1}{2}\,(\epsilon_{mni}\,\alpha_{mn,k} + \epsilon_{mnk}\,\alpha_{mn,i}) - \\ &- \frac{\rho}{p}\,\frac{\partial}{\partial t}\,\frac{1}{2}\,(J_{ik} + J_{ki}) + \frac{1}{2p}\,(f_{i,k} + f_{k,i})\ . \end{aligned} \tag{408}$$

In order to arrive at the equations for v_i, we construct from (384c)

$$v_{i,rr} + v_{r,ri} = 2\,\frac{\partial}{\partial t}\,\varepsilon_{ir,r} - J_{ir,r} - J_{ri,r}$$

as well as

$$v_{r,r} = \frac{\partial}{\partial t}\,\varepsilon_{rr,i} - J_{rr,i}\ .$$

Together this results in

$$v_{i,rr} = 2\,\frac{\partial}{\partial t}\,\varepsilon_{ir,r} - \frac{\partial}{\partial t}\,\varepsilon_{rr,i} - J_{ir,r} - J_{ri,r} + J_{rr,i}\ .$$

Here, we insert (407) and receive after simple calculation

$$\begin{aligned} v_{i,rr} = &\frac{\partial}{\partial t}\left[\frac{1}{p}\,\sigma_{ir,r}\right] + 2\frac{\partial}{\partial t}\,\frac{\partial}{\partial x_r}\left[\widetilde{S}_{irpq}\,C_{pqmn}\varepsilon_{mn}\right] - \\ &- \frac{\partial}{\partial t}\,\frac{\partial}{\partial x_i}\left[\widetilde{S}_{rrpq}\,C_{pqrs}\varepsilon_{mn}\right] - J_{ir,r} - J_{ri,r} + J_{rr,i}\ . \end{aligned}$$

We finally find using (403)

$$
\begin{aligned}
&\triangle v_i - \frac{\rho}{p}\frac{\partial^2}{\partial t^2} v_i + \\
&+ \frac{1}{p}\frac{\partial}{\partial t}\left[p\left(-2\frac{\partial}{\partial x_r}\left[\widetilde{S}_{irpq}\, C_{pqmn}\varepsilon_{mn}\right] + 2\frac{\partial}{\partial x_i}\left[\widetilde{S}_{rrpq}\, C_{pqrs}\varepsilon_{mn}\right]\right)\right] \\
&= -J_{ir\,,r} - J_{ri\,,r} + J_{rr\,,i} - \frac{1}{p}\frac{\partial}{\partial t} f_i \ .
\end{aligned}
\tag{409}
$$

Instead of Eqs. (389) and (390) for the isotropic continuum, Eqs. (408) and (409) apply for the general case of HOOKE's tensor $\mathbf{C}$ used in the determination of the fields $\boldsymbol{\varepsilon}$ and $\mathbf{v}$ for an arbitrary prescribed distribution of dislocations $\boldsymbol{\alpha}$ and their currents $\mathbf{J}$ as well as arbitrary volume forces $\mathbf{f}$.

The structure of the left-hand sides of these equations is somewhat complicated when dealing with the general case of HOOKE's tensor and its 21 independent parameters. The following decoupling of these equations can be achieved, see GÜNTHER [35]. We introduce an elastic displacement vector $\mathbf{s}$, so that the portions $\boldsymbol{\varepsilon}^I$ and $\mathbf{v}^I$ can be removed from the corresponding parts of the total elastic deformation $\boldsymbol{\varepsilon}$ and the matter velocity $\mathbf{v}$ according to

$$
\left.
\begin{aligned}
\varepsilon_{ik} &= \varepsilon^I_{ik} + \frac{1}{2}\left(s_{i\,,k} + s_{k\,,i}\right) \ , \\
v_i &= v^I_i + \frac{\partial}{\partial t} s_i \ .
\end{aligned}
\right\}
\tag{410}
$$

One can now show, for this we refer to GÜNTHER [35]: The quantities ε^I_{ik} and v^I_i simply fulfil the D'ALEMBERTian wave equations

$$
\begin{aligned}
\triangle \varepsilon^I_{ik} - \frac{\rho}{p}\frac{\partial^2}{\partial t^2}\varepsilon^I_{ik} = &-\frac{1}{2}\left(\epsilon_{mni}\,\alpha_{nk\,,m} + \epsilon_{mnk}\,\alpha_{ni\,,m}\right) \\
&-\frac{1}{2}\left(\epsilon_{mni}\,\alpha_{mn\,,k} + \epsilon_{mnk}\,\alpha_{mn\,,i}\right) - \frac{\rho}{p}\frac{\partial}{\partial t}\frac{1}{2}\left(J_{ik} + J_{ki}\right)
\end{aligned}
\tag{411}
$$

and

$$
\triangle v^I_i - \frac{\rho}{p}\frac{\partial^2}{\partial t^2} v^I_i = -J_{ir\,,r} - J_{ri\,,r} + J_{rr\,,i} \ , \tag{412}
$$

if the elastic vector $\mathbf{s}$ fulfils the equations of the classical theory of elasticity, thus equations of the type (342a) with a certain additional force created by the strain components ε^I_{mn},

$$
\begin{aligned}
\rho\frac{\partial^2 s_i}{\partial t^2} = &\frac{1}{2} C_{rimn}\left(s_{m\,,nr} + s_{n\,,mr}\right) + \\
&+ f_i + p\left(-2\frac{\partial}{\partial x_r}\left[\widetilde{S}_{irpq}\, C_{pqmn}\varepsilon_{mn}\right] + 2\frac{\partial}{\partial x_i}\left[\widetilde{S}_{rrpq}\, C_{pqrs}\varepsilon_{mn}\right]\right) \ .
\end{aligned}
\tag{413}
$$

The quantities ε^I_{ik} and v^I_i will be denoted here by us in this relationship as structural eigen strains, so-called incompatible eigen deformations that are completely independent from any elastic parameters or volume forces and that are thus only determined by the dislocations and their currents. We have thus achieved our goal, cf. also GÜNTHER [35]:

For every value of the decomposition parameter p, an elastic displacement field $\mathbf{s}$ exists according to (410), so that the structural eigen strains ε^I and $\mathbf{v}$ can be calculated from (411) and (412) independent of the elastic parameters of the medium.

We can then choose the parameter p, so that the signal velocity in the wave Eqs. (411) and (412) corresponds to the signal velocity c_o of the sine-GORDON equation or also with the transversal sound velocity c_T, namely

$$\frac{p}{\rho} = c_o^2 \approx c_T^2 \ . \tag{414}$$

In this case, Eqs. (411) and (412) fulfil the same LORENTZ symmetry as the sine-GORDON equation. It is the relativistic field equations for all internal observers moving uniformly with respect to each other that have one and the same mathematical form. The inertial mass of the field ε^I, $\mathbf{v}$, created according to Eqs. (411) and (412), also shows the same velocity dependency as the naked mass of a dislocation introduced in Chap. 7 and later examined in (295). We can therefore—and this situation is well known from the discussions on the mass of an electron—no longer distinguish the two parts of mass from one another. Measuring the inertial mass of a dislocation only gives the sum of the naked mass and the field mass defined by Eqs. (411) and (412).

We must assume that the consequences of a quantized theory of field Eqs. (411), (412) do in fact correspond to those of quantum electrodynamics. This means that we principally have to, based on these equations, expect effects that correspond to the creation and annihilation of electron–positron pairs. It is known that electrons in the limits of medium field strengths can only be accelerated, namely by the LORENTZ force. In an extremely high-energetic electromagnetic field, there are, however, certain supplementary specific relativistic effects. If the field energy suffices, according to the energy mass equivalence, then electron–positron pairs are spontaneously created. A single electron cannot be created this way because of the charge conservation.

Quantitatively, the relationship where dislocations are concerned is completely different. Due to the fact that dislocations have a finite extend, we are dealing with linear, thus one-dimensional objects in comparison with the 'zero-dimensional' point electrons, the inertial mass of one dislocation is enormously large. This means that a highly energetic, elastic field is needed in order to accelerate this dislocation. In comparison with this, the energy needed to create an elementary; thus, the smallest possible dislocation loop is minutely small. The equation of charge conservation of electrons corresponds to Eq. (371) in Chap. 24 which stated that a dislocation could not end inside of a volume. Instead of the electron–positron pairs, closed dislocation loops are created. These loops can be arbitrarily small, as small as the lattice structure allows. The energetic relationship is, therefore, when compared to those in the electromagnetic case virtually reversed. The process of elementary dislocation loop creation occurs whenever low-energetic elastic fields come into play, because of the relatively low creation energy required. However, the acceleration of a dislocation requires a high-energetic environment. This conclusion is completely proven correct by the experiment. Plastic deformations are practically accompanied by intensive dislocation creation processes from the very beginning. We can only expect micro-plastic processes in low-energetic regions; in other words, we can expect dislocation motions as described by the sine-GORDON equation, for example the propagation of a kink along a line of dislocation and also the creation and destruction of elementary dislocation loops that lead on the line of dislocation to the so-called kink pair creation, cf. GÜNTHER [35].

The formulation of phenomena in solids in the system of the Special Theory of Relativity as we found for the so-called structural eigen strain using Eqs. (411) and (412) is not only of pure academic interest. We expect, from the application of theoretical quantum field methods on these equations, an approach to the theory of plasticity. This theory thus depends, from the point of view developed here, on the secondary relativistic effects, i.e. with the relativistic consequences arising from quantisation. The use of the relativistic spacetime structure in a solid, as developed from the sine-GORDON equation is therefore no longer left to us. The relativistic time of the moving, internal observer in the crystal leads to far-reaching consequences, namely to a theoretical concept of elasto-plastic interactions based on the concept of the quantum field theory.

The internal observers of our crystal, that we approximate as our continuum, can include the structural elastic strain ε^I, $\mathbf{v}$ that are described by Eqs. (411) and (412) in their considerations without any deviation from their Special Theory of Relativity. Which role do Eq. (413) play? These equations break every LORENTZ symmetry. The displacement vector $\mathbf{s}$ therefore has to be excluded from such considerations. There is however a way out of this dilemma, but this cannot be discussed here. Such a displacement vector always defines a coordinate transformation—it can be 'transformed away'. The result of this would be special relativistic equations at arbitrary coordinates, even in accelerated reference systems. This leads us mathematically into the world of the General Theory of Relativity, a problematic situation for the solid that we do not wish to follow here.

Chapter 28
Particles and Tachyons

In Chap. 20, we examined solution (261), $q^T(x, t)$, of the sine-GORDON equation,

$$q^T(x,t) = \frac{2a}{\pi} \arctan \exp \left[\frac{-\pi(x - u\,t)}{L_o\,\kappa} \right] + \frac{a}{2} \tag{261}$$

with

$$\kappa = \frac{u}{c_o}\sqrt{1 - \frac{v^2}{c_o^2}} = \operatorname{sign} u \sqrt{\frac{u^2}{c_o^2} - 1}\ , \quad u = \frac{c_o^2}{v}\ , \quad |u| > c_o\ . \tag{258}$$

Here, with this solution, an energy density also moves through the crystal; however, here it moves with a velocity faster than the speed of sound, to be precise $|u| > c_o$. Solutions of this type are therefore defined by EILENBERGER [11], as tachyons, and we accepted in Chap. 20 that tachyons are indeed particles or quasi-particles in the crystal, respectively. Firstly, $q^T(x, t)$ is a field, just as much so as the kink solution (111), $q^I(x, t)$. Such a field creates an energy–momentum tensor (279). Only in certain cases is it possible to attribute this tensor the definite physical parameters of a particle, here a total energy E and a total momentum P. In Chap. 23, we saw how we could understand the kink $q^I(x, t)$, using this method as a relativistic particle, a quasi-particle in a crystal. But what of the tachyon? Is the tachyon's field (261) mechanically a particle, so that arbitrary collisions between particles and tachyons must underlie the mechanical laws of energy and momentum conservation? Mathematically, this leads to the following question: Can we attribute the tachyon's field (261) in any arbitrary reference system Σ an energy E^T and a momentum P^T in such a way that these quantities are linked together in different reference systems by the Eq. (298).

How does the mass m^T of a tachyon and its velocity change when the tachyon changes reference systems? There is no restmass here. The mass of a tachyon can only be determined using the equation $E^T = m^T c_o^2$. And even the term velocity

H. Günther, *Elementary Approach to Special Relativity*,
https://doi.org/10.1007/978-981-15-3168-2_28

of a tachyon has, as we will see below, a rather abstract meaning and can only be introduced using the momentum P^T of the tachyon. The relationships for 'normal' particles cannot just be carried over to tachyons.

We will, to start off with, examine the general EINSTEINian composition of velocities that we developed in Chap. 17, Eq. (194). A 'flying' object may have in the reference system $\Sigma'(x', t')$ the *arbitrary* velocity $u' = dx'/dt'$. The reference system Σ', which we should consider a 'normal object' such as a train for example, may have the velocity V measured from out the reference system Σ_o. $|V| < c_o$ always applies to the velocities $|V|$ of a reference system. The velocity $u = dx/dt$ that an observer in the reference system Σ_o measures for the same flying object must then be, according to the composition of velocities,

$$u = \frac{V + u'}{1 + V\,u'/c_o^2} \qquad \text{with the inversion} \qquad u' = \frac{-V + u}{1 - V\,u/c_o^2} \;. \tag{415}$$

The following can be deduced by simple calculation:
(a) For every velocity $|u'| < c_o$ it follows that $|u| < c_o$ and vice versa.
(b) For every velocity $|u'| > c_o$ it follows that $|u| > c_o$ and vice versa.
(c) The principle of the constancy of critical signal velocity c_o is reproduced, i.e. from $|u'| = c_o$ it follows that $|u| = c_o$ and vice versa.

The kinematics of the Special Theory of Relativity therefore principally allows for three types of particles:
(a) particles with $|u| < c_o$,
(b) particles with $|u| > c_o$,
(c) particles with $|u| = c_o$.

Each of these particles are defined independently from the reference system and can therefore not change if the reference system changes.

We will not deal with the particles of type (c) here, e.g. photons in electrodynamics in the case of our physical spacetime. Particles of type (b) have received the name tachyon. Particles of type (a) are distinguished by the fact that they can also have the velocity zero. Whether any of these particles actually exist, tachyons or 'normal' particles, is not discussed by the SRT. Only that the particles of type (a) exist is a conclusion drawn from our elementary everyday experiences.

The general mathematical framework that includes all of the characteristic properties of the three types of particles, namely MINKOWSKI space formalism, will not be discussed here. The mathematical realisation of the Special Theory of Relativity according to H. MINKOWSKI's method, which is of great importance for dealing with, and for the development of relativistic physics, can be found in all major physics course books, as well as in the appropriate representations, cf. e.g. A. PAPAPETROU [99], A. P. FRENCH [23] and in the original paper of MINKOWSKI [66] from 1908, printed in LORENTZ [60]. In our one-dimensional Special Theory of Relativity, we intend on solving this problem without the help of too much mathematics.

The energy E of a particle is equivalent to its mass m. This has been demonstrated in Chap. 23 using the kink as an example, see Eqs. (296) and (297). The mass of a particle is therefore defined by means of its energy. We found our further considerations on particles in SRT on the equation

$$E = m\, c_o^2 \ . \tag{416}$$

We further assume that the interaction between particles during collisions is based on the mechanical laws of energy and momentum conservation. We now begin with the dependence of the mass m of a particle on its velocity u. In other words, we look for the function $f(u)$ in equation

$$m = \xi_o\, f(u) \ . \tag{417}$$

For the velocity u we make no limitations, the constant ξ_o is a parameter that, from the start we do not see as a restmass.

We observe the total inelastic collision in the reference system Σ' of two particles that have the same parameter ξ_o and the same opposite momentum, so that the total momentum disappears according to momentum conservation. For the momentum conservation in Σ', which we would normally note as $\xi_o\, f(u')\, u' + \xi_o\, f(-u')\, (-u') = 0$ with an even function $f(-u) = f(u)$, we make a more general ansatz, which is justificated only later on,

$$\xi_o\, f(u')\, u' + \delta \cdot \xi_o\, f(-u')\, (-u') = 0 \ . \qquad \text{Momentum conservation in } \Sigma' \tag{418}$$

Here, we introduced a factor δ for the parameter of the second particle that can assume the values $+1$ or -1. We will make use of this later on. Furthermore, we do not automatically have the function $f(u)$ at our disposal. The ansatz (418) will prove very useful when intend on registering particles of type (b) during the process of collision.

According to our assumptions of a total inelastic collision, both particles should remain static after the collision fused to one single particle in Σ' with the mass parameter M_o. We can therefore write in Σ' for the constancy of the total energy before and after the collision,

$$\xi_o\, f(u')\, c_o^2 + \delta \cdot \xi_o\, f(-u')\, c_o^2 = M_o\, f(0)\, c_o^2 \ , \qquad \text{Energy conservation in } \Sigma' \tag{419}$$

so that

$$\xi_o \left[f(u') + \delta \cdot f(-u') \right] = M_o \ . \tag{420}$$

In the reference system, Σ_o the mass M_o has after the collision the velocity V. The velocities of the colliding particles before the collision occurs can be calculated in Σ_o using the composition of velocities (415). We insert u' in Σ' for the velocity of the first particle and for the second particle with the parameter $\delta \cdot \xi_o$ we insert $-u'$. We therefore receive for the energy conservation in Σ_o

$$
\begin{aligned}
&\xi_o\, f\Big(\frac{V+u'}{1+V\,u'/c_o^2}\Big)\,c_o^2\;+ \\
&+\delta\cdot\xi_o\, f\Big(\frac{V-u'}{1-V\,u'/c_o^2}\Big)\,c_o^2 = M_o\, f(V)\,c_o^2\ .
\end{aligned}
\qquad \text{Energy conservation in } \Sigma_o \tag{421}
$$

From (420) and (421), a functional equation for the function $f = f(u)$ follows,

$$
f\Big(\frac{V+u'}{1+V\,u'/c_o^2}\Big) + \delta\cdot f\Big(\frac{V-u'}{1-V\,u'/c_o^2}\Big) = \big[f(u') + \delta\cdot f(-u')\big]\cdot f(V)\ . \tag{422}
$$

For the general solution of this equation, we must distinguish between two cases.

(a) We are dealing with two particles of type (a), thus $|u'| < c_o \longrightarrow |u| < c_o$ with $\frac{V\pm u'}{1\pm V\,u'/c_o^2}$. In this case, we choose $\delta = 1$ and the solution of the functional equation is determined by the LORENTZ factor $\sqrt{1-u^2/c_o^2}$ according to the well known relation

$$
f(u) = \frac{1}{\sqrt{1-u^2/c_o^2}}\ , \quad |u| < c_o\ . \tag{423}
$$

Insertion of (423) in (422) gives us,

$$
\begin{aligned}
&\frac{1}{\sqrt{1-\frac{1}{c_o^2}\Big(\frac{V+u'}{1+V\,u'/c_o^2}\Big)^2}} + \frac{1}{\sqrt{1-\frac{1}{c_o^2}\Big(\frac{V-u'}{1-V\,u'/c_o^2}\Big)^2}} \\
&= \frac{1+V\,u'/c_o^2}{\sqrt{\big(1+V\,u'/c_o^2\big)^2 - \frac{1}{c_o^2}\,(V+u')^2}} + \frac{1-V\,u'/c_o^2}{\sqrt{\big(1-V\,u'/c_o^2\big)^2 - \frac{1}{c_o^2}\,(V-u')^2}} \\
&= \frac{1+V\,u'/c_o^2}{\sqrt{1-V/c_o^2}\,\sqrt{1-u'^2/c_o^2}} + \frac{1-V\,u'/c_o^2}{\sqrt{1-V/c_o^2}\,\sqrt{1-u'^2/c_o^2}} \\
&= 2\,\frac{1}{\sqrt{1-V/c_o^2}\,\sqrt{1-u'^2/c_o^2}} = 2\,f(V)\,f(u')\ , \qquad \Box
\end{aligned}
$$

(For 'normal' particles $\delta = 1$ and $f(u) = f(-u)$ must be observed).

(b) We are dealing with two particles of type (b), thus $|u'| > c_o \longrightarrow |u| > c_o$ where for u is once again $u = \frac{V\pm u'}{1\pm V\,u'/c_o^2}$. In order to receive a solution of the functional equation (422), we must now assume $\delta = -1$ and thus supplement our ansatz (423) in the following way:

$$
f(u) = \left.\begin{array}{lll} \dfrac{1}{\sqrt{1-u^2/c_o^2}} \,, & & |u| < c_o \,, \\ & \text{if} & \\ \dfrac{\operatorname{sign} u}{\sqrt{1-u^2/c_o^2}} \,, & & |u| > c_o \,. \end{array}\right\} \tag{423a}
$$

The sign inserted in the second part of the solution, $\operatorname{sign} u = u/|u|$, may seem somewhat arbitrary. We will come back to this in a moment and firstly verify that the functional equation (422) is fulfilled by this ansatz not only for a positive u, but also for a negative u. Here, one must observe that $\delta = -1$. For the first particle, we assume a velocity $u' > 0$ in Σ'. Then, $\operatorname{sign}(V - u') = +1$ and $\operatorname{sign}(V - u') = -1$. We find

$$
\frac{\operatorname{sign}\dfrac{V+u'}{1+V\,u'/c_o^2}}{\sqrt{\dfrac{1}{c_o^2}\left(\dfrac{V+u'}{1+V\,u'/c_o^2}\right)^2 - 1}} + \delta\,\frac{\operatorname{sign}\dfrac{V-u'}{1-V\,u'/c_o^2}}{\sqrt{\dfrac{1}{c_o^2}\left(\dfrac{V-u'}{1-V\,u'/c_o^2}\right)^2 - 1}}
$$

$$
= \frac{\left[\operatorname{sign}\dfrac{V+u'}{1+V\,u'/c_o^2}\right] t(1+V\,u'/c_o^2)\left[\operatorname{sign}(1+V\,u'/c_o^2)\right]}{\sqrt{\dfrac{1}{c_o^2}(V+u')^2 - \left(1+V\,u'/c_o^2\right)^2}} +
$$

$$
+\,\delta\,\frac{\left[\operatorname{sign}\dfrac{V-u'}{1-V\,u'/c_o^2}\right](1-V\,u'/c_o^2)\left[\operatorname{sign}(1-V\,u'/c_o^2)\right]}{\sqrt{\dfrac{1}{c_o^2}(V-u')^2 - \left(1-V\,u'/c_o^2\right)^2}}
$$

$$
= \frac{\left[\operatorname{sign}(V+u')\right](1+V\,u'/c_o^2)}{\sqrt{1-V^2/c_o^2}\,\sqrt{u'^2/c_o^2-1}} + \delta\,\frac{\left[\operatorname{sign}(V-u')\right](1-V\,u'/c_o^2)}{\sqrt{1-V^2/c_o^2}\,\sqrt{u'^2/c_o^2-1}}
$$

$$
= \frac{1}{\sqrt{1-V^2/c_o^2}}\,\frac{1-\delta}{\sqrt{u'^2/c_o^2-1}} = f(V)\left[f(u') + \delta\, f(-u')\right] \,, \qquad \square
$$

For both types of particles (a) and (b), we discover the following dependency of their momentum and energy to their velocity.

(a) For the momentum and the energy of a 'normal' type (a) particle with $v < c_o$ we write $\xi_o = m_o$ and the following is valid:

$$P = \frac{m_o}{\sqrt{1 - v^2/c_o^2}}\, v \ , \quad E = \frac{m_o}{\sqrt{1 - v^2/c_o^2}}\, c_o^2 \ . \tag{424}$$

The parameter ξ_o of this particle is called its restmass m_o and it is

$$P_o = 0 \ , \quad E_o = m_o\, c_o^2 \ . \tag{424a}$$

These are the familiar relationships that we proved for the kink solution of sine-GORDON equation in Chap. 23, see Eqs. (296) and (297).

(b) The general solution (423a) allows for a second type of particle that also fulfils the mechanical laws of energy and momentum conservation. For the energy E^T and the momentum P^T of this type (b) particle with $u > c_o$, the so-called tachyons, the following applies

$$P^T = \frac{m_*^T}{\operatorname{sign} u \sqrt{u^2/c_o^2 - 1}}\, u \ , \quad E^T = \frac{m_*^T}{\operatorname{sign} u \sqrt{u^2/c_o^2 - 1}}\, c_o^2 \ . \tag{425}$$

In this case, we have written $\xi_o = m_*^T$. This parameter m_*^T obviously cannot be a restmass. The physical meaning of m_*^T can be seen in the critical case of $u \longrightarrow \pm\infty$. As one can easily see, the following applies for both $u \longrightarrow +\infty$ and for $u \longrightarrow -\infty$, namely

$$\lim_{u\to\pm\infty} P^T = P_\infty^T = m_*^T\, c_o \ , \quad \lim_{u\to\pm\infty} E^T = E_\infty^T = 0 \ . \tag{425a}$$

If one should insert the forbidden velocities $u < c_o$ into Eq. (425), then imaginary values for the mass of a tachyon can be calculated, therefore also an imaginary restmass, so that momentum and energy remain real. Tachyons are therefore also classified as particles with imaginary restmasses. Such a classifications of tachyons is however misleading, as we will see below.

The physical parameter of a tachyon is the quantity m_*^T. This parameter takes the place of the restmass for particles of type (a). There is a fundamental difference between these parameters, which we will discuss later on.

The velocity u of a tachyon is defined by the quotient from momentum and energy according to

$$u := \frac{c_o^2\, P^T}{E^T} \ . \tag{426}$$

The mass of the tachyon defined using the energy $E^T = m^T\, c_o^2$ can be both negative and positive, as the energy E^T. With the help of the ansatz (423a) we were only able to fulfil the energy and the momentum conservation (421) and (418) resp. and therefore also the functional equation (422), because we assumed a negative mass, $\delta = -1$, for the second particle. This would not have been necessary without the factor $\operatorname{sign} u$ in Eq. (423a). However, this factor guarantees that the momentum P^T

of the tachyon has one and the same critical value $m_*^T\, c_o$ for both $u \longrightarrow +\infty$ and $u \longrightarrow -\infty$. This condition must be met for the following reason.

We can immediately see from the composition of velocities (415): Not only the tachyon with the velocity $u' = +\infty$, as seen from Σ', but also the tachyon with the velocity $u' = -\infty$ in Σ' has, when observed from Σ_o, one and the same velocity $u = c_o^2/V$, where V the velocity of Σ' is with respect to Σ_o. Seen from Σ_o, the tachyon has the same momentum for both critical values. It can therefore not possess two different critical values for the momentum in Σ'. The factor δ is therefore absolutely necessary in order to guarantee uniqueness for the momentum of the tachyon.

From Eq. (425) for the energy and the momentum of a tachyon, the physical particle properties of the tachyon follow: If the values E^T and P^T are measured for the energy and the momentum of a tachyon in the reference system Σ_o, as well as $E^{T\prime}$ and $P^{T\prime}$ in the reference system Σ' that moves with respect to Σ_o with the velocity V, then the following is valid:

$$\left.\begin{aligned}
P^{T\prime} &= \frac{P^T - V\,E^T/c_o^2}{\sqrt{1 - V^2/c_o^2}} \quad , \qquad E^{T\prime} = \frac{E^T - V\,P^T}{\sqrt{1 - V^2/c_o^2}} \\
&\text{with the inversion} \\
P^T &= \frac{P^{T\prime} + V\,E^{T\prime}/c_o^2}{\sqrt{1 - V^2/c_o^2}} \quad , \qquad E^T = \frac{E^{T\prime} + V\,P^{T\prime}}{\sqrt{1 - V^2/c_o^2}} \ .
\end{aligned}\right\} \tag{427}$$

These equations are identical to Eq. (298) that we verified in Chap. 23 for the kink solution of the sine-GORDON equation. Equations (427) and (298) are valid for both type (a) and type (b) particles, respectively. They are the paramount of the physical particle characteristics. The verification of these equations is based, even for tachyons, explicitly on the composition of velocities (415) and it is applied in the same way, as shown in Chap. 23 for the type (a) particle. We will be satisfied with applying this to the last equation of (427). According to the definition

$$v := \frac{c_o^2}{u} \qquad \longrightarrow \qquad \sqrt{\frac{u^2}{c_o^2} - 1} = \frac{|u|}{c_o}\sqrt{1 - \frac{v^2}{c_o^2}} \tag{428}$$

we attribute the velocities u and u' of the tachyon in Σ_o and Σ' the velocities v and v', with an amount smaller than c_o. Using the composition of velocities (415) for u and u', the composition is also valid for v and v', as one can easily examine (V is once again the velocity of Σ' in Σ_o),

$$u = \frac{V + v'}{1 + V\,v'/c_o^2} \qquad \text{with the inversion} \qquad v' = \frac{-V + v}{1 - V\,v/c_o^2} \ . \tag{429}$$

From this, we receive, as in Chap. 23, the following, which can be verified by squaring

$$\sqrt{1-\frac{v^2}{c_o^2}}=\frac{\sqrt{1-V^2/c_o^2}\,\sqrt{1-v'^2/c_o^2}}{1-V\,v'/c_o^2} \tag{430}$$

and then, with the primed quantities corresponding to (425), we find using (428), (429), (430) and $\operatorname{sign} x = x/|x|$,

$$\begin{aligned}
\frac{E^{T\prime}+V\,P^{T\prime}}{\sqrt{1-V^2/c_o^2}} &= \frac{\dfrac{m_*^T\,c_o^2}{\dfrac{u'}{|u'|}\dfrac{|u'|}{c_o}\sqrt{1-v'^2/c_o^2}}+V\,\dfrac{m_*^T\,u'}{\dfrac{u'}{|u'|}\dfrac{|u'|}{c_o}\sqrt{1-v'^2/c_o^2}}}{\sqrt{1-V^2/c_o^2}} \\
&= \frac{m_*^T\,c_o^2}{u'/c_o}\,\frac{1+V\,u'/c_o^2}{\sqrt{1-V^2/c_o^2}\,\sqrt{1-v'^2/c_o^2}} \\
&= m_*^T\,c_o^2\,\frac{\dfrac{V}{c_o}+\dfrac{v'}{c_o}}{\sqrt{1-V^2/c_o^2}\,\sqrt{1-v'^2/c_o^2}} \\
&= m_*^T\,c_o^2\,\frac{\dfrac{v}{c_o}\left(1+\dfrac{V\,v'}{c_o^2}\right)}{\sqrt{1-V^2/c_o^2}\,\sqrt{1-v'^2/c_o^2}} \quad = \quad m_*^T\,c_o^2\,\frac{v}{c_o}\,\frac{1}{\sqrt{1-v^2/c_o^2}} \\
&= \frac{m_*^T\,c_o^2}{u/c_o\sqrt{1-v^2/c_o^2}}
\end{aligned}$$

and finally

$$\frac{E^{T\prime}+V\,P^{T\prime}}{\sqrt{1-V^2/c_o^2}}=\frac{m_*^T}{\operatorname{sign} u\,\sqrt{u^2/c_o^2-1}}\,c_o^2=E^T\ .$$

This is however E^T according to (427), which is what we intended to show. The same calculation can be made for P^T.

Tachyons now have mysterious properties. The momentum of a tachyon has one and the same sign in all reference systems. This follows from the Eq. (427) if one chooses the reference system for Σ_o in which the tachyon has the critical value (425a): The energy is zero. The LORENTZ factor $\sqrt{1-V^2/c_o^2}$ is however always positive. Thus, $P^{T\prime}$ always retains the same sign as P^T. Based on this property, again with the composition of velocities (415) or even (429) we can see that the velocity u of a tachyon does not have to have the same sign as its momentum P^T.

With a positive v' (and also positive u'), v and therefore also u become negative if $-V > v'$ is chosen, which is always possible for the reference system.

We will now illustrate the peculiarities of tachyons using two numerical examples. An observer in Σ_o determines the velocity of a tachyon as $u = \frac{5}{4}\, c_o$. He sends this tachyon towards an observer moving with the velocity $V = \frac{1}{2}\, c_o$. Using (415) one can calculate the velocity u' that the observer (moving with V in Σ_o) determines for this tachyon, $u' = \frac{-\frac{1}{2}\, c_o + \frac{5}{4}\, c_o}{1 - \frac{c_o 5 c_o}{2 \cdot 4 c_o^2}} = 2c_o$. The observer running behind the tachyon registers that the tachyon is in fact moving away from him at a faster speed than bevor! We can find a tentative explanation in correspondence with Eq. (425). The observer following the tachyon assumes a portion of the tachyon's energy of motion. According to (425), the amount of the energy of a tachyon becomes smaller if the amount of its velocity increases. The tachyon loses energy when it gets faster. If the velocity increases infinitely, the tachyon completely loses its energy.

The following situation is even more peculiar. In Σ_o the velocity $u = 2c_o$ is measured for a tachyon. An observer follows this tachyon with the velocity $V = \frac{4}{5}\, c_o$. This observer then registers the velocity $u' = \frac{-\frac{4}{5} + 2c_o}{1 - \frac{4 c_o 2 c_o}{5 c_o^2}} = -2c_o$ for the tachyon. Is the tachyon in fact moving towards the observer, who was sent out to follow the tachyon? This would not be anything extraordinary for velocities in the region of $|u| < c_o$, a simple overtaking procedure, e.g. one car overtaking the other. A tachyon cannot however be overtaken! its velocity is always greater than c_o, a velocity that we can never achieve. Nevertheless, the velocity $u' = \Delta x'/\Delta t'$ of the tachyon in Σ' becomes negative, whilst its velocity $u = \Delta x/\Delta t$ in Σ_o is positive. The sign of the quantity Δx for the tachyon cannot change when the tachyon changes reference systems, just as the tachyon's momentum retains its sign as shown above (with exactly the same argumentation). For a positive u and a positive Δx, u' can only become negative if $\Delta t'$ becomes negative! The tachyon therefore does not spatially come closer to the observer in Σ'. Its velocity becomes negative, because it moves backwards through time! We will say that a tachyon moves backwards through time if its velocity and momentum have opposite signs. If we uphold the notion of tachyon velocity, we must then accept the fact that a tachyon, depending on the reference system in which it is observed, can move with the attributed velocity even backwards through time. This is where the causality problem connected to tachyons is anchored. We will return to this during the next two chapters.

When the sign of the tachyon's velocity changes, the sign of its energy E^T also changes according to (425) and thus also per definition that of its mass $m^T = E^T/c_o^2$.

An explanation for all these peculiarities can be found in the following considerations. We remain in one and the same reference system Σ_o and realise an inversion (a mirroring) of the spatial coordinates, see also Chap. 12. In other words, we describe all positions using new space coordinates $\overline{x}$, whilst retaining the time coordinates according to

$$\overline{x} = -x \ , \quad \overline{t} = t \ . \tag{431}$$

Then, per definition, the following applies to the velocity $\overline{u}$ of the tachyon (as for every other particle) in the new coordinates

$$\overline{u} = \frac{d\overline{x}}{d\overline{t}} = -u \ . \tag{432}$$

Furthermore, the momentum is, according to its physical attribution a spatial vector. The components of spatial vectors change their signs during inversion, in other words

$$\overline{P}^T = -\overline{P} \ . \tag{433}$$

Due to $\overline{u}/\mathrm{sign}\,\overline{u} = u/\mathrm{sign}\,u$ it follows from (425) that the tachyon parameter m_*^T must also change its sign so that (433) can be fulfilled,

$$\overline{m}_*^T = -\overline{m}_*^T \ . \tag{434}$$

We can make the following conclusion from this:

The tachyon parameter m_*^T is a spatial vector.

This is the key needed to unlock and understand the properties of tachyons. The quantity m_*^T has nothing at all in common with the restmass m_o of a normal particle of the type (a). The particle parameter m_o, the restmass, is a spatial scalar, therefore unchangeable with respect to any change of space coordinates. In comparison to this, m_*^T multiplied by the constant quantity c_o is a characteristic momentum value that is assumed by the tachyon in the reference system where its energy is zero. The occasional classification of the tachyon as a particle with an imaginary restmass is therefore misleading.

The quantity m_*^T is the momentum parameter of the tachyon.

Whilst the restmass m_o of a 'normal' type (a) particle is always positive, the momentum parameter m_*^T of a tachyon can be both positive and negative. In our one-dimensional Theory of Relativity, we have two types of tachyons, those with a positive and those with a negative momentum parameter. We will be able to observe these in the next chapter directly using the solutions of the sine-Gordon equation. If we change the direction of the x-axis defined as positive, then the signs of the tachyon's momentum parameters also change.

For all of these tachyon properties that have the transformation Eq. (427) in common with the 'normal' particles, but vary so distinctly in their kinematics, there is a simple mathematical frame. In the formalism of Minkowski geometry all particles are described using spacetime vectors. Our 'normal' particles of type (a) are described by so-called time-like vectors with the consequence of a preferred frame in which only the so-called time component $P_o = E_o/c_o = m_o c_o$ of this vector is different from zero. In this way, the restmass m_o is well-defined for all 'normal' particles of the type (a). In comparison to this, tachyons are described using this geometry by so-called space-like vectors with the consequence of a preferred frame in which now the time component disappears. In our one-dimensional Special Theory of Relativity, only one single component of the preferred space-like momentum vector P_{∞}^{T} remains, where we discovered $P_{\infty}^{T} = m_{+}^{T} c_o$ in (425a). Due to the fact that c_o is a constant, the spatial vector property is completely transferred to m_{*}^{T}. In our one-dimensional case only the change of the sign would remain after an inversion according to (433). For a detailed explanation of the mathematics of Minkowski space, we refer the reader to A. Papapetrou [99], A. P. French [23]. The mathematical treatment of tachyons in the framework of Minkowski space can be found in D. E. Liebscher [57].

Up to now, we have only shown that tachyons are compatible with the principles of the Special Theory of Relativity, and how these can be fitted into the general frame of this theory. If tachyons actually exist, and if their existence leads to experimentally proven consequences, then nothing has of yet been said. Only then when we are able to say more about this question, will we concern ourselves with the causality problem which is inevitably thrown up in the light of tachyons.

Chapter 29
Tachyons of Plastic Deformation

We now come back to the question that we formulated at the beginning of the last chapter: Is the field of the solution (261) of the sine-GORDON equation,

$$
\begin{aligned}
q^T(x,t) &= \frac{2a}{\pi}\arctan\exp\left[\frac{-\pi(x-u\,t)}{L_o\,\kappa}\right]+\frac{a}{2}\,, \\
\kappa &= \frac{u}{c_o}\sqrt{1-\frac{v^2}{c_o^2}} = \operatorname{sign} u\sqrt{\frac{u^2}{c_o^2}}\,, \qquad u=\frac{c_o^2}{v}\,, \qquad |u|>c_o\,,
\end{aligned}
\tag{435}
$$

mechanically seen a particle? Can we really attribute the field (435) an energy E^T and a momentum P^T in every reference system Σ, so that these quantities are related in various reference systems to each other via the Eq. (427)? Only then would we be able to state that the solution (435) is a tachyon, a particle of type (b) as defined in the last chapter.

First of all we formally construct, according to the rules (289), (290) the energy-momentum tensor (276) for the field $q^T = q^T(x,t)$. The calculations are done in the same manner as for the solution q^I in Chap. 23. A difference between the two consists in the exponents of the exponential function, where now a minus sign is present, and where γ is replaced with the quantity $\kappa = \gamma\,u/c_o$. Furthermore it must be observed that a velocity u with $|u| > c_o$ replaces the velocity v with $|v| < c_o$. The additive constant $a/2$ in q^T disappears during differentiation. According to this additive constant $a/2$ in q^T, in the cosine term of the LAGRANGian L in (287) the argument increases by π, which reverses the sign of the cosine term. We thus find by applying the results of the calculations made in Chap. 23,

H. Günther, *Elementary Approach to Special Relativity*,
https://doi.org/10.1007/978-981-15-3168-2_29

$$\cos\left(\frac{2\pi}{a}q^T\right) - 1 = -\cos\left(4\arctan\exp\left[\frac{\pi(x-ut)}{L_o\kappa}\right]\right) - 1$$
$$= -\left[\cos\left(4\arctan\exp\left[\frac{\pi(x-ut)}{L_o\kappa}\right]\right) - 1\right] - 2$$
$$= 8\frac{\exp\left[-2\frac{\pi(x-ut)}{L_o\kappa}\right]}{\left(1+\exp\left[-2\frac{\pi(x-ut)}{L_o\kappa}\right]\right)^2} - 2\ ,$$

$$\frac{\partial q^T}{\partial x} = -\frac{2a}{L_o\kappa}\frac{\exp\left[\frac{-\pi(x-ut)}{L_o\kappa}\right]}{1+\exp\left[2\frac{\pi(x-ut)}{L_o\kappa}\right]} \quad \text{as well as} \quad \frac{\partial q^T}{\partial t} = -u\frac{\partial q^T}{\partial x}\ .$$

This leads to

$$\frac{\sigma}{2}\left(\frac{\partial q^T}{\partial x}\frac{\partial q^T}{\partial x} + \frac{1}{c_o^2}\frac{\partial q^T}{\partial t}\frac{\partial q^T}{\partial t}\right) + \frac{\sigma a^2}{4L_o^2}\left(\cos\left(\frac{2\pi}{a}q^T\right) - 1\right)$$
$$= \frac{\sigma}{2}\frac{4a^2}{L_o^2\kappa^2}\left(1+\frac{u^2}{c_o^2}\right)\frac{\exp\left[-2\frac{\pi(x-ut)}{L_o\kappa}\right]}{\left(1+\exp\left[-2\frac{\pi(x-ut)}{L_o\kappa}\right]\right)^2} + \frac{\sigma a^2}{4L_o^2}\left(8\frac{\exp\left[-2\frac{\pi(x-ut)}{L_o\kappa}\right]}{\left(1+\exp\left[-2\frac{\pi(x-ut)}{L_o\kappa}\right]\right)^2} - 2\right)$$
$$= \frac{2\sigma a^2}{L_o^2\kappa^2}\left[2+\left(\frac{u^2}{c_o^2}-1\right)\right]\frac{\exp\left[-2\frac{\pi(x-ut)}{L_o\kappa}\right]}{\left(1+\exp\left[-2\frac{\pi(x-ut)}{L_o\kappa}\right]\right)^2} - \frac{\sigma a^2}{2L_o^2}\frac{\left(1-\exp\left[-2\frac{\pi(x-ut)}{L_o\kappa}\right]\right)^2}{\left(1+\exp\left[-2\frac{\pi(x-ut)}{L_o\kappa}\right]\right)^2}$$
$$= \frac{2\sigma a^2}{L_o^2\kappa^2}\left[2+2\left(\frac{u^2}{c_o^2}-1\right)\right]\frac{\exp\left[-2\frac{\pi(x-ut)}{L_o\kappa}\right]}{\left(1+\exp\left[-2\frac{\pi(x-ut)}{L_o\kappa}\right]\right)^2} - \frac{\sigma a^2}{2L_o^2}\frac{4+\left(1+\exp\left[-2\frac{\pi(x-ut)}{L_o\kappa}\right]\right)^2}{\left(1+\exp\left[-2\frac{\pi(x-ut)}{L_o\kappa}\right]\right)^2}$$
$$= \left(\frac{1}{\kappa^2}+1\right)\frac{4\sigma a^2}{L_o^2}\frac{\exp\left[-2\frac{\pi(x-ut)}{L_o\kappa}\right]}{\left(1+\exp\left[-2\frac{\pi(x-ut)}{L_o\kappa}\right]\right)^2} - \frac{\sigma a^2}{2L_o^2}\frac{\left(1+\exp\left[-2\frac{\pi(x-ut)}{L_o\kappa}\right]\right)^2}{\left(1+\exp\left[-2\frac{\pi(x-ut)}{L_o\kappa}\right]\right)^2}$$
$$= u^2\frac{1}{c_o^2}\frac{1}{\kappa^2}\frac{4\sigma a^2}{L_o^2}\frac{\exp\left[-2\frac{\pi(x-ut)}{L_o\kappa}\right]}{\left(1+\exp\left[-2\frac{\pi(x-ut)}{L_o\kappa}\right]\right)^2} - \frac{\sigma a^2}{2L_o^2}\ .$$

If we introduce another function $\rho^T = \rho^T(x - v\,t)$ according to

$$\rho^T(x-ut) = \frac{1}{c_o^2}\frac{1}{\kappa^2}\frac{4\,\sigma a^2}{L_o^2}\frac{\exp\left[-2\dfrac{\pi(x-ut)}{L_o\kappa}\right]}{\left(1+\exp\left[-2\dfrac{\pi(x-ut)}{L_o\kappa}\right]\right)^2} \tag{436}$$

as well as a parameter Θ_o according to

$$\Theta_o = \frac{\sigma\,a^2}{2Ł_o^2}\ , \tag{437}$$

we receive for the first component $-t$ of the energy-momentum tensor

$$-\,t = u^2\rho^T(x-ut)\ . \tag{438}$$

We also find

$$-\frac{\sigma}{c_o^2}\frac{\partial q^T}{\partial x}\frac{\partial q^T}{\partial t} = -\frac{1}{c_o^2}(-u)\frac{4\,\sigma a^2}{L_o^2\kappa^2}\frac{\exp\left[-\dfrac{\pi(x-ut)}{L_o\kappa}\right]}{\left(1+\exp\left[-2\dfrac{\pi(x-ut)}{L_o\kappa}\right]\right)^2}$$

and thus

$$p = u\,\rho^T(x-ut) \tag{438a}$$

and also, according to (289),

$$-\,s = -\,c_o\,u\,\rho^T(x-ut)\ . \tag{438b}$$

Finally it follows that

$$-\frac{\sigma}{2}\left(\frac{\partial q^T}{\partial x}\frac{\partial q^T}{\partial x}+\frac{1}{c_o^2}\frac{\partial q^T}{\partial t}\frac{\partial q^T}{\partial t}\right)+\frac{\sigma a^2}{4L_o^2}\left(\cos(\frac{2\pi}{a}q^T)-1\right)$$

$$= -\frac{\sigma}{2}\frac{4a^2}{L_o^2\kappa^2}\left(1+\frac{u^2}{c_o^2}\right)\frac{\exp\left[-2\dfrac{\pi(x-ut)}{L_o\kappa}\right]}{\left(1+\exp\left[-2\dfrac{\pi(x-ut)}{L_o\kappa}\right]\right)^2}+\frac{\sigma a^2}{4L_o^2}\left(8\frac{\exp\left[-2\dfrac{\pi(x-ut)}{L_o\kappa}\right]}{\left(1+\exp\left[-2\dfrac{\pi(x-ut)}{L_o\kappa}\right]\right)^2}-2\right)$$

$$= -\frac{2\,\sigma a^2}{L_o^2\kappa^2}\left[2+\left(\frac{u^2}{c_o^2}-1\right)\right]\frac{\exp\left[-2\frac{\pi(x-ut)}{L_o\kappa}\right]}{\left(1+\exp\left[-2\frac{\pi(x-ut)}{L_o\kappa}\right]\right)^2} - \frac{\sigma a^2}{2L_o^2}\frac{\left(1-\exp\left[-2\frac{\pi(x-ut)}{L_o\kappa}\right]\right)^2}{\left(1+\exp\left[-2\frac{\pi(x-ut)}{L_o\kappa}\right]\right)^2}$$

$$= -\frac{4\,\sigma a^2}{L_o^2\kappa^2}\frac{\exp\left[-2\frac{\pi(x-ut)}{L_o\kappa}\right]}{\left(1+\exp\left[-2\frac{\pi(x-ut)}{L_o\kappa}\right]\right)^2} - \frac{\sigma a^2}{2L_o^2}\frac{\left(1+\exp\left[-2\frac{\pi(x-ut)}{L_o\kappa}\right]\right)^2}{\left(1+\exp\left[-2\frac{\pi(x-ut)}{L_o\kappa}\right]\right)^2}$$

$$= -\frac{1}{\kappa^2}\frac{4\,\sigma a^2}{L_o^2}\frac{\exp\left[-2\frac{\pi(x-ut)}{L_o\kappa}\right]}{\left(1+\exp\left[-2\frac{\pi(x-ut)}{L_o\kappa}\right]\right)^2} - \frac{\sigma a^2}{2L_o^2}$$

and therefore, once again with the function $\rho^T(x-ut)$ according to (436), as well as Θ_o according to (437),

$$-\,e = -\,c_o^2\rho^T(x-ut) - \Theta_o\ . \tag{438c}$$

From (438) to (438c) we can construct the following energy-momentum tensor $\widetilde{\mathbf{T}}^T$ for our field q^T according to the rule (276),

$$\widetilde{\mathbf{T}}^T = \begin{pmatrix} u^2\rho^T & u\rho^T \\ -u\,c_o^2\,\rho^T & -c_o^2\,\rho^T \end{pmatrix} - \begin{pmatrix} \Theta_o & 0 \\ 0 & \Theta_o \end{pmatrix}\ . \tag{439}$$

Here the second term is an additive constant, independent of both spacetime as well as the velocity u. Such a constant, defined alone by the parameters of an ideal lattice, does not affect the transformations of energy as a result of mechanical processes in this lattice, these processes only being observable by internal observers. We can therefore just ignore this constant and assume that we have a measurable energy-momentum tensor T^T for our field q^T according to

$$\mathbf{T}^T = \begin{pmatrix} u^2\rho^T & u\rho^T \\ -u\,c_o^2\,\rho^T & -c_o^2\,\rho^T \end{pmatrix} \quad \text{with} \quad |u| > c_o\ . \qquad \begin{matrix}\text{Energy-momentum tensor}\\ \text{of the field } q^T(x,t)\end{matrix} \tag{439a}$$

Here, according to (436) $\rho^T = \rho^T(x-ut)$, and therefore $\operatorname{div}\mathbf{T}^T = 0$ is valid.

In Chap. 22 we saw that: If the energy-momentum tensor of a field $q(x,t)$ has the structure (279), then we can appoint this field to a particle, to be more precise, to a particle of type (a), whereby the physical particle parameters for the field q, its energy E and its momentum P, can be calculated according to the Eqs. (281) and (282). This is a well known fact in field theory. The kink solution (111) with its energy-momentum tensor (294) was identified as a particle in Chap. 23 using these facts.

There is a further relationship less well known in the field theory: The energy-momentum tensor $\mathbf{T}$ once again has the structure (279),

$$\mathbf{T} = \begin{pmatrix} u^2\rho & u\rho \\ -u\, c_o^2\, \rho & -c_o^2\, \rho \end{pmatrix} \;, \quad \operatorname{div}\mathbf{T} = 0 \;. \tag{279}$$

This time however, the condition $|u| > c_o$ applies to the velocity u. The velocity u in (279) cannot therefore stand for the velocity of a reference system. Our tensor (439a) has just this structure. Such a tensor can be associated with a particle of type (b), a tachyon with the velocity u according to the following rule,

$$m^T = \operatorname{sign} u \int\limits_{-\infty}^{+\infty} \rho^T dx \;, \tag{440}$$

and

$$\left.\begin{aligned} P^T &= m^T u \;, \\ E^T &= m^T c_o^2 \;. \end{aligned}\right\} \tag{441}$$

According to these equations we can calculate the momentum and the energy of a tachyon (435). This would be

$$\begin{aligned}
\int\limits_{-\infty}^{+\infty} \rho^T dx &= \frac{1}{c_o^2}\,\frac{1}{\kappa^2}\,\frac{4\,\sigma a^2}{L_o^2} \int\limits_{-\infty}^{+\infty} \frac{\exp\left[-2\dfrac{\pi(x-ut)}{\text{Ł}_o\kappa}\right]}{\left(1+\exp\left[-2\dfrac{\pi(x-ut)}{L_o\kappa}\right]\right)^2} \\
&= \frac{1}{c_o^2}\,\frac{1}{\kappa^2}\,\frac{4\,\sigma a^2}{L_o^2}\,\frac{L_o\kappa}{2\pi}\,\operatorname{sign} u \int\limits_{-\infty}^{+\infty} \frac{e^{-x}}{(1+e^{-x})^2}\,dx \\
&= \frac{1}{c_o^2}\,\frac{1}{\kappa^2}\,\frac{2a^2\sigma}{\pi L_o}\left[\frac{1}{1+e^{-x}}\right]_{-\infty}^{+\infty} \\
&= \frac{\operatorname{sign} u}{\kappa}\,\frac{1}{c_o^2}\,\frac{2a^2 E}{\pi L_o} \;.
\end{aligned}$$

Due to the fact that we introduce the new integration variable $2\frac{\pi(x-ut)}{L_o\kappa}$ into the second line the direction of integration changes, depending on the sign of u in κ. We kept the direction of integration from $-\infty$ to $+\infty$ and introduced the factor $\operatorname{sign}\mathrm{u}$.

We therefore precisely find, by observing $(\text{sign})^2 = 1$,

$$P^T = \frac{m_*^T}{\operatorname{sign} u \sqrt{u^2/c_o^2 - 1}} u \ , \quad E^T = \frac{m_*^T}{\operatorname{sign} u \sqrt{u^2/c_o^2 - 1}} c_o^2 \ , \tag{442}$$

a tachyon according to (425), however with a momentum parameter m_*^T specified by the parameters of the lattice,

$$m_*^T = f\, m_a \ , \quad f = \frac{2a}{\pi L_o} \ , \tag{443}$$

with the mass $m_a = a\,\sigma/c_o^2$ of a dislocation on one lattice distance a according to (295). We therefore discover a remarkable symmetry between the tachyon belonging to (435) and the type (a) particle associated with the kink (111). This symmetry relies on the fact that the tachyon solution (435) arises out of the so-called π-transformation of the kink solution, see SEEGER [86].

For the critical case of an infinite velocity u we once again get (425a),

$$\lim_{u \to \pm\infty} P^T = P_\infty^T = m_*^T\, c_o = \frac{2a^2\sigma}{\pi L_o c_o} \ , \quad \lim_{u \to \pm\infty} E^T = E_\infty^T = 0 \ . \tag{444}$$

Here we received, according to (383), a positive momentum parameter m_*^T for the solution (435). For the critical case $u \longrightarrow \infty$ of the tachyon solution (435) we have the special solution (265),

$$q_\infty^T(t) = \lim_{u \to \infty} q^T(x,t) = \frac{2a}{\pi} \arctan \exp \Big[\frac{\pi c_o}{L_o} t \Big] + \frac{a}{2} \ . \tag{265}$$

We have illustrated this tachyon in Fig. 20.2. The momentum P^T of the tachyon (435) is positive in every reference system. This tachyon's momentum tends with increasing velocity to the smallest possible value. The tachyon, if it actually exists, cannot loose it.

A negative momentum parameter of the same amount we get for the following solution $\widetilde{q}^{\,T}(x,t)$ of the sine-GORDON equation,

$$\widetilde{q}^{\,T}(x,t) = \frac{2a}{\pi} \arctan \exp \Big[\frac{-\pi(x+ut)}{L_o \kappa} \Big] + \frac{a}{2} \ . \tag{445}$$

This tachyon solution of the sine-GORDON equation represents a particle of type (b) according to

$$\widetilde{P}^T = \frac{\widetilde{m}_*^T}{\operatorname{sign} u \sqrt{u^2/c_o^2 - 1}} u \ , \quad \widetilde{E}^T = \frac{\widetilde{m}_*^T}{\operatorname{sign} u \sqrt{u^2/c_o^2 - 1}} c_o^2 \tag{446}$$

with

$$\widetilde{m}_*^T = -f\, m_a \ , \quad f = \frac{2a}{\pi L_o} \ . \tag{447}$$

In the critical case $u \longrightarrow \pm\infty$ we now get

$$\lim_{u\to\pm\infty} \widetilde{P}^T = \widetilde{P}_\infty^T = \widetilde{m}_*^T\, c_o = -\frac{2a^2\sigma}{\pi L_o c_o} \ , \quad \lim_{u\to\pm\infty} \widetilde{E}^T = \widetilde{E}_\infty^T = 0 \ . \tag{448}$$

For the tachyon solution (445) we get the following special solution for the critical case $u \longrightarrow \infty$

$$\widetilde{q}_\infty^T(t) = \lim_{u\to\infty} \widetilde{q}^T(x,t) = \frac{2a}{\pi} \arctan\exp\left[-\frac{\pi c_o}{L_o}\, t\right] + \frac{a}{2} \ . \tag{449}$$

The negative sign of the momentum parameter for the tachyon $\widetilde{q}^T(x,t)$ belonging to (445) simply follows from the validity of

$$\frac{\partial \widetilde{q}^T(x,t)}{\partial t} = +u\, \frac{\partial \widetilde{q}^T(x,t)}{\partial x} \ . \tag{450}$$

However, for the tachyon solution (435) the following applies,

$$\frac{\partial q^T(x,t)}{\partial t} = -u\, \frac{\partial q^T(x,t)}{\partial x} \ . \tag{450a}$$

This results in a difference of sign for the components p and $-s$ of the energy-momentum tensor according to (289) and (290), and we get the following equation (451) instead of (439a),

$$\mathbf{T}^T = \begin{pmatrix} u^2 \rho^T & -u\, \rho^T \\ u\, c_o^2\, \rho^T & -c_o^2\, \rho^T \end{pmatrix} \quad \text{with } |u| > c_o \ , \tag{451}$$

as well as a function (436), $\rho^T(x+ut)$ for the solution $\widetilde{q}^T(x,t)$. With (279), (440) and (441) we get a tachyon with the negative momentum parameter $\widetilde{m}_*^T$.

The function $\widetilde{q}^T(x,t)$ according to (445) differs from the function $q^T(x,t)$ according to (435) alone in the direction of motion. We get the same snapshot as shown in Fig. 20.1, but the direction of motion is reversed, see Fig. 29.1. The same situation occurs for the tachyon $\widetilde{q}_\infty^T(t)$ belonging to (449), whose graphic representation corresponds to that in Fig. 20.2 with the exception of the direction of motion, see Fig. 29.2.

Of the tachyon we can say: it moves into the past if its velocity u and its momentum P^T have opposite signs. If and whether such a situation is present depends on the reference system. In every reference system the momentum P^T of a tachyon has one and the same sign defined by its momentum parameter m_*^T. The sign of the tachyon's velocity u can change during the transition into another reference system,

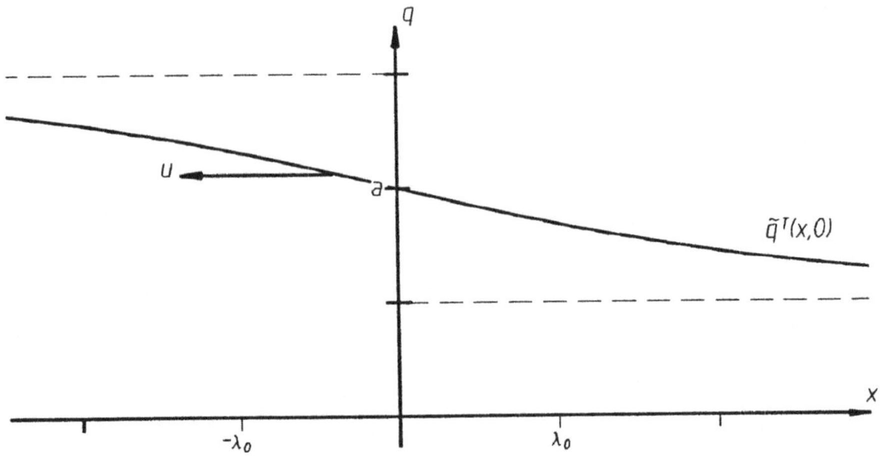

Fig. 29.1 The tachyon $\widetilde{q}^T(x,t) = \frac{2a}{\pi} \arctan \exp\left[\frac{-\pi(x+ut)}{L_o\kappa}\right] + \frac{a}{2}$ with negative momentum parameter $\widetilde{m}_*^T$ at $t = 0$ and $u = -2c_o$

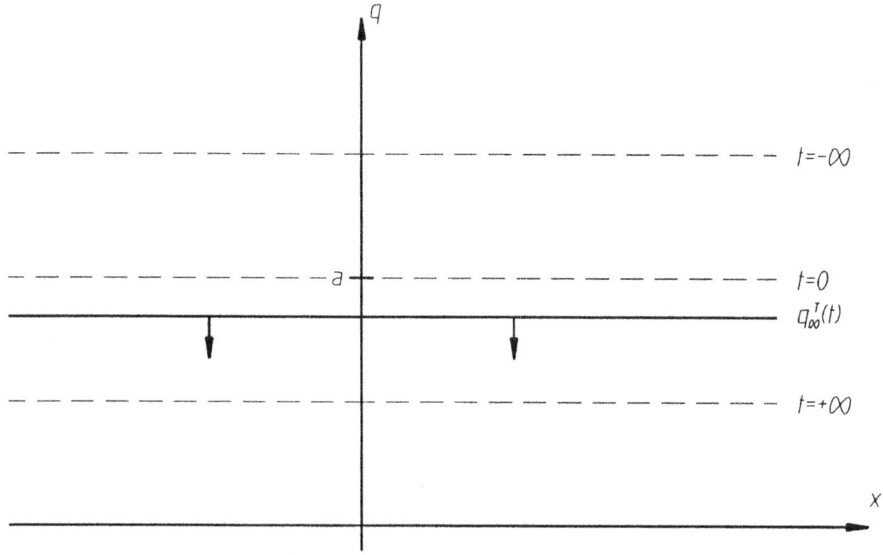

Fig. 29.2 The tachyon solution $\widetilde{q}^T_\infty(t) = \frac{2a}{\pi} \arctan \exp\left[-\frac{\pi c_o}{L_o} t\right] + \frac{a}{2}$. The limiting positions at $t = -\infty$ and $t = +\infty$ as well as the position at $t = 0$ are plotted as dashed lines. The arrows indicate the direction of motion

as we have shown in the last chapter with the help of the composition of velocities, see the discussion after the examples in Chap. 28.

We will now discuss a paradox that can be used against the particle properties of the solution (435) of the sine-GORDON equation, therefore against the particle properties of tachyons. The density function (436) disappears for $u \longrightarrow \infty$ as $1/u^2$, and the following is valid,

$$\lim_{u\to\infty} \rho^T(x-ut) = 0 \ , \tag{452}$$

$$\lim_{u\to\infty} \left[u\,\rho^T(x-ut)\right] = 0 \ , \tag{452a}$$

$$\lim_{u\to\infty} \left[u^2\,\rho^T(x-ut)\right] = \frac{4\sigma a^2}{c_o^2}\,\frac{\exp\left[\frac{2\pi}{L_o}t\right]}{\left(1+\exp\left[\frac{2\pi}{L_o}t\right]\right)^2} := -t_\infty^T \tag{452b}$$

and thus for the critical case for the energy-momentum tensor $\mathbf{T}_\infty^T$,

$$\mathbf{T}_\infty^T = \begin{pmatrix} -t_\infty^T & 0 \\ 0 & 0 \end{pmatrix} \ . \tag{453}$$

This is however not the special structure (439a) that we discovered above. If we integrate this tensor over the momentum density and energy density, then it immediately follows that

$$P_\infty^T = 0 \ , \quad E_\infty^T = 0 \ , \tag{454}$$

which contradicts (444). According to (454) the tachyon's momentum and energy would disappear in every reference system. The tachyon would not be a particle as can be immediately seen in the structure of the energy-momentum tensor (453).

How can this contradiction be solved? For a velocity u increasing towards infinity, the energy and momentum densities become infinitely minute according to (452) and (452a). We can also see from the density function (436) that this propagates with the velocity $u \longrightarrow \infty$, namely proportionally to $L_o|u| = \sqrt{u^2/c_o^2 - 1}$. We must observe the correct order of the limiting values! The contradictory result (454) was caused by the fact that energy and momentum were at first distributed throughout infinite one dimensional space and then that we then constructed the integrals according to the rule

$$\int_{-\infty}^{+\infty} \lim_{u\to\infty} [\cdots \rho^T]\,dx = \lim_{r\to\infty} \int_{-r}^{+r} \lim_{u\to\infty} [\cdots \rho^T]\,dx \ . \tag{455}$$

These integrals disappear for (452) and (452a) by definition. We must firstly, where finite velocities are concerned, calculate the integrals and only then let the velocity tend to infinity. The rule (455) gives us a false result, namely (454). We cannot let the momentum move towards infinity and then state that the total momentum is zero, because its density is zero. The correct rule is therefore

$$\lim_{u\to\infty}\left[\lim_{r\to\infty}\int_{-r}^{+r}\cdots\rho^T\right]dx\ . \tag{456}$$

Using this procedure we get the correct solution (444) and thus the correct tachyons with all their peculiar properties.

One final objection could be made against the physical reality of the tachyon solution (435). This solution does not have the same physical importance as the kink solution (111). Whilst the kink realises a deviation from the stabile equilibrium, the tachyon solution (435) is an instabile solution—a dislocation on a potential barrier, as we saw in Chap. 20 during the discussion concerned with Fig.20.1. It will therefore be difficult to verify such conditions. This does not however make their physical reality uncertain. The instabile position of a mechanical pendulum above its suspension is not something one sees everyday. No human would thus come to the conclusion that such a position does not exist—the artists in a circus earn their money by showing such positions.

The existence of tachyons can therefore not be questioned. It is of no relevance that we concern ourselves with the tachyons of internal observers of an infinitely extended crystal. The important factor is that tachyons, tachyons with both negative and positive momentum parameters actually exist inside of the world of our internal observers, for whom the Special Theory of Relativity is valid. Let us now take another closer look at these tachyons.

The tachyon $q^T_\infty(t)$, the special solution (265) of the sine-GORDON equation, describes the exact simultaneous gliding of a straight dislocation along the x-axis from one potential barrier of the lattice to a neighbouring. During the process of a plastic deformation the motion of the dislocation elements, arbitrarily placed at large distances from each other on the x-axis, occurs in exactly the same timed sequence. A motion of the complete dislocation line, embedded in the crystalline lattice, which is taken into account using the sine-GORDON equation, can only occur if its line tension can be sustained for arbitrary distances whilst upholding the correlations caused by the lattice. Using the expression 'tachyon' we can state for this procedure that: The stress state of a moving straight dislocation line is realised by a tachyon, which moves in the x-direction with infinite velocity with a finite momentum, but with vanishing energy. We can also interpret the tachyon belonging to $q^T(x,t)$ according to (435). One easily realises that its initial state at $t=-\infty$ and its final state at $t=+\infty$ coincide with the corresponding states of the tachyon belonging to $q^T_\infty(t)$. Hence, a dislocation line moved from one potential barrier to the neighbouring. Even this motion occurs strictly correlated, but not in one and the same timed procedure for all dislocation segments. The dislocation pieces with large x-coordinates limp behind if u is positive. The inevitable creation of inhomogeneity in the stress state can be interpreted using the momentum and the energy of a tachyon, which moves with $|u|>c_o$ along the x-axis.

For every solution $q^T(x,t)$ there is always a preferred frame of the internal observer, in which this tachyon has the special form $q^T_\infty(t)$. For the internal observer in this frame, the gliding of the dislocation line occurs simultaneously. Mathematically, the dependency of x and t in the solution $q^T(x,t)$ follows simply from an adapted LORENTZ transformation of the time coordinate, thus arises from a state of simultaneity in a special frame. Physically, such a dislocation motion results in this special frame by a simultaneous sounding along the complete dislocation line. For the plastic tachyon, which we external experimenters observe in our laboratory, e.g. as $q^T_\infty(t)$, the internal observer registers a velocity $u = -c_o^2/V$ from out of a reference system, opposite to the laboratory, moving with the velocity $+V$. Because the momentum P of the tachyon remains positive, if we started from solution (435) with a positive momentum parameter m^T_*, the internal observer states, referring to the tachyon, that it 'moves into the past'. Another internal observer moving with the velocity $-V$, observes the velocity $u = +c_o^2/V$ for the same tachyon. For both observers, the elements of the plastic deformation, which move the dislocation from one potential barrier to the neighbouring, occur in reverse timed order. For the first observer the procedure with the larger space coordinates occur first, for the second those with the smaller coordinates occur first. Important is that both solutions of the internal observers occur according to a corresponding LORENTZ transformation won from $q^T_\infty(t)$, and that these physical solutions also exist for us in the laboratory. What remains of the 'tachyon moving into the past', as registered by some internal observers in certain reference systems, for the external experimenter? The latter registers not only a simultaneous gliding of the complete line of dislocation from one potential barrier to the neighbouring, but also such motions where the dislocation elements with positive x-coordinates move ahead faster, and other motions where those with negative x-coordinates are faster. All three motions are based, for us observing from the outside, on a simultaneous sounding of the lattice and not on a localised sounding, whose results would lead us to expect particle solutions of type (a). In order to mathematically find the formation of the various solutions, we need to incorporate the external forces $\mathbf{F}_A$ into our starting equation (79), as done in (53). This results in an additive stress term on the right hand side of the sine-GORDON equation. The propagation of a localised sounding, the transmission of a signal remains the privilege of type (a) particles. We can therefore state:

Plastic tachyons cannot transmit signals inside of a crystal.

These situations leads us, in our considerations on tachyons, to ask the question about the possibility, according to the laws of mechanics, of a collision between a 'normal' particle and a tachyon, so that the particle either transfers energy to the tachyon, or receives energy from it. Only in such a case would it be principally possible to use a tachyon to transmit a signal or message with its own, extremely large velocity $|u| > c_o$. We will come back to this question during our discussion on the problems of causality in the next chapter.

Chapter 30
On the Causality Problem: Particle–Tachyon Collisions

Just merely changing the time order when observing two events E_1 and E_2 from different reference systems doesn't break causality. If the one event E_2 was caused by the event E_1, then E_2 must always occur after E_1, regardless of the reference system from which these events are observed. A reversal of the time order of these events would severely confuse our causal view of the world. One can however ask the question, when does a break of causality inside of the smallest spatial distances not contradict our view of the world? Here, quantum theoretical deliberations play a role. We refer the reader to the discussions of H. TREDER [94].

The mere existence of tachyons as particles is not enough. The all decisive question is, can we use tachyons to transmit signals, meaning:

Can we transmit energy using tachyons?

This is what it is all about. To begin with, we will examine, in full generality, elastic collisions between a particle and a tachyon. We use the term 'particle' as defined in Chap. 28, as a type (a) particle and the term 'tachyon' as a type (b) particle.

A particle with restmass m_o shall collide with a tachyon with the momentum parameter m_*^T. The velocities of the particle before and after the collision are w and w', the corresponding velocities of the tachyon are u and u', respectively. The momenta and energies of particles and tachyons are defined by the formulas (297) and (442). Due to the presumed 'elasticity' of the collision process, the energy and momentum conservation is valid, the LORENTZ invariant parameters, the mass parameter m_o for the particle, and the momentum parameter m_*^T for the tachyon are retained, no 'other' particles or tachyons are created. The following is valid for the momenta and energies before and after the collision,

$$\left.\begin{aligned} \frac{m_o\, w}{\sqrt{1-w^2/c_o^2}} + \frac{m_*^T\, u}{\operatorname{sign} u\sqrt{u^2/c_o^2-1}} &= \frac{m_o\, w'}{\sqrt{1-w'^2/c_o^2}} + \frac{m_*^T\, u'}{\operatorname{sign} u'\sqrt{u'^2/c_o^2-1}}\,, \quad \text{Momentum} \\ \frac{m_o\, c_o^2}{\sqrt{1-w^2/c_o^2}} + \frac{m_*^T\, c_o^2}{\operatorname{sign} u\sqrt{u^2/c_o^2-1}} &= \frac{m_o\, c_o^2}{\sqrt{1-w'^2/c_o^2}} + \frac{m_*^T\, c_o^2}{\operatorname{sign} u'\sqrt{u'^2/c_o^2-1}}\,. \quad \text{Energy} \end{aligned}\right\} \quad (457)$$

H. Günther, *Elementary Approach to Special Relativity*,
https://doi.org/10.1007/978-981-15-3168-2_30

We now replace the tachyon velocities u and u' with their respective allocated velocities v and v' from $u = c_o^2/v$ and $u' = c_o^2/v'$. Furthermore, we replace the corresponding square root factors according to the rule $\kappa = \gamma\, u/c_o$. The conservation laws (457) then have the following form:

$$\left.\begin{aligned}
\frac{m_o\, w}{\sqrt{1-w^2/c_o^2}} + \frac{m_*^T\, c_o}{\sqrt{1-v^2/c_o^2}} &= \frac{m_o\, w'}{\sqrt{1-w'^2/c_o^2}} + \frac{m_*^T\, c_o}{\sqrt{1-v'^2/c_o^2}}\,, && \text{Momentum}\\
\frac{m_o\, c_o^2}{\sqrt{1-w^2/c_o^2}} + \frac{m_*^T\, c_o\, v}{\sqrt{1-v^2/c_o^2}} &= \frac{m_o\, c_o^2}{\sqrt{1-w'^2/c_o^2}} + \frac{m_*^T\, c_o\, v'}{\sqrt{1-v'^2/c_o^2}}\,. && \text{Energy}
\end{aligned}\right\} \quad (457a)$$

In order to simplify Eq. (457a), we use to our advantage the fact that we may freely select the reference system $\overline{\Sigma}$ in which we want to describe the occurring collision process. Furthermore, we also have the positive-valued direction of the x-axis at our disposition. We can therefore assume that the particle in the laboratory, in which we describe its collision with the tachyon, is at rest before the collision, and furthermore, that the momentum of the colliding tachyon points in the direction of the positive x-axis, the momentum parameter m_*^T being therefore positive. We then introduce a positive parameter μ according to

$$\mu := \frac{m_o}{m_*^T}\,. \quad (458)$$

We again refer the reader to the curiosities of tachyon kinematics. Even though the momentum of our colliding tachyon must always remain positive, its velocity can change its sign. Equation (457a) becomes, by dividing by m_*^T,

$$\overline{\Sigma}:\quad \left.\begin{aligned}
\frac{c_o}{\sqrt{1-v^2/c_o^2}} &= \frac{\mu\, w'}{\sqrt{1-w'^2/c_o^2}} + \frac{c_o}{\sqrt{1-v'^2/c_o^2}}\,, && \text{Momentum}\\
\mu\, c_o^2 + \frac{c_o\, v}{\sqrt{1-v^2/c_o^2}} &= \frac{\mu\, c_o^2}{\sqrt{1-w'^2/c_o^2}} + \frac{c_o\, v'}{\sqrt{1-v'^2/c_o^2}}\,. && \text{Energy}
\end{aligned}\right\} \quad (457b)$$

In order to get a clearer overview of the possible solutions of these equations, we now introduce hyperbolic functions according to,

$$\left.\begin{aligned}
&\frac{v'}{c_o} = \tanh\alpha'\,,\ \frac{v}{c_o} = \tanh\alpha\,,\\
&\frac{w'}{c_o} = \tanh\beta'.
\end{aligned}\right\} \quad (459)$$

Because $w = 0$, we get with $w/c_o = \tanh\beta = 0$, also $\beta = 0$.

Under observance of $1/\sqrt{1-\tanh^2 x} = \cosh x$, $\tanh x/\sqrt{1-\tanh^2 x} = \sinh x$ and simple rearrangements, Eq. (457b) is followed by

$$\left.\begin{aligned} \mu \sinh \beta' &= \cosh \alpha - \cosh \alpha' \ , \\ \mu \cosh \beta' &= \sinh \alpha - \sinh \alpha' + \mu \ . \end{aligned}\right\} \tag{460}$$

Due to $\mu > 0$ according to (458), $\cosh \beta' - 1 > 0$ and the monotony of the hyperbolic sine,

$$\sinh \alpha' < \sinh \alpha \quad \longleftrightarrow \quad \alpha' < \alpha \ , \tag{461}$$

from the second Eq. (460) according to

$$0 \leq \sinh \alpha - \sinh \alpha' = \mu \left[\cosh \beta' - 1\right] \tag{462}$$

it results the condition

$$\alpha' \leq \alpha \ . \tag{463}$$

With $\cosh^2 \beta' - \sinh^2 \beta' = 1$ we eliminate, from Eq. (460), the velocity $w = c_o \tanh \beta'$ of the particle after the collision and find after short calculations, keeping the rule $\cosh(\alpha - \alpha') = \cosh \alpha \cosh \alpha' - \sinh \alpha \sinh \alpha'$ in mind,

$$0 \leq \frac{1}{\mu}[\cosh(\alpha - \alpha')] = \sinh \alpha' - \sinh \alpha \ . \tag{464}$$

However, according to the same conclusion as above, it follows now that

$$\alpha \leq \alpha' \ . \tag{465}$$

Eqs. (463) and (465) result in

$$\alpha' = \alpha \quad \longleftrightarrow \quad u' = u \ . \qquad \text{No signal transmission by the tachyon} \tag{466}$$

The tachyon must pass the particle without changing its velocity. It can therefore not transmit even the slightest amount of energy and thus also not transmit a signal. Causality cannot be broken this way. We discover:

A breaking of causality caused by tachyons cannot occur within the limits of elastic collision processes.

If the tachyon cannot transmit energy how can we even realise that the tachyon exists? Well, in the above presumption, we restricted ourselves to the elastic particle-tachyon collision. Inelastic collisions completely and fundamentally change the situation, see Liebscher [59].

As an initial state, we now solely observe a resting particle of the restmass m_o. The aim is a final state in which this particle, with conservation of its restmass, has received a velocity w' different from zero by the emission of a tachyon with the velocity u'. One can easily understand that a particle at rest can never emit a particle without changing its restmass due to energy conservation, because the emitted particle would have a positive energy. A particle at rest with $E_o = m_o c_o^2$ cannot however emit energy without changing its restmass. This is different for tachyons. Tachyons can have negative energy. Using the same notation as above, the following equations result from the conservation laws of energy and momentum in place of those in (460),

$$\left.\begin{aligned} \mu \sinh \beta' &= -\cosh \alpha' \ , \\ \mu \cosh \beta' &= -\sinh \alpha' + \mu \ . \end{aligned}\right\} \qquad \text{No tachyon before the collision} \tag{467}$$

The difference between the two Eqs. (460) and (467) is just the missing incoming tachyon in (467).

Equation (467) leads us, by eliminating β', after simple calculation, to,

$$\begin{aligned} \sinh \alpha' &= -\frac{1}{2\mu} \ , \\ \frac{v'}{c_o} &= \tanh \alpha' = \frac{\sinh \alpha'}{\sqrt{1+\sinh^2 \alpha'}} = -\frac{1}{2\mu} \frac{1}{\sqrt{1+1/4\mu^2}} \ , \\ \frac{u'}{c_o} &= \frac{c_o}{v'} = -\sqrt{4\mu^2+1} \ , \end{aligned}$$

$$u' = -c_o \frac{\sqrt{m_*^{T\,2} + 4m_o^2}}{m_*^T} \tag{468}$$

and thus from (467) to

$$\begin{aligned} \sinh \beta' &= -\frac{1}{\mu} \cosh \alpha' = -\frac{1}{\mu}\sqrt{1+\sinh^2 \alpha'} = -\frac{\sqrt{1+4\mu^2}}{2\mu^2} \ , \\ \frac{w'}{c_o} &= \tanh \beta' = \frac{\sinh \beta'}{\sqrt{1+\sinh^2 \beta'}} = -\frac{\sqrt{1+4\mu^2}}{2\mu^2} \frac{1}{\sqrt{1+(1+4\mu^2)/(4\mu^4)}} \\ &= -\frac{\sqrt{1+4\mu^2}}{\sqrt{(1+2\mu^2)^2}} \ , \end{aligned}$$

$$w' = -c_o \frac{\sqrt{m_*^{T\,2} + 4m_o^2}}{1+2m_o^2} \ . \tag{469}$$

With a given restmass m_o of the particle, Eq. (467) thus also have a non-trivial solution w', the velocity of the particle after the collision, for any arbitrary momentum parameter m_*^T of the tachyon to be emitted. This would mean that the particle could emit an arbitrary number of tachyons, whereby this particle would continuously change its velocity—without changing its restmass m_o—without any external influences exerted on it. A devastating result! Who has ever seen a particle start to move without any recognisable reason, just because it emitted a tachyon. Moreover, the tachyon that was emitted could also be absorbed. So, the deadly shot using a tachyon and the end of causality a reality?

Well, in fact, the emission of a particle by a tachyon can be compatible to the conservation laws. If we just remove the particle before the collision in Eq. (460), we receive

$$\left.\begin{aligned} \mu \sinh \beta' &= \cosh \alpha - \cosh \alpha' \ , \\ \mu \cosh \beta' &= \sinh \alpha - \sinh \alpha' \ . \end{aligned}\right\} \qquad \text{No particle before the collision} \tag{470}$$

From $0 \leq \mu \cosh \beta' = \sinh \alpha - \sinh \alpha'$ and the monotony of the hyperbolic sine follows $\alpha' \leq \alpha$. Using $\cosh^2 \beta' - \sinh^2 \beta' = 1$ we eliminate β' by squaring and subtraction of both Eq. (470) with the result

$$\alpha' = \alpha - \operatorname{arcosh}\left(1 + \frac{\mu^2}{2}\right) \ . \tag{471}$$

According to the above, an arbitrary arriving tachyon, in other words α and m_*^T are arbitrarily given, can emit a particle with an arbitrary restmass $m_o = \mu \, m_*^T$. The tachyon velocity u' after emission can be calculated from (471) keeping $c_o/u' = v'/c_o = \tanh \alpha'$ in mind, and the velocity w' of the emitted particle can be calculated from (470) and (471) keeping $w'/c_o = \tanh \beta'$ in mind.

All of these processes are formally possible according to the conservation laws of energy and momentum. Do they however actually exist? Physically speaking, do these processes exist as solutions of the fundamental equations of physics known to us? If so, then we could only save our causal view of the world by referring back to TREDER's [94] proposition of banning these processes into sub-microscopic regions, or by simply formulating an additional condition stating that these solutions are excluded—which would of course only be a stopgap used to bridge the holes in the theory. This is the general situation in which we cannot discuss these problems without actually coming to a consensus about which elementary particle theory is the best to use in modern physics. Let us remain with that what we can elementarily comprehend, the sine-GORDON equation with its particle and tachyon solutions inside of a crystal.

The preceding considerations do not lead us to the conclusion that the particle $q^I(x,t)$, the solution (111) of the sine-GORDON equation can in fact emit a tachyon $q^T(x,t)$, the solution (261) of the sine-GORDON equation. Absolutely not! Due to the non-linearity of the sine-GORDON equation, the sum of both solutions is not another solution of the sine-GORDON equation. We cannot even state that the sum of the states defined by the tachyon and the particle for $t \longrightarrow \infty$ results in a solution of the sine-GORDON equation. The sine-GORDON equation therefore does not allow for such an emission of tachyons.

Basically, the question with regards to causality in a non-linear field theory cannot be formulated in its original form. In the original form, the action of one particle on another particle was asked for. Such a situation however no longer exists, at least not as strictly as before. We now have a composed state consisting of particle and tachyon parts, and that from the very beginning. In a non-linear field theory, one would generally have to re-consider which initial condition basically approximates that of an incoming tachyon colliding with a particle. The question with respect to the transmission of an amount of energy ΔE by the tachyon to the particle can be seen as a solution in a non-linear field theory of an initial value problem of a corresponding partial differential equation. If exact calculations are not supplied, we drift into the realm of speculation with respect to the comprehensive question about causality. As we have seen above with respect to the plastic tachyons of the sine-GORDON equation, they represent states which exist at any time along the complete x-axis, thus also at the position of an arbitrary, isolated 'normal' particle. The requirements for the collision of two particles, used by us for the above calculations, cannot be fulfilled.

There are however solutions to the sine-GORDON equation that can be interpreted as a superposition of particle and tachyon components. We use as an example the so-called general, velocity $0 < v < c_o$ propagating breather solution (117), see SEEGER [86],

$$q^{III}(x,t) = \frac{2a}{\pi} \arctan \frac{\sin \dfrac{-\pi c_o (x - u\,t)}{\sqrt{2}\,L_o \kappa}}{\cosh \dfrac{\pi (x - v\,t)}{\sqrt{2}\,L_o \gamma}} \,. \tag{472}$$

Here, both the 'normal' particle velocity v and the tachyon velocity $u = c_o^2/v > c_o$ occur. The complete object can be seen and comprehended as a mixture of particles and tachyons. The particle velocity controls the propagation of the solution as a whole, whilst the tachyons realise the correlated oscillations over an arbitrarily large distance. Only the regulated phase relations with the velocity $|u| > 0$ between the oscillations remain left over in the composed state (472) from the energy transport carried out with velocity $|u| > 0$ by the isolated tachyon (261). If one constructs the energy–momentum tensor of the breather $q^{III}(x,t)$, then we arrive at a type (a) particle. This is of course only possible because tachyons have the same transformation properties (427) for energy and momentum as normal particles, (298).

We see, tachyons give us a hard time whenever they can be superimposed with normal particles in the framework of a linear theory. Inside of a non-linear theory, however, we seem to be able to solve the problems that tachyons cause:

> Tachyons do not realise signals; they realise correlations.

In order to illustrate the causality problems in connection to tachyons, we will refer back to our detective story from Chap. 23. The suspect DR FAST, accused of murdering the lawyer MR STUS, was irrevocably exonerated by the observations made by SHERLOCK HOLMES. HOLMES, sitting in his specially reserved cabin in the MERCURY Express had measured that MR STUS was already dead before DR FAST had even left prison, therefore DR FAST had no access to any type of equipment with which to kill MR STUS. These two events, event E_o: release of DR FAST from prison, and event E_1: death of MR STUS occurred, when observed from the MERCURY Express, in reversed timed order. DR FAST was therefore exonerated and found not guilty. A, from DR

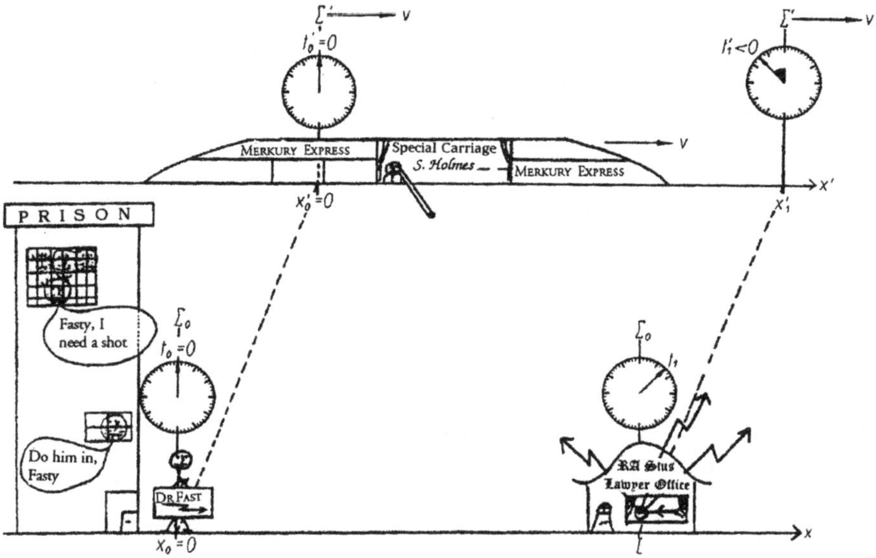

Fig. 30.1 SHERLOCK HOLMES observes from his special carriage in the MERCURY Express how DR FAST is released from prison and the sudden death of the lawyer MR STUS. The prison is located in the reference system Σ_o at $x_o = 0$. DR FAST leaves this prison at time $t_o = 0$. This is event E_o. The reference system Σ' is realised by the MERCURY Express. This reference system moves with the positive velocity of $v = 0,8\,c_o$ with respect to Σ_o. The initial values for the time and space coordinates in Σ' were chosen so that HOLMES also registered the values $x'_o = 0$ and $t'_o = 0$ for the event E_o. MR STUS' office is at rest in Σ_o at $x_1 = L$. MR STUS suddenly dies at time $t_1 = L/2c_o$, which, in the reference system Σ_o, is after the release of DR FAST from prison. This is event E_1. Using our Eq. (151), Chap. 13, we can calculate the time and space coordinates registered by HOLMES in the MERCURY Express for event E_1. We find, see also (476) or (253), $x'_1 = L$ and $t'_1 = -L/2c_o$ (the same value for the space coordinates of the event E_1 in both reference systems is due to our special choice of example). The time t'_1 is negative! DR FAST however leaves prison at time $t'_o = 0$. HOLMES thus observes that MR STUS was already dead before DR FAST left prison and thus proved the innocence of his client. We will keep the following in mind: The positive velocity v of the MERCURY Express with respect to Σ_o causes the event with the larger space coordinates in Σ_o, which is registered from Σ', to be observed to occur earlier than in Σ_o. The dotted lines connect one and the same event

FAST, possibly sent out tachyon could not have been the cause of death of STUS, see Fig. 30.1.

There was however a second suspect, DR FAST's old laboratory assistant MR WACKER who was also sentenced. MR WACKER however was sent, on security reasons, to a different prison, which seen from the reference system Σ_o has the position $x_2 = 2L$. MR WACKER was too released from prison at time $t_2 = 0$ and was arrested again after the death of MR STUS, which occurred at $t_1 = \frac{1}{2}L/c_o$. MR WACKER was in possession of a suspicious tachyon machine, and it was believed that this machine may have been used to send out a second tachyon. We therefore have a second event, E_2, the release of laboratory assistant MR WACKER from prison, which is described in the reference system Σ_o as $E_2(x_2 = 2L,\ t_2 = 0)$. This event is also registered by

HOLMES in his reference system Σ'. We calculate the time t_2' that HOLMES registered in his reference system Σ' at which WACKER was released from prison.

Using the LORENTZ transformation (151), Chap. 13, we calculate for $t_2 = 0$, $x_2 = 2L$ and again with $v = \frac{4}{5}\,c_o$,

$$t_2' = \frac{t_2 - v x_2/c_o^2}{\sqrt{1 - v^2/c_o^2}} = \frac{0 - \frac{4}{5}\,c_o\,2L/c_o^2}{\sqrt{1 - \frac{16}{25}\,c_o^2/c_o^2}} = -\frac{8}{5}\,\frac{5}{3}\,\frac{L}{c_o}\ ,$$

thus

$$t_2' = -\frac{8}{3}\,\frac{L}{c_o}\ . \tag{473}$$

We also calculate

$$x_2' = \frac{x_2 - v t_2}{\sqrt{1 - v^2/c_o^2}} = \frac{5}{3}\,2L\ ,$$

thus

$$x_2' = \frac{10}{3}\,L\ . \tag{474}$$

This time, measurement (473) cannot exonerate MR WACKER. We saw above, cf. Eq. (253), that HOLMES registered the time $t_1' = -\frac{1}{2}\,L/c_o$, so that HOLMES observes $t_2' < t_1'$. This means that, from the reference system Σ', WACKER was already free before MR STUS was killed and could rightly be held suspect. The measurements made by MR HOLMES can only exonerate DR FAST. The occurring event E_2, to the right-hand side of E_1, is seen to occur even earlier to the approaching MERCURY Express than the event E_o, see Fig. 30.2. HOLMES, anticipating this result, sent his friend DR WATSON to travel with the HERMES Courier, a train travelling with the same velocity as the MERCURY express, but in the opposite direction, and asked him to take several measurements using the equipment he had deposited there. DR WATSON was thus in the reference system Σ'', which seen from Σ_o had a velocity of $v'' = -\frac{4}{5}\,c_o$.[1] The instruments were set up in such a fashion that DR WATSON registered DR FAST's release from prison at the coordinates $x_o'' = 0$, $t_o'' = 0$, thus

$$E_o:\quad \left.\begin{array}{l} \Sigma_o : x_o = 0\,,\ t_o = 0\ , \\ \Sigma' : x_o' = 0\,,\ t_o' = 0\ , \\ \Sigma'' : x_o'' = 0\,,\ t_o'' = 0\ . \end{array}\right\} \tag{475}$$

[1]This reference system Σ'' is not identical to the introduced reference system Σ'' in Chap. 17 dealing with the twin paradox.

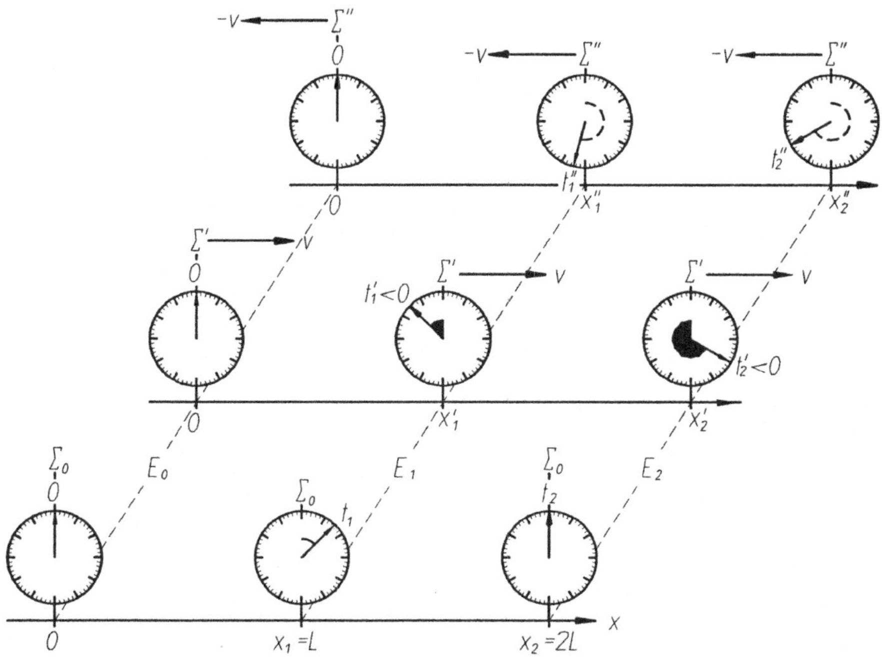

Fig. 30.2 The timed order of events E_o, E_1 and E_2 in dependence of the reference system. The coordinate origin is determined in all three reference systems by event E_o, so that $t_o = t'_o = t''_o = 0$ and $x_o = x'_o = x''_o = 0$ are chosen. The events E_o and E_2 occur, according to supposition, simultaneously in Σ_o, whilst event E_1 occurs time $t_1 = \frac{1}{2}\,L/c_o$ later. In the illustration, all clocks are synchronised so that a hand displays the time L/c_o at the quarter-past position, thus $L/c_o = 15$ scale marks, whereby 60 scale marks would mean a complete rotation of the clock hand. This results in $t_1 = \frac{1}{2}L/c_o = 7,5$ scale marks. The reference system Σ', realised by the MERCURY Express, has the velocity $v = 0,8\,c_o$ with respect to the reference system Σ_o. Using (151), one can calculate the time $t'_1 = -\frac{1}{2}\,L/c_o = -7,5$ scale marks and $t'_2 = -\frac{8}{3}\,L/c_o = -40$ scale marks for $x_1 = L$, $t_1 = \frac{1}{2}\,L/c_o$ and $x_2 = 2L$, $t_2 = 0$, respectively. In Σ' the event E_2 occurs before E_1 and E_1 occurs before E_o. The reference system Σ'', realised by the HERMES Courier, has the velocity $v = -0,8\,c_o$ with respect to the reference system Σ_o. One thus calculates $t''_1 = \frac{13}{6}\,L/c_o = 32,5$ scale marks and $t''_2 = \frac{8}{3}\,L/c_o = 40$ scale marks. In Σ'' the event E_o occurs before E_1 followed by E_2. The opposite direction of movement of the train causes the order of events to be reversed with respect to Σ'

We once again calculate the times t''_1 and t''_2 using LORENTZ transformation (151) that DR WATSON registered for the release of both DR FAST and MR WACKER from prison. Using $t_1 = \frac{1}{2}L/c_o$ and $v = v'' = -\frac{4}{5}\,c_o$, WATSON discovered for MR STUS the time of death t''_1 according to

$$t''_1 = \frac{t_1 - v''x_1/c_o^2}{\sqrt{1 - v^2/c_o^2}} = \frac{5}{3}\left(\frac{L}{2c_o} + \frac{\frac{4}{5}\,c_o\,L}{c_o^2}\right) = \frac{5}{3}\frac{L}{c_o}\left(\frac{1}{2} + \frac{4}{5}\right) = \frac{5}{3}\frac{L}{c_o}\frac{13}{10}\;,$$

thus

$$t_1'' = \frac{13}{16}\frac{L}{c_o} \ . \tag{476}$$

We also find

$$x_1'' = \frac{5}{3}\left(x_1 - v''t_1\right) = \frac{5}{3}\left(L + \frac{4}{5}c_o\frac{L}{2c_o}\right) = \frac{5}{3}\frac{14}{10}L \ ,$$

thus

$$x_1'' = \frac{14}{15}L \ . \tag{477}$$

The following observations are thus valid for event E_1, see Fig. 30.2,

$$E_1: \quad \left.\begin{aligned} \Sigma_o &: x_1 = L \ , \quad & t_1 &= \frac{1}{2}\frac{L}{c_o} \ , \\ \Sigma' &: x_1' = L \ , \quad & t_1' &= -\frac{1}{2}\frac{L}{c_o} \ , \\ \Sigma'' &: x_1'' = \frac{14}{15}L \ , & t_1'' &= \frac{13}{6}\frac{L}{c_o} \ . \end{aligned}\right\} \tag{478}$$

Together with the corresponding coordinate origins according to (475), the quotient of the time and space coordinates from (478) can be used to determine the velocities of the tachyon emitted by Dr Fast as seen in all three reference systems, namely

$$\left.\begin{aligned} \Sigma_o &: u_1 = \frac{x_1}{t_1} = 2c_o \ , \\ \Sigma' &: u_1' = \frac{x_1'}{t_1'} = -2c_o \ , \\ \Sigma'' &: u_1'' = \frac{x_1''}{t_1''} = \frac{28}{65}c_o \ . \end{aligned}\right\} \quad \text{Velocities of the first tachyon} \tag{479}$$

We will assume that the tachyon emitted by Dr Fast has a positive momentum parameter m_*^T. The tachyons velocity therefore has the same sign as its momentum in both Σ_o and Σ''. This is situation that fits into our conception based on 'normal' particles.

The situation however is completely different for the observer in Σ'. The velocity of this tachyon is negative and has the opposite direction with respect to its own momentum. We remind the reader that the momentum of a tachyon has the same sign in every reference system. One could therefore state that the tachyon is travelling backwards through time! This peculiarity of the tachyon being able to travel backwards through time completely breakes our view of the world. We are however, based on this, not allowed to come to the conclusion that tachyons do not exist. This peculiarity only means that we must be very careful when dealing with tachyons so as to not get caught up in contradictions, also see the discussion based on the examples shown in Chap. 28. Let us now more closely examine the observations made by Dr Watson.

With $t_2 = 0$, $x_2 = L$ and again $v = v'' = -\frac{4}{5}\,c_o$, Dr Watson observes for t_2'', the time of release of Mr Wacker,

$$t_2'' = \frac{t_2 - v''x_2/c_o^2}{\sqrt{1 - v^2/c_o^2}} = \frac{0 + \frac{4}{5}\,c_o\,2L/c_o^2}{\sqrt{1 - \frac{16}{25}\,c_o^2/c_o^2}} = \frac{8}{5}\,\frac{5}{3}\,\frac{L}{c_o} \ ,$$

thus

$$t_2'' = \frac{8}{3}\,\frac{L}{c_o} \ . \tag{480}$$

He also observes for x_2'',

$$x_2'' = \frac{x_2 - v''t_2}{\sqrt{1 - v^2/c_o^2}} = \frac{5}{3}\,2L \ ,$$

thus

$$x_2'' = \frac{10}{3}\,L \ . \tag{481}$$

The following observations were made for the event E_2, see also Fig. 30.2,

$$E_2 : \left.\begin{array}{l} \Sigma_o : x_2 = 2L \ , \quad t_2 = 0 \ , \\ \Sigma' : x_2' = \dfrac{10}{3}\,L \ , \ t_2' = -\dfrac{8}{3}\,\dfrac{L}{c_o} \ , \\ \Sigma'' : x_2'' = \dfrac{10}{3}\,L \ , \ t_2'' = \dfrac{8}{3}\,\dfrac{L}{c_o} \ . \end{array}\right\} \tag{482}$$

We see $t_1'' < t_2''$ from (478) and (482). Observed from the reference system Σ'' the event E_1 occurs before the event E_2. The more spatially right-hand side events do occur, from the Σ'' point of view, at a later time point than in Σ_o due to the opposite direction of motion of the Hermes Courier. This can cause the reversal of the order of events in time, which itself is reversed from the viewpoint of the Mercury

> Express. Here, we once again refer the reader to our theorem on the twin paradox in Chap. 17: "The time dilatation of a moving clock is solely dependent on the square of its velocity. The synchronisation regulation of a clock changes its sign when the direction of velocity changes."

According to the measurements made by Dr Watson, Dr Fast's assistant was still in prison after Mr Stus had already died. Thus, Mr Wacker also had an alibi and had to be set free. Although the measurements made by Dr Watson showed Dr Fast to be outside of prison before Mr Stus was killed, the measurements made by Holmes exonerated him. Both suspects had to be set free because of their respective alibis.

Fig. 30.3 The confession

According to our above calculations on the collision between particles and tachyons, we can only agree to this evaluation if all tachyon energy transfers covered by the energy–momentum conservation can be ruled out using the governing physical equations. We therefore must be certain that inelastic collisions using tachyons, as discussed by us according to Eq. (467) with the result (469), are ruled out. We have seen, using the sine-Gordon equation, how this can be realised when based on the non-linearity of equations. We will therefore presuppose, that the totality of physical equations, which must fulfil all occurring interactions, exclude any possibility of non-causal action using tachyons. The acquittal of both Dr Fast and WackerMr Wacker are established facts.

Our detective story now takes a turn. Dr Fast, now, an old man on his deathbed, makes a confession to a priest, see Fig. 30.3, and admits murdering the lawyer Mr Stus.

What really happened back then? Dr Fast and his assistant Mr Wacker were both charged of a murder using a tachyon machine and were both sentenced to long-term prison based on circumstantial evidence, even though they in fact had nothing to do with the murder. Dr Fast knew that his assistant, Mr Wacker, was going to be released from prison at the same time, and he knew that Wacker would immediately aim his tachyon machine at Mr Stus. Dr Fast also knew that Wacker would not be able to do any damage at all. Those tachyons would not be able to 'damage' anything or anyone. They would just go right 'through' MR STUS without a trace, just as we saw in Eq. (466) concerning the elastic collision process.

DR FAST however also wanted revenge and calculated the following experiment: The device in the possession of his assistant can only create tachyons with a negative momentum parameter. He himself adjusts his device to create tachyons with a positive momentum parameter. He now aims his device at STUS, who is in his office located at $x_1 = L$, simultaneously together with WACKER's, the later located at $x_2 = 2L$ and his device at $x_o = 0$. What happens, or what could happen if the tachyons inelastically collide at the location where MR STUS is standing?

This situation is exactly the same situation that we had when we started to discuss tachyons in Chap. 28. In both situations, we observe the inelastic collision of two tachyons. However, we now assume that the rest frame Σ_o of DR FAST is the system in which the total momentum disappears and where the restmass M_o is created after the collision. The tachyon with a positive velocity u created by DR FAST has the momentum parameter m_*^T. The momentum parameter of the tachyon with the negative velocity $-u$ created by WACKER is $\delta \cdot m_*^T = -m_*^T$, so that the total momentum disappears. As we know, the momentum Eq. (418) and the energy equation (419) with $f(u)$ according to (419a) and $|u| > c_o$ are both fulfilled, whereby the Eq. (423a) determines M_o. This means,

$$M_o = \frac{2m_*^T}{\sqrt{u^2/c_o^2 - 1}} \, . \tag{483}$$

If sufficiently 'slow' tachyons could be created, meaning tachyons whose velocities are not much greater than c_o, then a particle with an arbitrarily large mass M_o could be created by the collision. This particle would of course destroy anything at the location of its creation.

In order to prohibit such an inelastic tachyon reaction resulting in the creation of a restmass M_o all the above-mentioned general objections of a non-linear field theory can be applied. A particle with the mass M_o can only be created if such an 'elementary particle' actually exists. We have seen that, e.g. the restmass of the 'elementary particle' kink, in other words the solution (111) of the sine-GORDON equation (88) can only assume a certain value m_o according to (296). A kink with an arbitrary restmass M_o cannot be created in our crystal, because there is no such thing as a kink with an arbitrary restmass in our crystal. In our depiction above, we will have to trust the competent explanations made by the physicist DR FAST and accept that the described process, the creation of a particle with a certain restmass M_o from two tachyons, is possible.

In the framework of the sine-GORDON equation, it is a mathematically important question whether the two tachyons with opposing momentum parameters, i.e. the tachyons belonging to the solutions (261) and (445), coming from $t \longrightarrow -\infty$, for $t \longrightarrow +\infty$ can be composed into a particle solution, a kink of the form (111). The answer to this question cannot be given here.

If the constructed case above were true, what would be the effects on causality? Would not we have proven the following above: If DR FAST had caused the death of MR STUS with the help of his tachyon device then MR HOLMES, sitting in the MERCURY Express, would have observed the effect (the death of STUS) occurred before the cause (shooting the tachyon). Causality broken?

In this case, the answer must in fact be no. A further piece of information is to be added to the facts concerning DR FAST's action, namely that his assistant, MR WACKER, who was positioned at a distance of $2L$ away from FAST, simultaneously emitted a tachyon, simultaneously in Σ_o. The cause of death must therefore include that MR WACKER indeed knew the exact time point when to fire the tachyon. We always assumed that this was somehow known to both FAST and WACKER. This assumption however is not exact enough when investigating the question of causality. DR FAST would have to have sent WACKER a message containing the information about the time point, here $t_o = 0$ in the reference system Σ_o, when WACKER should fire the tachyon. This information cannot be sent faster than the velocity c_o. The respective signal to WACKER must have been sent no later than $t^* = -2L/c_o$ in order for WACKER to be able to simultaneously, at $t_o = 0$, shoot the tachyon at MR STUS. The cause of death thus begins in the reference system Σ_o at time $t^* = -2L/c_o$ and not $t_o = 0$. For this time point, HOLMES in his MERCURY Express (reference system Σ') calculates

$$t^{*\,\prime} = \frac{t^* - v''x_o/c_o^2}{\sqrt{1 - v^2/c_o^2}} = \frac{5}{3}\left(\frac{2L}{c_o} - 0\right) ,$$

thus

$$t^{*\,\prime} = -\frac{10}{3}\frac{L}{c_o} . \tag{484}$$

This time point is however before STUS' time of death $t_1' = -\frac{1}{2}L/c_o$, (478).

The effect must be seen in the simultaneous firing of arbitrarily distanced tachyons in the reference system Σ_o. In order to create this simultaneous action, a signal is required which first must traverse this distance. The time required for this signal to arrive guarantees causality. Cause and effect cannot be reversed in their timed order, not even if tachyons can travel at infinite velocity.

Let us summarise. The existence of type (b) particles, tachyons, whose velocities u are greater than c_o, can be explained with the help of the Special Theory of Relativity. In a relativistic theory, one must expect solutions that represent such tachyons. One will always find, in the totality of possible reactions between particles and tachyons, which alone underlay the conservation laws of energy and momentum, situations that lead to a breaking of causality. This means that the order in time of two events, of which one event must be seen as the cause of the other event, would be reversed for such non-causal reactions when observing from a different, specially selected reference system. In order to guarantee causality, certain selection criteria may be formulated that forbid such reactions, or one postulates that those reactions which lead to a breaking of causality only take place in sub-microscopic regions in which our knowledge of physical reality is low, see TREDER [94].

There is also a third possibility, which, inside the framework of macroscopic theories, would be the best, namely that we have a 'good' relativistic field theory in which we do not have to postulate any additional exception conditions in order to secure causality. This could be, for example the non-linearity of the theory which only allows for solutions that fulfil our experiences and views of causality, so that we can better understand how our world is causally constructed. We believe that the non-linear field theory, as formulated by the sine-GORDON equation, could indeed be such a 'good's relativistic field theory without actually being able to prove this. If we accept this assumption, we come to the following statement at the end of our discussion:

> The totality of the solutions of the sine-GORDON equation defines the principles of a Special Theory of Relativity with causality preserved.

Glossary of Symbols

a	lattice parameter
A, α	coefficients in the LAGRANGian of sine-GORDON field
$\boldsymbol{\alpha}$, α_{ik}	tensor of dislocation density
a_o	amplitude of the fundamental oscillation for a system of two masses
$\mathbf{b}$	BURGERS vector
$\boldsymbol{\beta}$	tensor of distorsion
c	sound velocity, critical velocity of the wave equation
$\mathbf{C}$, C_{ikrs}	HOOKE's tensor
c_L	vacuum light velocity
c_l	longitudinal sound velocity
c_L	transversal sound velocity
c_o	critical velocity of sine-GORDON equation
d	interlattice-plane distance
D	force constant of a dislocation
D_m	force constant of a dislocation
D	force constant, spring stiffness
D_N	force constant, spring stiffness
δ	non-integrable increment of a function
$\delta(x)$	DIRAC's function
δ_{ik}	KRONECKER symbol
$\triangle$	LAPLACian
Δ	increment of a physical quantity
e	energy density of a field
E	Energy of a mass
E	modulus of elasticity of
E	modulus of elasticity of a dislocation
E, E'	energy of a mass in the reference systems Σ_o and Σ'
E_{eff}	effective modulus of elasticity
E_{kin}	kinetic energy

H. Günther, *Elementary Approach to Special Relativity*,
https://doi.org/10.1007/978-981-15-3168-2

E^T	energy of a tachyon
$E^T, E^{T\prime}$	energy of a tachyon in the reference systems Σ_o and Σ'
E_o	rest energy of a kink
ε	relative elastic strain
$\varepsilon, \varepsilon_{rr}$	dilatation
$\boldsymbol{\varepsilon}, \varepsilon_{ik}$	strain tensor
$\mathbf{F}, F$	force vector, force, resulting total force
F	tension of a one dimensional field
$\mathbf{F}_a, F_a$	single force vector, single force
$\mathbf{f}_{ab}, \mathbf{F_{AB}}$	interacting forces
$\mathbf{f}_{ik}$	interacting forces
ϕ	phase angle of an oscillation
G	modulus of shear
k	wave number
L	length of a rod
L	LAGRANGian
L_o	natural unit of measure for length in a lattice
L_o, L'	length of a static and a moving rod resp.
λ	wave length
λ	LAMé's constant
λ_o	length composed of lattice parameters
m	(inertial) mass
m, m_o	moving mass and restmass of a particle
m, m_o	moving mass and restmass of a kink
m_a	mass of a dislocation of the length a of a lattice distance
m_a, M_A	single masses of a system
Δm_α	inertial masses of a dislocation
m_*^T	momentum parameter of a tachyon
μ	LAMé's constant
μ	parameter for the particle tachyon collision
ν	frequency
ν	POISSON's ratio
ω	circular frequency
Ω_o	circular frequency of an oscillating line
P	momentum of a mass
P, P'	momentum of a mass in the reference systems Σ_o and Σ'
P^T	momentum of a tachyon
$P^T, P^{T\prime}$	momentum of a tachyon in the reference systems Σ_o and Σ'
P_∞^T	momentum of a tachyon with 'infinite' velocity
$\mathbf{p}$	momentum vector of a partricle
p	momentum density of a one dimensional field
ΔP	momentum of a one dimensional field
$q, q(x,t)$	displacement of a dislocation transversal to the line direction
q	dimensionless measure for displacement of a dislocation
q_a, q_i	coordinates of masses and their centers of inertia

q_α coordinates of dislocations
r, r_o amplitude of the fundamental oscillation of a system of two masses
ρ density of masses and the centers of inertia
ρ_{eff} effective mass density of a linear chain
ρ_o mass density of a sine-GORDON field
ρ_o mass density of a dislocation
s, s_o elastic displacement and maximal elastic displacement
s density of energy current of a one dimensional field
σ line tension of a linear chain
σ line tension of a dislocation
$\boldsymbol{\sigma}, \sigma_{ik}$ stress tensor
Σ_o preferred frame, defined with the help of a lattice
$\Sigma, \Sigma', \Sigma''$ reference systems
t time, coefficient of measure for time, time coordinate
t dimensioneless coefficient of measure for time
t, t' hand settings of the static and moving clocks
t stress of the one dimensional field
$\mathbf{t}$ vector of line direction of a dislocation
$\mathbf{T}$ energy-momentum tensor of a field
T density of kinetic energy of a field
$\mathbf{T}$ physical time
T_o natural unit of measure for time in a crystalline lattice
T_o, T' period of a static and a moving clock
τ one dimensional stress
τ_o time composed of lattice parameters
u velocity
u velocity of a tachyon
U total energy of an oscillating particle
U_A, U_B, U_V, U_1,...clocks
v, v_o velocities
w velocity
x space, coefficient of measure for space, space coordinate
$\mathbf{X}$ physical length
x dimensionles coefficient of measure for space
ξ_o mass parameter

Notice that in this book we use γ for the square root, $\gamma = \sqrt{1 - v^2/c^2}$, and not for the inverse quantity. This is in accordance with the previous German text "Grenzgeschwindigkeiten und ihre Paradoxa", Springer Fachmedien Wiesbaden 1996.

Units of Measurement

Å	ÅNGSTRÖM
°C	degree centigrade
K	degree KELVIN
Hz	HERTZ
J	JOULE
kcal	kilogram-calorie
kg	kilogram
m	meter
cm	centimeter
gm	'lattice meter'
s	second
gs	'lattice second'
N	NEWTON

H. Günther, *Elementary Approach to Special Relativity*,
https://doi.org/10.1007/978-981-15-3168-2

Remarks to the Notation

All calculations refer to CARTESIAN coordinates.
The indices $1, 2, 3$ stand for $x,\ y,\ z$, so $a_1 = a_x,\ a_2 = a_y,\ a_3 = a_z$.
Multicomponent quantities are characterised by bold face typing,

$$\mathbf{a} = (a_1,\ a_2,\ a_3)\,, \quad \mathbf{F} = (F_1,\ F_2,\ F_3)\ ,$$

$$\boldsymbol{\varepsilon} = \begin{pmatrix} \varepsilon_{11} & \varepsilon_{12} & \varepsilon_{13} \\ \varepsilon_{21} & \varepsilon_{22} & \varepsilon_{23} \\ \varepsilon_{31} & \varepsilon_{32} & \varepsilon_{33} \end{pmatrix}\,, \quad \boldsymbol{\sigma} = \begin{pmatrix} \sigma_{11} & \sigma_{12} & \sigma_{13} \\ \sigma_{21} & \sigma_{22} & \sigma_{23} \\ \sigma_{31} & \sigma_{32} & \sigma_{33} \end{pmatrix}\ .$$

Bold face typing is not used for vectors and tensors if these are used for one dimensional problems where one component quantities are concerned, see Chap. 22.
A dot '$\cdot$' between two quantities describes the scalar product.
As usual '$:=$' denotes equal by definition.
The scalar product of two vectors is a number,

$$\mathbf{a} \cdot \mathbf{b} = \sum_{r=1}^{3} a_r b_r\ .$$

The point can be dropped for products with numbers,

$$\lambda \cdot \nu = \lambda \nu\ .$$

The quantity $\mathbf{v}\,\mathbf{v} \cdot d\mathbf{A}$ created from vector $\mathbf{v}$ and vector $d\mathbf{A}$ is therefore a vector according to

$$\mathbf{v}\,\mathbf{v} \cdot \mathbf{A} = \Big(v_1 \sum_{r=1}^{3} v_r\, dA_r\,,\ v_2 \sum_{r=1}^{3} v_r\, dA_r\,,\ v_3 \sum_{r=1}^{3} v_r\, dA_r\Big)\ .$$

H. Günther, *Elementary Approach to Special Relativity*,
https://doi.org/10.1007/978-981-15-3168-2

The scalar product of a vector and a tensor (matrix) is a vector,

$$\mathbf{a}\cdot\varepsilon = \Big(\sum_{r=1}^{3} a_r\,\varepsilon_{r1}\,,\ \sum_{r=1}^{3} a_r\,\varepsilon_{r2}\,,\ \sum_{r=1}^{3} a_r\,\varepsilon_{r3}\Big)\ ,$$

$$\varepsilon\cdot\mathbf{b} = \Big(\sum_{r=1}^{3} \varepsilon_{1r}\,b_r\,,\ \sum_{r=1}^{3} \varepsilon_{2r}\,b_r\,,\ \sum_{r=1}^{3} \varepsilon_{3r}\,b_r\Big)\ .$$

The scalar product of two tensors is a tensor,

$$\varepsilon\cdot\boldsymbol{\sigma} = \begin{pmatrix} \sum_{r=1}^{3} \varepsilon_{1r}\,\sigma_{r1} & \sum_{r=1}^{3} \varepsilon_{1r}\,\sigma_{r2} & \sum_{r=1}^{3} \varepsilon_{1r}\,\sigma_{r3} \\ \sum_{r=1}^{3} \varepsilon_{2r}\,\sigma_{r1} & \sum_{r=1}^{3} \varepsilon_{2r}\,\sigma_{r2} & \sum_{r=1}^{3} \varepsilon_{2r}\,\sigma_{r3} \\ \sum_{r=1}^{3} \varepsilon_{3r}\,\sigma_{r1} & \sum_{r=1}^{3} \varepsilon_{3r}\,\sigma_{r2} & \sum_{r=1}^{3} \varepsilon_{3r}\,\sigma_{r3} \end{pmatrix}\ .$$

Should the scalar product sign '·' be set twice then summation must be done twice correspondingly,

$$\varepsilon\cdot\cdot\,\boldsymbol{\sigma} = \sum_{r=1}^{3}\sum_{s=1}^{3} \varepsilon_{rs}\sigma_{sr}\ .$$

The KRONECKER symbol δ_{ik} is defined according to

$$\delta_{ik} = \begin{cases} 1 & \\ & \text{for} \\ 0 & \end{cases} \quad \begin{matrix} i = k\ . \\ \\ i \neq k\ . \end{matrix}$$

The LEVI- CIVITA symbol ϵ_{ijk} is defined according to

$$\epsilon_{123} = \epsilon_{231} = \epsilon_{312} = +1\ ,$$
$$\epsilon_{213} = \epsilon_{132} = \epsilon_{321} = -1\ ,$$
$$\epsilon_{ijk} = 0\quad \text{otherweise}\ .$$

A cross $\times$ respresents the vector product,

$$\big(\mathbf{a}\times\mathbf{b}\big)_i = \sum_{r,s=1}^{3} \epsilon_{irs}\,a_r b_s\ ,$$

$$\big(\mathbf{a}\times\varepsilon\big)_{ik} = \sum_{r,s=1}^{3} \epsilon_{irs}\,a_r\varepsilon_{sk}\ ,\quad \big(\boldsymbol{\sigma}\times\mathbf{b}\big)_{ik} = \sum_{r,s=1}^{3} \epsilon_{krs}\,\sigma_{ir}b_s\ .$$

From Chap. 25 onwards multiple component quantities are summed using identical indices according to the EINSTEINian summation convention,

$$\mathbf{a} \cdot \mathbf{b} = a_r\, b_r \ , \quad \left(\mathbf{a} \cdot \varepsilon\right)_k = a_r\, \varepsilon_{rk} \ , \quad \boldsymbol{\varepsilon} \cdot\cdot\, \boldsymbol{\sigma} = \varepsilon_{rs}\, \sigma_{sr} \ , \quad \left(\mathbf{a} \times \mathbf{b}\right)_i = \epsilon_{irs}\, a_r\, b_s \ .$$

From Chap. 23 onwards, a comma is used to describe the partial derivative,

$$\frac{\partial f}{\partial x_k} = f_{,k} \ .$$

The vectorial operator $\boldsymbol{\nabla} = (\nabla_1, \nabla_2, \nabla_3)$ is also written as grad and defined by

$$\nabla_k = \frac{\partial}{\partial x_k} \ .$$

One thus notes

$$\operatorname{div} \mathbf{a} = \boldsymbol{\nabla} \cdot \mathbf{a} \ , \quad (\operatorname{curl} \mathbf{a})_k = (\boldsymbol{\nabla} \times \mathbf{a})_k = \epsilon_{krs}\, a_{s,r} \ , \quad \operatorname{grad} f = \boldsymbol{\nabla} f = (f_{,1}\, ,\, f_{,2}\, ,\, f_{,3}) \ .$$

The LAPLACian $\triangle$ is defined as

$$\triangle = \operatorname{div} \operatorname{grad} = \boldsymbol{\nabla} \cdot \boldsymbol{\nabla} = \frac{\partial^2}{\partial x^2} + \frac{\partial^2}{\partial y^2} + \frac{\partial^2}{\partial z^2} \ .$$

The notation $\exp[x]$ is used for the function e^x if complicated exponents are included.

Curriculum Vitae

Helmut Günther

CV: Born 1940 in Bochum. 1963 diploma in physics at the Humboldt-University Berlin. 1966 and 1972 first and second DSc degrees. 1972-1962 lectures on Theoretical Physics at the Humboldt-University Berlin.
Academic positions: 1963–1969 Institut f. Reine Mathematik, Dt. Akad. Wiss. Berlin. 1969–1982 Zentralinstitut f. Astrophysik, Potsdam-Babelsberg. 1982–1986 Einstein-Laboratorium f. Theoretische Physik, Potsdam-Babelsberg.
1986 escape from the GDR.
1987–1989 Max-Planck-Institut f. Metallforschung, Stutgart. 1989-1990 Institut f. Theoretische & Angewandte Physik, Universität Stuttgart.
1990–2005 Prof. for Mathematics and Physics at the University of Applied Sciences, Bielefeld.

An easily comprehensible, new approach is developed for deriving Einstein's Special Theory of Relativity. In this context, the universal constancy of the speed of light, Einstein's special principle of relativity will be explained. Furthermore, the complete framework of special relativity—including the paradoxa and philosophical topics, tachyons and causality and even the limits of validity—can be understood with the help of a surprising physical model.

H. Günther, *Elementary Approach to Special Relativity*,
https://doi.org/10.1007/978-981-15-3168-2

References

1. R. Becker, *Electromagnetic Fields Interactions and Chap.DIV* (Dover Publications Inc., NY, 1982)
2. V. Berzi, V. Gorini, J. Math. Phys. **10**, 1518 (1969)
3. V. Berzi, V. Gorini, Ann. Inst. HENRI POINCARé, XVI **1**, 1–22 (1972)
4. V. Berzi, Rep. Math. Phys. **II**(3), 297 (1976)
5. U. Bleyer, W. Haubold (ed.), MICHELSON-Ausstellung des Zentralinst. f. Astrophys. der Akad. Wiss. DDR (Potsdam-Babelsberg, 1979)
6. I.N. Bronstein, K.A. Semendjajew, *Handbook of Mathematics* (Springer, Berlin, 1997)
7. I.M. Burgers, I. Proc, Kon. Nederl. Akad. Wetesch. **42**, 243 (1939)
8. I.M. Burgers, I. Proc, Kon. Nederl. Akad. Wetesch. **42**, 378 (1939)
9. D.C. Champeney, G.R. Isaac, A.M. Khan, Phys. lett. **7**, 241 (1963)
10. U. Dehlinger, Ann. Phys. (Lpz.) **2**, 749 (1929)
11. E. Eilenberger, *Solitons. Mathematical Methods for Physicists* (Springer, Berlin, 1981)
12. A. Einstein, *Ausgewählte Texte* (Meister (Goldmann Verlag, München, Herausgegeben von H.C, 1986)
13. A. Einstein, *Sidelights on Relativity. I. Ether and Relativity. II. Geometry and Experience.* Translated by G.B. JEFFERY and W. PERRETT (Methuen & Co., London, 1922)
14. A. Einstein, Ann. Phys. (Lpz.) **17**, 891 (1905)
15. A. Einstein, see H. A. LORENTZ [60]
16. A. Einstein, Ann. Phys. (Lpz.) **18**, 639 (1905)
17. A. Einstein, *Relativity. The Special and the General Theory* (University Paperbacks. Methuen, London, 1960)
18. A. Enneper, Nachr. Königl. Gesellsch. d. Wiss. Göttingen **493**, (1870)
19. J.D. Eshelby, Proc. R. Soc. A **266**, 222 (1962)
20. G.F. FitzGerald, Science **13**, 390 (1889)
21. F.C. Frank, Phil. Mag., VII. Ser., **42**, 809 (1951)
22. F.C. Frank, J.H. van der Merwe, Proc. R. Soc. Lond., Ser. A, **201**, 261 (1950)
23. A.P. French, *Special Relativity* (Massachusetts Institute of Technology, Massachusetts, 1968)
24. J. Frenkel, Z. Phys. **37**, 572 (1926)
25. J. Frenkel, T. Kontorova, J. Phys. Acad. Sci. USSR **1**, 362 (1939)
26. H. Genz, R. Decker, *Symmetrie und Symmetriebrechung in der Physik* (Vieweg-Verlag, Braunschweig, 1991)
27. H.F. Goenner, *Einführung in die spezielle und allgemeine Relativitätstheorie* (Spektrum Akademischer Verlag, Heidelberg, 1996)

H. Günther, *Elementary Approach to Special Relativity*,
https://doi.org/10.1007/978-981-15-3168-2

28. H.F. Goenner, *Spezielle Relativitätstheorie und die klassische Feldtheorie* (Spektrum Akademischer Verlag, München, 2004)
29. H. Goldstein, *Classical Mechanics*, 2nd edn. (Addison-Wesley Publishing Company, Mass & London, 1981)
30. H. Günther, Phys. Stat. Sol. (b) **185**, 335 (1994)
31. H. Günther, *Zur nichtlinearen Kontinuumstheorie bewegter Versetzungen* (Akademie-Verlag, Berlin, 1967)
32. H. Günther, Ann. Phys. (Lpz.) **40**, 220 (1983)
33. H. Günther, Z. angew. Math. Mech. **74**(6), T603 (1994)
34. H. Günther, Phys. Stat. Sol. (b) **149**, 101 (1988)
35. H. Günther, Z. Phys, B **76**, 89 (1989)
36. H. Günther, Euromech Colloquium 348, Tallin (1996)
37. H. Günther, Phys. Stat. Sol. (b) **161**, 179 (1990)
38. H. Günther, V. Müller, *The Special Theory of Relativity - Einsteins World in New Axiomatics* (Springer Nature, Berlin, 2019)
39. M.P. Haugan, C.M. Will, Phys. Today **69**, (1987)
40. F.W. Hehl, E. Kröner, Z. Naturf. **20**a, Heft 3, 336 (1965)
41. F.W. Hehl, Y.N. Obukhov, *Foundations of Classical Electrodynamics*. Progress in Mathematical Physics, vol. 33 (Birkhäuser Verlag, Basel, 2003)
42. L. Infeld, *Albert Einstein. His Work and Its Influence on Our World* (Charles Scribner's Sons, New York, 1950)
43. D. Ivanenko, A. Sokolow, *Klassische Feldtheorie* (Akademie-Verlag, Berlin, 1953)
44. H.J. Ives, G.J. Stillwell, J. Opt. Soc. **28**, 215 (1938)
45. H.J. Ives, J. Opt. Soc. **29**(183), 294 (1939)
46. C. Kittel, *Introduction to Solid State Physics*, 5th edn. (Wiley Inc, New York, 1967)
47. E. Kröner, *Kontinuumstheorie der Versetzungen und Eigenspannungen* (Springer, Berlin, 1958)
48. E. Kröner, Continuum theory of defects, in *Les Houches, Session 35, 1980, Physics of Defects*, ed. by R. Balian, et al. (North-Holland, Amsterdam, 1981), pp. 215–315
49. E. Kröner, G. Rieder, Z. Phys. **145**, 424 (1956)
50. E. Kröner, *Plastizität und Versetzungen*. In: Sommerfeld, A.: *Vorlesungen über Theoretische Physik*, Bd.**2** (Akademische Verlagsgesellschaft Geest & Portig, Leipzig, 1964), S.363
51. E. Kröner, ZAMM 66, 5, T **284**, (1968)
52. E. Kröner, *Symposium on Materials Modelling: From Theory to Technology. 1991 Part I* (IOP Publishing Ltd, Oxford, 1992)
53. E. Kröner, in *Proceedings of the 2nd International Conference on Nonlinear Mechanics*, ed. by C. Weizang (Peking University, Beijing, 1993)
54. L.D. Landau, E.M. Lifschitz, *Theory of Elasticity* (Pergamon Press, Oxford, 1970)
55. G. Leibbrandt, J. Math. Phys. **21**(10), 1613 (1980)
56. G. Leibbrandt, Phys. Rev. (D) **23**, 2484 (1981)
57. D.E. Liebscher, *Relativitätstheorie mit Zirkel und Lineal* (Akademie-Verlag, Berlin, 1991)
58. D.E. Liebscher, B. Brosche, Astron. Nachr. **319**, (1998)
59. D.E. Liebscher, Ann. Phys. (Lpz) **32**, 363 (1975)
60. H.A. Lorentz, A. Einstein, H. Minkowski, H. Weyl, *The Principle of Relativity* (Jeffery (Dover Publications Inc, New York, Translated by W. Perret and G.B, 1952)
61. H.A. Lorentz, K. Versl, Ak. Amsterdam **1**, 74; Coll. Papers **4**, 219 (1892)
62. H.A. Lorentz, Proc. Acad. Sc. Amsterdam **6**, 809 (1904)
63. H.A. Lorentz, Revue génerale des Sciences **25**, 179 (1914)
64. H.A. Lorentz, *The Theory of Electrons* (Teubner-Verlag, Leipzig, 1916)
65. A.A. Michelson, E.W. Morley, Am. Sci. **34**, 333 (1887)
66. H. Minkowski, *Vortrag auf der 80* (Versammlung Deutscher Naturforscher und Ärzte, Köln, 1908)
67. G.L. Naber, *The Geometry of* MINKOWSKI *Spacetime. An Introduction to the Mathematics of the Special Theory of Relativity* (Springer, New York, 1992)
68. J.F. Nye, Acta metallurgica **1**, 153 (1953)

69. E. Orowan, Z. Phys. **89**, 605 (1934)
70. G. Otting, Dissertation München. Phys. Z. **40**, 681 (1939)
71. A. Pais, *Subtle is the Lord* (The Science and the Life of Albert Einstein (Oxford University Press, Oxford, 1982)
72. R.K. Pathria, *The Theory of Relativity*, 2nd edn. (Pergamon Press, Oxford, 1974)
73. H. Poincaré, Rev. Métaphys. Morale **6**, 1 (1898)
74. H. Poincaré, *The Value of Science* (Dover Publications Inc., New York, 1958)
75. H. Poincaré, *Sechs Vorträge aus der Reinen Mathematik und Mathematischen Physik* (Teubner-Verlag, Leipzig, 1910)
76. M. Polanyi, Z. Phys. **89**, 660 (1934)
77. K.R. Popper, *The Logic of Scientific Discovery* (Harper & Row, New York, 1976)
78. L. Prandtl, Z. angew. Math. Mech. **8**, 85 (1928)
79. S.P. Puri, *Special Theory of Relativity* (Asia Publishing House, London, 1972)
80. C. Rebbi, G. Soliani, *Solitons and Particles (World Scientific Publishing Co* (Pte. Ltd., Singapore, 1984)
81. H.F.H.G. Reichenbach, *Philosophie der Raum-Zeit-Lehre*, Gesammelte Werke, Bd.2 (Vieweg-Verlag, Braunschweig, 1972)
82. H.F.H.G. Reichenbach, *Die philosophische Bedeutung der Relativitätstheorie*, Gesammelte Werke, Bd.3 (Vieweg-Verlag, Braunschweig, 1973)
83. H.F.H.G. Reichenbach, *The Philosophy of Space and Time* (Dover Publications Inc, New York, 1958)
84. J. Rubinstein, Math. Phys. **11**, 258 (1970)
85. J.A. Schouten, *Ricci-Calculus* (Springer, Berlin, 1954)
86. A. Seeger, *Solitons in crystals, in Continuum Models of Discrete Systems* (University of Waterloo Press, Waterloo, 1980)
87. A. Seeger, *Diplomarbeit Universität Stuttgart* (1948/49)
88. A. Seeger, P. Schiller, in *Physical Acoustics*, ed. by W.P. Mason, vol. IIIA (Academic, New York, 1966)
89. A. Seeger, *Solitons and statistical thermodynamics*, vol. 249, Lecture Notes in Physics (Springer, Berlin, 1986)
90. A. Seeger, F. Ackermann, H. Mugrabi, A. Seeger, Acta metall. **31**, 1353 (1953)
91. T.H.R. Skyrme, Proc. R. Soc. Lond. Ser. A **261**, 237 (1961)
92. G.I. Taylor, Proc. R. Soc. Lond. Ser. A, **145**, 362 (1934)
93. J. Tiele, *Wissenschaftliche Kommunikation. Die Korrespondenz* ERNST MACH*s*. (OTTO NEURATH an ERNST MACH) (A. Henn - Verlag, Kastellaun, 1978)
94. H. Treder, *Elementarlänge, Tachyonen und Kausalität*, in *Philosophische Probleme des physikalischen Raumes* (Akademie-Verlag, Berlin, 1974)
95. H. Treder, *Das* EINSTEIN - BOHR*sche Kasten - Experiment*, in *Philosophische Probleme des physikalischen Raumes* (Akademie-Verlag, Berlin, 1974)
96. W. Voigt, Nachr. Königl. Gesellsch. d. Wiss. Göttingen **41**, (1887)
97. W. Voigt, Phys. Zeitschr. **9**, 762 (1908)
98. C.F.V. Weizsäcker, *Materie, energie, information*, in *Die Einheit der Natur* (Deutscher Verlag der Wiss., Berlin, 1957)
99. A. Papapetrou, *Spezielle Relativitätstheorie* (Deutscher Verlag der Wiss., Berlin, 1957)

Index

H. Günther, *Elementary Approach to Special Relativity*,
https://doi.org/10.1007/978-981-15-3168-2

M

N

O

T

The manufacturer's authorised representative in the EU is Springer Nature Customer Service Centre GmbH, Europaplatz 3, 69115 Heidelberg, Germany. If you have any concerns regarding our products, please contact ProductSafety@springernature.com

Printed and bound by CPI Group (UK) Ltd, Croydon, CR0 4YY
30/09/2026
02225510-0014